W9-COX-954

House, Home, and Community:
Progress in Housing Canadians, 1945-1986

GEORGIAN COLLEGE LIBRARY

GEBL-BK
#29.20

House, Home, and Community: Progress in Housing Canadians, 1945-1986

Edited by

John R. Miron

with

Larry S. Bourne, George Fallis, A. Skaburskis,

Marion Steele, and Patricia A. Streich

DISCARD

Library Commons
Georgian College
One Georgian Drive
Barrie, ON
L4M 3X9

McGill-Queen's University Press
Montreal and Kingston · London · Buffalo

Canada Mortgage and Housing Corporation (CMHC)
Ottawa

CMHC SCHL

Copyright © Canada Mortgage and Housing Corporation 1993

ISBN 0-7735-0995-X

Legal deposit first quarter 1993
Bibliothèque nationale du Québec

Printed in Canada on acid-free paper

Canadian Cataloguing in Publication Data

House, home and community: progress in
housing Canadians, 1945-86

Includes bibliographical references and index.
Co-published by: Canada Mortgage and Housing Corporation.

ISBN 0-7735-0995-X

1. Housing – Canada – History. I. Miron, John R., 1947-
II. Canada Mortgage and Housing Corporation.

HD7305.A3H85 1993   363.5'0971'0904   C92-090723-7

The Editors gratefully acknowledge the support of the
Centre for Urban and Community Studies,
University of Toronto.

The authors gratefully acknowledge the financial support of
Canada Mortgage and Housing Corporation. The views expressed
in this book are the personal views of the authors and the
Corporation accepts no responsibility for them.

Cette publication est aussi disponible en français
sous le titre: Maison, foyer, et collectivité:
évolution du logement au Canada, 1945 à 1986.

ISBN 0-7735-1121-0

IN MEMORY OF

Joan Simon and John Bradbury,
whose work symbolizes the determination of
social scientists to understand housing
as both consequence and cause of the social and
economic behaviour of individuals and the
institutions within which we live.

Contributors

John Bossons, Department of Economics, University of Toronto

Larry S. Bourne, Centre for Urban and Community Studies, and Department of Geography and Planning, University of Toronto

John H. Bradbury (Deceased), Formerly Department of Geography, McGill University

Francine Dansereau, Institut National de la Recherche Scientifique – Urbanisation, Montreal, Quebec

George Fallis, Department of Economics, York University

Lynn Hannley, The Communitas Group Ltd., Edmonton, Alberta

Richard Harris, Department of Geography, McMaster University

John Hitchcock, Department of Geography and Planning, University of Toronto

Deryck Holdsworth, Department of Geography, Pennsylvania State University

J. David Hulchanski, Faculty of Social Work, University of Toronto

Janet McClain, Department of Political Science, University of Western Ontario

James McKellar, Program in Real Property Development, York University

John R. Miron, Centre for Urban and Community Studies, and Division of Social Sciences, Scarborough College, University of Toronto

Eric G. Moore, Department of Geography, Queen's University

Jeffrey Patterson, Institute for Urban Studies, University of Winnipeg

James V. Poapst, Faculty of Management, University of Toronto

Damaris Rose, Institut National de la Recherche Scientifique – Urbanisation, Montreal, Quebec

Joan Simon, (Deceased) Formerly Department of Consumer Studies, University of Guelph

A. Skaburskis, School of Urban and Regional Planning, Queen's University

Marion Steele, Department of Economics, University of Guelph

Patricia A. Streich, Forma Consulting, Kingston, Ontario

Martin Wexler, Service de l'habitation et du développement urbain, Ville de Montréal

Table of Contents

Prologue

THIS MONOGRAPH is the result of a project first conceived at Canada Mortgage and Housing Corporation in the early 1980s. The idea was to commission a pair of monographs, one on the housing industry (Clayton Research Associates Limited 1988) and the other – this volume – on progress in the housing of Canadians. The purposes of this monograph were:

- to benchmark the evolution of housing in Canada since 1945,
- to address important dimensions of the current and future housing situation,
- to identify housing research issues and priorities in Canada over the mid to long term, and
- to identify important housing policy issues likely to emerge in Canada over the mid to long term.

A consortium of scholars coordinated by the Centre for Urban and Community Studies of the University of Toronto was chosen by CMHC to write this monograph. The consortium proposed an unusual – some critics said unworkable – plan in which the monograph was to be written by twenty-one authors from across Canada, including both scholars and housing advocates/practitioners, and from a variety of academic disciplines. While housing anthologies are commonplace, what was seen as novel about this monograph was that it would be a "seamless" monograph: written as if by one hand, yet reflecting the richness of diverse views of the Canadian housing experience.

Work on this project began in January 1986, and the text was completed in the form of a Final Report in August 1989. The Final Report consisted of twenty-three chapters. This book is an abridged version wherein each chapter of the Final Report has been substantially reduced in length. To conserve space, Chapter 12 of the Final Report, entitled "Indicators of Housing Adequacy" (authored by Streich and Hannley) is not included in this book; subsequent chapters of the Final Report are renumbered as a consequence.

In drafting the monograph, the authors benefited from two parallel papers

1

commissioned as part of this project: "Evolution de la situation du logement locatif à Montréal," by Professor Marc Choko (Université de Québec à Montréal) and "Housing in Rural Areas and Small Towns," by Andy Rowe (NORDCO Ltd.). Material from these papers has been integrated into various chapters in this monograph. The authors also benefited from the comments of many reviewers, including nine internal reviewers who met with the authors at two workshops in 1986 during the writing of the first draft: Mr. R. Adamson (CMHC, retired), Dr. S. Carreau (Ville de Montréal), Professor T. Carter (University of Winnipeg), Dr. F. Clayton (Clayton Research Associates), Professor W. Grigsby (University of Pennsylvania), Dr. A. McAfee (City of Vancouver), Dr. P. Tomlinson (City of Toronto), Professor G. Wanzel (Technical University of Nova Scotia), and Mr. R. Langlais (Langlais, Hurtibise). In addition, eleven external reviewers were commissioned to comment on the completed first draft, including Dr. M. Audain (Polygon Properties), Professor R. Bellan (University of Manitoba), Professor B. Carroll (McMaster University), Professor F. DesRosiers (Université Laval), Mr. J. Friedlander (Alcan Ltd., retired), Professor M. Goldberg (University of British Columbia), Mr A. Hansen (National Research Council, retired), Professor J. Mercer (University of British Columbia), Professor M. Qadeer (Queen's University), Mr. J. Todd (Ecoanalysis Consulting Services), and Professor W. Michelson (University of Toronto). In addition, CMHC commented extensively on each draft of the manuscript and sought input from various provincial/territorial housing agencies.

When perusing this text, the reader should keep the following in mind:

- the coverage largely includes events and data available up to 1986 (when the first draft of the monograph was completed);

- while the monograph covers housing issues faced by various levels of government in Canada, it emphasizes those which have concerned the federal government;

- the monograph attempts to capture the many faces of Canada and the different housing experiences (for example, heartland *versus* hinterland, metropolitan *versus* non-metropolitan, rural *versus* urban, Aboriginal *versus* non-Aboriginal, north *versus* south, and haves *versus* have-nots);

- the views expressed are those of the authors who, as housing researchers and advocates, generally share a belief that adequate and affordable housing for everyone is an important goal for Canadian society;

- the monograph focuses more on events, policy initiatives, and data sources than on theory, although the explanations given for the housing condition of Canadians provide much that would be valuable in a university-level course on housing; and

- to minimize overlap, relevant details that are in common to several Chapters have been moved to two Appendices which readers may benefit from reading in conjunction with the Chapters.

This monograph is intended to be a comprehensive overview of the housing situation in Canada. In many respects, it is successful. Arguably, this is one of the most important volumes on housing ever produced in Canada. The volume is comprehensive in scope. The chapters are informative, thoughtful, and provocative. The book will stand as the principal record of Canada's post-war housing experience (1945-86) well into the next century.

At the same time, in retrospect, some issues and questions might have received more attention. The book focuses, for example, on the stock of private (that is, non-collective) dwellings. More attention could have been paid to the small, but often disadvantaged, group of Canadians who live in group homes, residential hotels, institutions, and other collective dwellings. The book might also have better distinguished between the stock of housing built to be rented (the conventional rental stock) and the stock of housing that was built to be owner-occupied but is currently rented (the non-conventional rental stock): for example, some condominium units, flat conversions in single-detached structures, and rented houses. The non-conventional rental stock has become more important in the overall supply of rental accommodation in recent years; we need to better understand the factors that motivate suppliers of this stock. Finally, the monograph could have paid more attention to the issue of dwelling maintenance. Repair, replacement, and additions have been a rising share of total new investment in housing in Canada. We need to better understand what determines the level of this investment, and what impact it has on the adequacy and affordability of housing.

Although this book was initially drafted in 1986 and revised over the following three years, the principal messages in this book remain relevant. Competitive and efficient markets continue to be important in meeting the housing needs of many Canadians. Unaffordable housing and poor access to the benefits of home ownership among disadvantaged Canadians, are still key concerns in housing policy today, and likely will remain so over the foreseeable future. Nationally, the incomes of consumers have continued to grow sluggishly at best since 1986 (even before the recession that began in 1990), and net new household formation has remained low at the level of the early 1980s. At the local level, the debate over accessory apartments, in-law suites, and other types of affordable housing continues to rage. And, because interrelationships between housing and the well-being of individuals and families have become more complex over the years, disparate perspectives remain useful in examining housing questions.

At the same time, the authors of this monograph were not omniscient. They did not foresee, or misread the significance of, important social and economic changes that have occurred since 1986. For example, the monograph does not extensively consider the problem of homelessness, a problem of which public awareness has risen in recent years. The authors also did not anticipate the severity and length of the recession that began in 1990 and the effects that it would have on housing consumption and social well-being. Furthermore, the monograph did not anticipate the repeal of the Federal Sales Tax (FST) in January 1991 and its replacement by a broadly-based Goods and Services Tax. Finally, the house price slump that afflicted many Canadian housing markets in 1989 was also unanticipated by the authors.

In addition, the authors did not fully anticipate several important public debates in the past six years on issues that have implications for housing. Among these was the re-emergence of environmental issues. Issues of soil contamination, radon gas, solid waste disposal, water and air pollution, interior air quality, sustainable development, and reduction in energy use, for example, have direct impacts on housing and housing policy. Another set of issues concerns heightened awareness of systemic violence against women and children, and personal security in general. The range of housing concerns that arise from this – from shelters for battered women to neighbourhood safety – receive less attention in this book than might be the case in a more contemporary study. As well, the authors did not foresee the significance of Canada's on-again off-again constitutional discussions. The Meech Lake accord and subsequent events have raised public awareness of issues surrounding the appropriate division of powers among federal and provincial governments and Aboriginal self-government. This book says relatively little about housing in the context of these issues. As well, the monograph was drafted before the Free Trade Agreement was struck between Canada and the United States of America (not to mention the recently-negotiated NAFTA), and hence does not discuss the effects that trade liberalization will have on the housing industry or on consumer prices.

Finally, since the Final Report was written, there have also been noteworthy changes in the kinds of data available to governments and the measures they use to identify housing needs and problems. For example, the federal and provincial governments (using information from the 1988 HIFE survey) switched to a Core Need measure based on the National Occupancy Standard for suitable housing. Another is the development in 1989 by CMHC of an index of the affordability of home ownership for renter households. Still another is the initiation in 1989 of a new annual report by Statistics Canada on repair and replacement expenditures by home owners.

On the policy front, governments did not stand still after 1986. For example, the ACT (Affordability and Choice Today) program – delivered by the Federation of Canadian Municipalities, the Canadian Home Builders Association, and the Canadian Housing and Renewal Association, and sponsored by CMHC – has, in recent years, encouraged improvements to regulation and procedures that affect the production and affordability of housing. The federal government has also:

1987 Broadened the benefits available under the Canada Pension Plan.

1988 Repealed the Home Improvement Loan Guarantee Program, permitted chattel mortgage insurance for non-long term tenure mobile homes, and delegated certain powers under NHA to the provinces and provincial agencies. On a minor note, NHA sections were renumbered (*Revised Statutes of Canada,* 1985); this book uses the old numbering scheme

1989 Increased the amounts available to non-profit social housing and cooperative housing groups to develop proposals for projects; clawed back Family Allowances and Old Age Security from higher-income taxpayers.

1990 Introduced under NHA new portfolio insurance, moveable home loan insurance, varying maturity mortgage-backed securities, and six-month mortgage products; imposed a 5% ceiling on growth in CAP expenditures to Ontario, Alberta, and British Columbia.

1991 Enhanced maternity, parental, and sickness benefits under Unemployment Insurance; increased UI qualifying period and reduced benefit period; reduced UI benefits for workers quitting without just cause.

1992 Eased downpayment requirements for NHA-insured mortgages; stopped funding of new projects under the federal cooperative housing program.

In addition, there has been a growing public debate about the nature, scope, size, and use of lot levies and development charges. Some argue that lot levies are exorbitant and serve only to unduly restrict development; others argue that levies are necessary to avoid punishing tax increases on existent residents.

Finally, the authors say little about the desirability or effectiveness of what in 1986 was a new innovation in housing policy: the social housing agreements struck between the federal government and each of the provinces. At the time that this book was drafted, we had simply not had enough time to evaluate the functioning of these agreements.

The shortcomings are mild, however, when viewed against the strengths of the monograph; the authors are to be commended for having constructed, overall, a milestone overview of progress in housing Canadians during those four decades. That this book is unique in its combination of historical sweep, range of disciplines, and focus on the breadth of issues of concern in housing policy is also due in part to the energy, efforts, comments, and enthusiastic support of CMHC, and especially of the program managers, first Philip Brown then Peter Spurr.

John R. Miron
Toronto
September 1992

CHAPTER ONE

On Progress in Housing Canadians

John R. Miron

HOUSING IS important to Canadians. For most of us, the purchase of a home is our single largest capital expenditure. And, whether owning or renting, shelter costs typically are a large component of the household budget. Housing is such an important component of consumer spending that, worldwide, it is used as an indicator of the standard of living. Also, from lumber, bricks, and nails to bathroom fixtures to carpeting to appliances, each new dwelling built has important effects throughout the economy. These effects flowed, in the 1980s, at the rate of about one new dwelling in Canada every 2.5 minutes. When so much is spent on housing, it is only natural to ask: " Has the money been spent wisely?" Have we spent too much on housing, at the expense of other forms of consumption or investment? Or, alternatively, should we have spent even more?

The Canadian concern with housing also has partly to do with climate. From the time of French explorer Samuel de Champlain's earliest settlements in the early seventeenth century, written records detail the harsh winters and the consequent need for adequate, durable housing. Early structures typically had little insulation, poor foundations, few windows, dirt floors or wood planking, no inside water or toilet facilities, and no central heating. As building materials were bulky to transport, indigenous materials were employed resulting in regional variations in building forms. Throughout much of Canada, however, the problem has been the same: to protect households adequately from long and inhospitable winters. To be poorly housed is to invite discomfort, ill health, the wrath of the elements, the spoiling of one's possessions, or personal injury.

In addition to providing protection from the elements, housing is important in the broader sense of determining an individual's quality of life and the achievement of various social goals. Satisfactory housing can make a vital contribution to equality of opportunity, the redistribution of wealth, and the nurturing of individual dignity and freedom of choice. Housing also fulfils our need for privacy. Home is the place where we usually sleep, prepare and eat food, attend to physical and emotional needs, and engage in family life. It is a place where we can be with our family or friends, a place where we can get away from the rest of society and be free from intrusion or observation. Indeed, housing

John R. Miron

Table 1.1

Comparing the housing conditions of Canadians, 1941 and 1986

	1941	1986
Total population ('ooos)	11,507	25,354
In urban areas	6,252	19,392
In collective dwellings	368	434
Occupied private dwellings ('ooos)	2,573	8,992
Rooms per dwelling	5.3	5.8
Persons per dwelling	4.5	2.8
Persons per room	0.8	0.5
Owned homes (%)	57	62
In urban areas	40	57
Single detached dwellings (%)	71	58
In urban areas	49	49
Dwellings (%)		
In need of major repair	27	7†
Using stove or space heater	61	7†
Using coal, coke, or wood fuel	93	4†
With refrigerator	21	98‡
With piped running water	61	96‡
With inside flush toilet	56	94‡
With installed bath or shower	45	91‡

SOURCE Taken from *Census of Canada 1941 and 1986*. Total population for 1986 includes estimates on incompletely enumerated Indian reserves and Indian settlements.
† Data shown are from the *Census of Canada 1981* and are latest available.
‡ Data shown are from the *Census of Canada 1971* and are latest available.

needs and circumstances figure, directly or indirectly, in most aspects of every-day life.

Setting the Historical Stage

The year 1945 is an appropriate point to begin the modern history of housing in Canada. The end of World War II was a time when Canadians were bracing themselves for an uncertain future as the memory of the hardship and misfortune generated by the Great Depression of the 1930s lingered. Although employment rose during the War, wages remained low and consumer goods were rationed. Fears were expressed of a return to a stagnant economy upon post-war demobilization. At the same time, concern was expressed for Canada's declining birth rate. After having dropped steadily in the 1920s and 1930s, fertility was approaching a level that would lead to absolute population decline, which was

expected to contribute further to economic stagnation by reducing aggregate demand.

THE HOUSING STOCK AROUND 1945

The 1941 Census found that just over half of all Canadians lived in urban areas (Table 1.1). A few of these individuals (368,000) lived in "collective" dwellings such as hospitals, nursing homes, hotels, tourist homes, lodging houses, work camps, staff or student residences, or barracks. The vast majority (11.1 million persons) lived in about 2.6 million "private" dwellings – an average of 4.5 persons per dwelling. About 40% of the private dwellings in urban areas were owner occupied, compared to 75% in rural areas. Just over 70% of all dwellings in Canada were single detached structures, but in urban areas, the figure was just 50%.[1] Private dwellings averaged 5.3 rooms.[2]

Such aggregate figures do not indicate much about the condition or quality of this housing stock. Census enumerators found that 27% of all private dwellings in 1941 were in need of major repair.[3] Urban dwellings, and those located in more prosperous Ontario, Quebec, and British Columbia were generally larger or better equipped than those in the rest of Canada. About 60% of all dwellings relied on stoves or space heaters and 93% used coal, coke, or wood as their heating fuel.[4] Although almost all urban dwellings had electricity, only 20% of rural dwellings were so equipped, and many had only a 25 hertz supply. Only 21% of all private dwellings had a mechanical refrigerator; most of the rest relied on ice boxes. Only 60% of dwellings had piped running water; 56% had an inside flush toilet; and 45% had a private installed bath or shower.[5] It has also been estimated that the average age of a private dwelling in Canada at that time was about thirty years (Firestone 1951, 49).

Many households, especially those with low incomes, were living in housing that was inadequate for their needs or too costly given their incomes.[6] The Curtis Report (Canada 1944, 110-22) estimated that, among the bottom third of metropolitan renter households by income in 1941, 89% paid more than one-fifth of earnings on rent and that 28% lived in quarters with more than one person per room. Among the middle third of renters, the corresponding figures were 51% and 21%. Overall, the Curtis Report (Canada 1944, 12-3) saw a need for 230,000 more urban dwellings, 23,000 more rural non-farm dwellings, 125,000 more rural farm dwellings as of 1946 – almost 15% of 1941 stock – to replace substandard and overcrowded housing in Canada.

Housing in Canada has traditionally been produced largely within the private sector, albeit with significant public regulation, implicit and explicit subsidization, and direct government involvement. Prior to 1945 involvement by the public sector was relatively limited. Nonetheless, by 1945 the principal elements of federal post-war housing policy had already been tried out. The first modern instance of a housing program was the $25 million loan program of 1918 that made mortgage money available for the construction of new-owned homes. The program provided low-interest mortgage loans with small downpayments

and a long amortization period. The target of this policy was the young returning soldier of modest income who needed help to purchase a small home. A similar program was enacted during World War II. In between came the Dominion
Housing Act of 1935. To speed recovery from the Great Depression, this Act provided for cheap and flexible first mortgage loans to buyers of new, moderately-
priced dwellings. The target was the young home buyer with a modest income.
The Canadian Farm Loan Act of 1927 provided similar assistance (for the construction of farm homes) as did the 1938 Full Recovery Low Rental Housing Program (although implementation of the latter was halted because of World War
II). In addition, during World War II the federal government introduced a number of programs designed to accommodate war production workers in low-cost
rental units (for example, the Home Conversion Plan, the Home Extension
Plan, and the Emergency Shelter Program). In focusing on assistance to the
home buyer of modest income and the low-income renter, pre-war and wartime policies foreshadowed the two key target groups of post-war housing policy.

PROGRESS SINCE 1945

In the early 1940s few people foresaw the demographic and economic booms
that were about to sweep Canada. Demographically, a new wave of immigration,
set off by the post-war resettlement of European refugees, marked a dramatic
change from the preceding few decades; prior to that, the last big wave of immigration had been in the first decade of this century. This time, the impact of rising immigration was augmented by a surge in fertility, a drop in infant mortality, and generally increasing life expectancy. The consequence was that Canada's
population more than doubled from 1941 to 1986. Accompanying this growth
was a great shift in composition. With continuing improvement in longevity,
the number of elderly (especially widows) proliferated. So too did the number
of unmarried adults after the late 1960s, partly an effect of the sheer number of
baby boomers and partly the result of an upturn in divorce and a downturn in
nuptiality. For various reasons, more and more individuals came to live outside
the nuclear family unit. Nonetheless, the number of families also grew rapidly in
the 1960s and 1970s, a consequence of the entry of the baby boomers into adulthood and marriage. While a smaller proportion chose to marry, the total number of families actually increased.

Economically, a rapid post-war expansion of employment was made possible
by much new investment. From 1945 to 1985 total employment rose by almost
150% (half again as fast as population), and per capita disposable income went
up by about 200% even after discounting for inflation. Income-support
schemes (such as unemployment insurance, old age security and guaranteed
income supplements, and public and private pension plans) spread this affluence over a wider group of Canadians. Also important was the post-war spread
of subsidized consumption (health care, higher education, and public housing,
for instance) that raised effective incomes, particularly among the poor.

These demographic and economic changes boosted the demand for housing in three important ways. First, in terms of sheer numbers, a growing population requires more housing. Second, the changing composition of Canada's population contributed still more to the demand for housing. Those young singles and families, the divorced, and the widowed all contributed to rapid household formation. Third, a growing affluence that lasted until the late 1970s combined with modest increases in the cost of housing made possible an undoubling of families and the separate accommodation of non-family adults. While Canada's population doubled, the number of households surged from 2.6 million in 1941 to 9.0 million in 1986. New dwelling completions totalled in excess of 6 million units. These figures largely count just the construction of principal residences. In post-war Canada, the stock of seasonal and second homes also surged.

Accompanying this remarkable growth were important geographic shifts. By 1986 Canada's population had become largely urban. Also important was the migration of population from the Atlantic and Prairie provinces and Quebec to Ontario, Alberta, and British Columbia – flows that would have added to housing demand nationally even in the absence of overall population growth.

Another important shift was in type of dwelling. The number of persons living in collective dwellings declined as a percentage of Canada's total population between 1941 and 1986. The net result of population increasing slower than the number of private dwellings is that average household size had declined to 2.8 persons by 1986. In addition, by 1986, 56% of urban dwellings were owner occupied – a rise that reflected growing affluence, the attraction of capital gains and its favourable taxation, and the emergence of new forms of ownership, such as condominiums and co-ownership. The incidence of ownership rose in rural areas as well. The 1960s and early 1970s were also characterized by the rise of the large, privately-owned apartment building in urban areas. As urbanization and metropolitanization proceeded, single detached structures dropped to just 57% of all dwellings in Canada. However, within urban areas the shifts were principally among other types of dwellings (for example, double, row, duplex, and low-rise and high-rise apartment structures); as a percentage of the total urban housing stock, the detached dwelling declined only modestly over this period.

The post-war period also saw improvement in the quality of housing. In 1986 just 7% of all private dwellings were in need of major repair.[7] Few relied on stoves or space heaters or used coal, coke, or wood as their heating fuel. By 1971 rural electrification and a 60 hertz supply were almost universal, as were refrigerators, piped running water, inside flush toilets, and private installed baths and/or showers. However, the stock is now aging. Only 25% of the stock in 1986 had been built in the previous decade, down from 33% in 1981 and 29% in 1971. Furthermore, the average Canadian became better housed in a structural sense. Builders gradually incorporated new materials and technologies in housing construction and renovation. These included plywood sheathing, steel girders, new fire-resistant materials, grounded wiring systems, higher insulation standards, and prefabricated components such as roof trusses, double/triple glazed

John R. Miron

FIGURE 1.1 All dwellings by condition, Native and non-Native
            households: Canada, 1981.

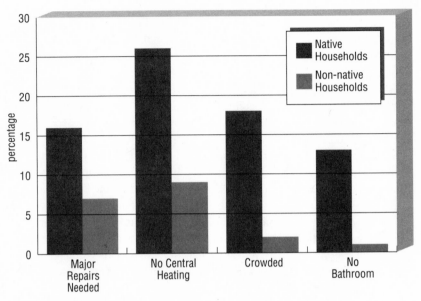

SOURCE: Statistics Canada (1984). *Canada's Native People.* Table 7.

insulated windows, and kitchen cupboard components. At the same time, the
incidence of crowding declined. For example, by 1986 the average number of
persons per room was only 0.5, and fewer than 3% of all private dwellings contained more than one person per room.

Against this backdrop of overall improvement, it is important to keep in
mind that not all Canadians have become equally well housed. Significant regional differences remain. Substandard housing is still generally more common in
rural areas and the less-affluent Atlantic provinces and to a lesser extent the Prairies. Canada's Aboriginal peoples, whether Inuit, Indian, or Métis, also tend to
be housed less adequately (Figure 1.1).

Since 1945 the targets of housing policy have shifted. In broad terms, the
objectives remained the same: to ensure that all Canadians were decently housed
and that this housing was affordable. In the early post-war years, the focus was
on improving the access of moderate-income families to owned homes. Governments then focused on better rental housing for low-income families, the
elderly, students, Aboriginal peoples, and the disabled.

Over the years households with affordability problems remained a primary
target of housing policy at all levels of government. However, it has proven

FIGURE 1.2 Real incomes of individuals and households: Canada,
1941-1986.

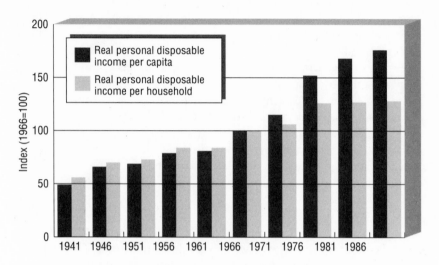

SOURCE: Personal disposable income taken from Statistics Canada. 1988.
         *National Income and Expenditure Accounts: Annual Estimates,
         1926-1986*. Catalogue 13-531. Income data deflated by CPI.
NOTE: Households in 1946 interpolated from 1941 and 1951 figures.

difficult to identify the target. In the case of home owners, the value of housing
as an asset (that is, taking into account capital gains and imputed rents) makes it
difficult simply to measure the cost of shelter. Among both renters and owners,
it is difficult to separate temporal changes in housing costs into quality and price
components.

Post-war changes in house prices, in part, mirrored the buoyancy of the Canadian economy. In the early post-war years, house prices escalated quickly as the
economy boomed. However, by about 1957 the economic boom had died, and
house prices slumped. From the late 1950s through the early 1960s, Canada's
economy was in a recession and the price of housing declined relative to other
consumer goods. Prices picked up again in the economic boom of the mid 1960s
and increased at a frantic pace in the 1970s before slumping again in the recession of the early 1980s. The latter boom was also in part demographically driven
as baby boomers swelled the ranks of young home buyers.

Most of the changes in the housing condition of Canadians were accommodated within private markets for both rental and owner-occupied housing. In
that sense, we can think of them as outcomes of shifts in the demand for, and

supply of, housing. By looking at population growth, we have already begun to consider demand factors. It has been estimated that about two-thirds of post-war household formation was directly attributable to changes in the size and demographic mix of population (Miron 1988, 119). However, the remaining one-third was the result of a greater propensity to live alone, especially among non-family individuals. In part, this was attributable to rising incomes (Figure 1.2), abetted by moderate increases in the price of housing (at least for renters) relative to other consumer prices. Canadians used part of their growing prosperity to purchase control of their living arrangement. Typically, this meant separate accommodation. In 1986, 88% of nuclear families in Canada lived by themselves, up from 79% in 1961. Among non-family persons the corresponding figure was 47%, up from 14% in 1951. The impact of rising real income on household formation in Canada has been large, accounting for about one-ninth to one-sixth of post-war household formation (Miron 1988, Chapter 6).

WHAT WAS TRIED

Also important in the post-war period have been the impacts of social policy initiatives, some of which were not even directly linked to housing. These include new income maintenance programs for the elderly,[8] the unemployed,[9] and low-income families and individuals and subsidy/insurance programs for basic services such as education and medical care. These programs made it possible for low-income individuals and families to devote more income to housing consumption. In subtle ways, income maintenance programs for the elderly also served to undermine the traditional role of the owned home as a nest egg for retirement.

Post-war social policy initiatives also included a concerted effort by governments to reduce the size of Canada's institutionalized population.[10] A public debate about deinstitutionalization began in the 1950s and 1960s. Some argued that inmates would be better housed in the private sector. Others argued that the special services and conditions found in institutional housing could not be effectively provided outside. In many cases, the advocates of private sector housing won out. Whatever the merits of deinstitutionalization, the effect was to push many non-family individuals, often marginal in income or wage-earning potential, into the private housing market.

Also important have been the tax expenditures implicit in the relative treatment of owners and renters under Canadian income tax legislation. Governments in Canada have never taxed the capital gains on sales of principal residences; however, neither have they allowed mortgage interest to be deducted from taxable income as occurs in the United States. Since 1972 capital gains on broad classes of other assets have been taxable. This increased the attractiveness of owning a home relative to owning these other taxable assets. Moreover, governments have never attempted to tax the imputed return on equity in an owner-occupied dwelling, which further increases the attractiveness of owning over renting.

This discussion serves to introduce broad questions about the appropriate instruments of housing policy. Over the post-war period, five main categories of instruments were used:

- tax expenditures (for example, MURBs, other income tax deductions and exemptions, and property tax abatements and credits);
- direct expenditures by all governments for the production and consumption of housing, including direct mortgage lending;
- regulation (for example, municipal zoning and development controls, provincial building codes, rent controls, landlord-tenant and new home warranty legislation, land subdivision restrictions, and federal regulation of financing, building materials, and building standards re NHA-produced housing);
- the sundry activities of crown corporations and government departments (for example, the federal BETT and R-2000 programs to promote energy efficiency and financial support for CCURR, ICURR, and other standard setting, research and development, and coordination activities); and
- loan guarantees (such as provided under NHA insurance provisions).

Over the post-war period, there have been shifts in the relative use of different instruments. What were the best tools and what caused governments to change the tools they use are important questions addressed in subsequent chapters.

Post-war Canadian housing policies fall into two broad categories.[11] One was concerned with eliminating perceived inefficiencies in the housing market. For one reason or another, the private market was seen to be incapable of producing an adequate supply of housing, and policies were implemented to regulate, assist, or encourage suppliers to compensate for the inefficiency. A second category of policy was concerned with issues of equity and social justice in housing. For various reasons, it was believed that the private market, even if operating efficiently, was incapable of providing adequate housing at an affordable cost for every Canadian.

Policies to improve market efficiency

In part, federal post-war housing policy concerned itself with the supply of mortgage funding. It was felt that potential lenders might be discouraged by the riskiness, lumpiness, or illiquidity of mortgages. Around 1945, there were few large lending institutions,[12] no secondary market for mortgages, and no private mortgage insurance companies. Small lenders were thought to be reluctant to enter the market and to be overly cautious when they did. This was seen to be injurious both to the production of an adequate amount of housing in particular and to overall economic growth in general.[13] After 1945 the federal government began to build on approaches used in the 1918 Loan Program, the 1935

Dominion Housing Act, and the 1938 NHA. Up until 1954 this approach involved making joint loans. Later, joint loans were replaced by a self-funding mortgage insurance scheme for modestly-priced housing. Since the mid 1980s the federal government has further expanded the supply of mortgage funds by introducing mortgage-backed securities that make investment mortgages more liquid. Grants were also given to help owners make their housing operate more efficiently. Included here over the years have been the Housing Improvement Loans (HIL) Program, the Canadian Home Insulation Program (CHIP), and the Canada Oil Substitution Program (COSP).

While the Canadian home-building industry has seen the emergence of a few large companies, it has been made up mainly of small firms. At the federal and provincial levels, there have been concerns that these firms are simply too small or too fragmented to encourage the kind of long-term research and development that is needed to spur innovation and new efficiencies. At the outset of the post-war period, the National Research Council and CMHC were mandated to develop appropriate standards for new construction and to explore new technologies and approaches. [14] By the 1970s part of this activity had been taken up by provincial housing agencies. [15]

This raises another concern of post-war housing policy. Housing construction is a complex technology. When consumers purchase housing, whether as owners or renters, they typically cannot be expected to know if it is well built. In a private market, they have to rely for the most part on the integrity of the builder and/or landlord. In economic terms, this imperfect information creates a risk that reduces the efficiency of the market, not unlike the risk-taking problems of small lenders. Post-war housing policy, in part, was an attempt to reduce this uncertainty. The spread of building code legislation throughout Canada assured the occupants of new housing that their housing was soundly built. In a similar vein, one can point to the spread of new home warranty legislation for owners and property maintenance legislation for renters. More generally, one can also cite the spread of regulation at all levels of government in regard to land subdivision, zoning, and development that helps assure occupants or neighbours that certain standards of good planning would be honoured.

A related policy issue concerns the income tax treatment of losses from rental property. Before 1972 a landlord could charge losses in the operation of rental property against other income. This included rental losses arising because of depreciation. After 1972 most landlords could not claim losses against other income if they were created by depreciation. Arguably, the Income Tax Act of 1972 made rental housing less attractive to small investors. In 1974 the federal government tried to compensate for this by introducing MURBs, under which small landlords could fully claim depreciation costs, a program that was terminated in the early 1980s.

Policies to promote equity and social justice
Canadian post-war housing policy has also been extensively concerned with an equitable allocation of housing. Much of this concern has been in the form of assistance for low-rental housing for needy Canadians. The 1938 NHA enabled subsidized joint mortgage loans for the construction of public housing by local housing agencies. The 1944 NHA amendments introduced the possibility of "limited dividend" housing, urban renewal programs, and rents geared to income. However, public housing did not get going in a major way until the 1949 amendment that enabled the Federal-Provincial Public Housing Program. In 1969 rent supplements were introduced to subsidize low-income households living in private rental accommodation. Both federal and provincial governments offered further subsidies to encourage new construction of private rental accommodation in the 1970s and 1980s.

In addition, several policies provided subsidies to low-income home owners. At the federal level, these included RRAP, started in 1973, which provided small, forgivable loans for upgrading substandard housing, the CMRP and MRPP (1981-3) that assisted home owners with mortgage renewal, and the Rural and Native Housing (RNH) program. In addition, the federal government has experimented with encouraging modest-income families to switch from renting to owning. At the provincial level, subsidies were introduced including property tax deferrals, abatements, and credits for the elderly.

In the 1970s there was a rising concern with consumer sovereignty and protection. In part, the concern was with improving the efficiency of the market by reducing imperfect information. In other respects, however, the concern was ultimately with the notion of fairness or social justice. In the housing area, this was perhaps best manifested in the emergence of security of tenure legislation at the provincial level. This legislation restricted the rights of landlords to specify onerous lease provisions, to evict tenants, and (in some provinces) to determine rents. The new home warranty legislation introduced at about the same time can also be seen partly in a similar vein. Still another example might be the federal mortgage rate protection plan under which borrowers are protected against large increases in interest rate upon mortgage renewal.

WHAT ABOUT THE FUTURE?
Thus, the period since 1945 represents an abrupt change from what had preceded it. At the same time, as we now look to the future, we may wonder if the 1980s mark the end of one era and the start of another. Rapid population growth is now history. Canadians now generally look ahead to a period of slow growth over the next few decades. Starting in the late 1970s, the growth in real incomes has also been sluggish at times. Of course, we should keep in mind the contrast between the dismal projections around 1945 and what actually happened. Nonetheless, the period between 1945 and 1985 forms an interesting unit of analysis. How portable are lessons learned during that period when we turn our attention to the future? How different will the future be, and what does this

imply for housing? Will housing policies that were successful from 1945 to 1985 also work in the future? What other policies should we be considering?

The Approach to Housing Progress

This is a book about housing progress, a phrase that admits of two meanings. One notion is simply of motion, direction, or change. The other is of movement or change for the better. In this book, the latter meaning is employed. However, improvement is in the eye of the beholder. One person might enjoy the spaciousness of 3.5 metre ceilings in a Victorian townhouse, while another dislikes the narrow windows. The scheme that we use to weight attributes of housing is part of our "perspective."

Over the post-war period, housing programs were used to promote social goals such as the redistribution of income, equality of opportunity, or social justice. In other words, Canadians as a whole were seen to benefit from an allocation of housing that promoted such goals. Of course, the extent of such benefits depend on the importance attached by a citizen to each goal. This can, and does, vary from one person to the next. One person might think that equality of opportunity is important, while another might not. Such differences in opinion about social goals are another element of one's perspective. The quality, quantity, location, and cost of our housing shapes our state of health and well-being, our sense of place and community, our self-esteem, and our access to public facilities, services, educational and job opportunities. It is an integral part of our standard of living and our view of how society should operate. In a pluralistic society, it is not surprising, therefore, to find differences in perspective.

Most of us have views on the extent to which the housing of Canadians has or has not progressed since 1945 and the impacts of government policies. It is always a surprise to the uninitiated to discover that someone else, equally informed, holds an opposite view. Such disagreements usually originate in one of two places: a disagreement about facts or a difference in perspective.

This book has been written by a group of authors – twenty-two in all. Each is expert in some aspect of housing. The group includes planners, architects, economists, geographers, and sociologists. They come from a variety of backgrounds: universities, planning practice, and housing consultancy. They come from various regions of Canada and from both large cities and smaller towns. Also, each of them is experienced in housing research, policy, and/or program delivery. As a consequence, they do not share a common viewpoint. Important differences among them are evidenced in subsequent chapters.

This book does not impose a single perspective on its authors. The authors present their viewpoints and show how their conclusions depend on them. The aim is to clarify the debate and to make the reader more aware of different views on housing progress and the impacts and desirability of particular housing policies. At the same time, the book also outlines what we understand about the demand, supply, and allocation of housing. It does more than simply

mirror the debate. It looks for the common ground – in terms of facts and perspective – and the bases for differences of viewpoint.

Structure of the Book

This book is divided into six sections. The first section looks at economic, demographic, and institutional factors underlying the post-war demand for housing. In Chapter 2, John R. Miron examines post-war patterns of household formation and housing consumption and estimates the importance of contributing demographic and economic factors. Marion Steele describes post-war changes in tenure and their causes, the links between economic well-being and tenure choice, and the rationale for government support of home ownership in Chapter 3. J. David Hulchanski traces the evolving legal basis of fee simple ownership and private renting in Chapter 4.

The second section discusses the principal aspects of the supply side of housing: housing finance, economics, technology, and regulation. In Chapter 5, George Fallis identifies the suppliers of housing and factors influencing their decisions. James V. Poapst examines post-war government regulation of mortgage markets, institutional participation, sources of demand for residential mortgage debt, use of equity financing and provisions of loan instruments in Chapter 6. John Bossons shows the range, scope, and rationale of regulation in the production of housing in Chapter 7. James McKellar, in Chapter 8, assesses the impact of building technology and the organization of the production process on housing form, cost, and quality.

The third section traces the implications of shifting demand and supply curves for housing stock growth and quality. A. Skaburskis considers in Chapter 9 how the components of housing stock change might be measured, and discusses the role of government regulation and stimulative programs on housing stock change. In Chapter 10, Skaburskis and E.G. Moore establish an accounting scheme for describing post-war transitions in the existing housing stock, identify known pressure points for policy intervention, and identify interactions between housing stock transitions and public policy that require more research. Joan Simon and Deryck W. Holdsworth in Chapter 11 document post-war changes in the design of housing forms and consider the emerging housing needs of families and changing concepts of neighbourhood.

The fourth section considers how changes in supply have matched shifts in demand. In Chapter 12, Lynn Hannley reviews indicators of substandard housing, documents the magnitude of the problem prior to 1945, explores post-war policy approaches, and suggests substandard housing indicators and policy approaches for the future. Janet McClain in Chapter 13 describes and discusses post-war changes in the scope of the need for supportive housing and related services, elaborates on the nature of consumer demand and the role of housing policy, and reviews the location and supply of special needs housing by type, level of care, funding, and sponsor. Damaris Rose and Martin Wexler examine

the adequacy of the post-war housing stock in light of important social and economic changes in Chapter 14. Patricia A. Streich examines the concept of housing affordability and the extent to which the magnitude of the affordability problem is conditioned by its definition in Chapter 15.

The fifth section examines housing stock change in its locational setting. In Chapter 16, L.S. Bourne demonstrates how the settlement environment of housing changed in post-war Canada and examines how the community development process contributed to this change and to housing progress. Francine Dansereau documents in Chapter 17 the evolution of intraurban disparities in socio-economic, ethnic, and physical terms, and considers implications for neighbourhood quality and housing progress. Richard Harris examines social mix as a goal of post-war community development and housing policy in Chapter 18. In Chapter 19, Jeffrey Patterson considers the relationships between housing and community development, focusing on the origins of the modern urban reform movement and the post-war shift in policy from urban renewal to neighbourhood improvement. John H. Bradbury considers housing needs and policies in single-enterprise communities in Chapter 20.

The chapters of the sixth section suggest what lessons might be learned from the post-war experience, what the next decade or two might hold for us, and the challenges and issues posed by these scenarios. Miron suggests important ideas about meeting future needs based on past experience in Chapter 21. John Hitchcock discusses future directions and challenges for consumers, industry, and governments in Chapter 22.

At the end of this book are two appendices: a Glossary and a Key Event Chronology. The Glossary will be of assistance to readers who are unfamiliar with proper names and acronyms that are used throughout the monograph. The Key Event Chronology gives a historical listing of major events in Canada's post-war housing progress.

Notes

1 Averages can mislead. In Montreal, single detached structures made up less than 7% of all dwellings in 1941, compared to 37% in Toronto and 75% in Vancouver.

2 In counting rooms in private dwellings, censuses exclude halls, bathrooms, closets, pantries and alcoves, attics and basements unless finished off for living purposes, and sunrooms and verandahs unless suitably enclosed for occupancy during all seasons.

3 In the Census, a dwelling was defined to be in need of major repair if it had a sagging or rotting foundation, a faulty roof or chimney, an unsafe outside steps or stairways, or an interior badly in need of repair (for example, large chunks of plaster missing from walls or ceiling).

4 In large cities, reliance on stove heating was minimal except in Montreal (62% of households) and Quebec City (64%). In Alberta, which has extensive gas and oil reserves, the use of natural gas as a fuel was greater.

5 Such amenities were more commonplace in urban dwellings. Among large cities, only in Edmonton did less than 80% of dwellings have flush toilets. Only in Edmonton and Quebec City did less than 75% of dwellings have an installed bath or shower.

6 Carver (1948, 74) reports that Toronto households with annual incomes below $1,000 spent on average about 40% of their income on shelter in 1941. Among households with incomes of $1,500 to $2,000, the figure was just 21%.

7 In the 1981 Census, respondents were asked to indicate whether their dwelling was in need of repair, excluding desirable remodelling or additions. Possible responses included "needs only regular maintenance," "needs minor repairs," and "needs major repairs." Respondents were advised that major repairs included defective plumbing or electrical wiring, structural repairs to walls, floor, or ceilings.

8 Old Age Security, Guaranteed Income Supplement, and Canada/Quebec Pension Plans.

9 Including Unemployment Insurance, work programs, a variety of provincial, local and third sector welfare programs.

10 Dear and Wolch (1987) further describe the deinstitutionalization process and the contemporary problem of homelessness that it engendered.

11 Traditionally, federal housing policy in Canada has served a third purpose: to spur economic growth or recovery from a recession. The construction of new housing creates jobs in the homebuilding industry directly, in addition to having positive effects on employment in the household furnishings industry and other related industries. Examples of policies that were primarily concerned with job creation are the Winter House-building Incentive Program of 1963-5, the Canada Rental Supply Program (CRSP) of 1981-4, and the Canada Home Renovation Plan (CHRP) and Canadian Home Ownership Stimulation Plan (CHOSP) of 1982-3.

12 Prior to 1954, chartered banks were prohibited from originating residential mortgages. Even after 1954, they were restricted to NHA-insured mortgages. Not until 1967 were chartered banks were allowed to originate conventional residential mortgages.

13 CMHC has always had a small residual lending role where private lenders were unprepared to cooperate. In 1957, when market interest rates rose above the ceiling set for NHA mortgages, the federal government briefly flirted with widespread direct lending. In all, 17,000 dwellings were financed in 1957 and 27,000 each in 1958 and 1959.

14 Of special note here is the model National Building Code developed by NRC, a code that has been widely adopted by the provinces as a standard for new construction.

15 The first contemporary use of minimum standards in housing may have been by a local government agency – the Spruce Court Housing Project of the Toronto Housing Company in 1914.

==

CHAPTER TWO

Demographic and Economic Factors in Housing Demand

John R. Miron

HOW DID post-war housing market outcomes differ from pre-war experience? One difference was in terms of living arrangements. In 1941, 11.5 million people in Canada were housed in 2.6 million "usual" residences (that is, households or, equivalently, occupied private dwellings), excluding seasonal and secondary homes. By 1986 the population had burgeoned to 25.2 million, but the number of usual residences surged even faster to 9.0 million units. The average number of persons in a usual residence fell by one-third (from 4.5 to 2.8 persons) between 1941 and 1986, and the proportion of households consisting of one person living alone trebled. When secondary and seasonal dwellings are added in, the growth of the housing stock is even more remarkable. [1]

A second difference was the surge in housing expenditures. Canadians spent $67.5 billion for rent (including imputed rents on owner-occupied housing), fuel, and power in 1986 – up from just $1.2 billion in 1946 (Table 2.1). Even after allowing for inflation, total real expenditure increased sevenfold, and per capita real expenditure threefold. At least through the 1960s, housing expenditure had just kept pace with the growth of income. As a percentage of total consumer expenditures, housing costs began to rise in the 1970s and 1980s.

How housing is produced in Canada, its price, its allocation among consumers, and its consumption are largely outcomes of the marketplace. Did these changed post-war outcomes simply reflect the changing preferences or incomes of consumers? Or, were the choices of consumers altered, broadened, or narrowed by supply factors?

Concepts, Definitions, and Data

The term "living arrangement" is not easily defined or measured. To minimize omissions and double-counting, census-takers survey all known places of residence and enumerate anyone usually resident there, including persons temporarily absent. Since everyone is assigned to their usual place of residence, census-takers tend to ignore dwellings at which no one is usually resident, for example, seasonal and vacant dwellings. In so doing, the amount of housing consumption and the size of the stock are underestimated.

Table 2.1

Gross domestic product and selected expenditure components:
Canada, 1946-1986
(billions of current dollars; unadjusted for inflation)

	Gross domestic product	Total gross rent, fuel, and power	Gross imputed rent	Gross rent paid
1946	12.2	1.2	0.5	0.3
1956	32.9	3.4	1.7	0.8
1966	64.3	6.8	3.7	1.7
1976	197.9	20.9	11.8	5.0
1986	509.9	67.5	40.1	15.4

SOURCE Statistics Canada (1988).

Another problem with census data lies in the definition of a household. In Census terms, a household is the individual or group of individuals living in a dwelling. In turn, a dwelling is a structurally separate set of living quarters with a private entrance. Being "structurally separate" means that the occupants of a dwelling do not have to pass through the living quarters of others to get to their own dwelling. With whom one shares a "usual place of residence" depends on how the dwelling is defined. Since 1945 many persons have switched from lodging within a larger household to living on one's own. Some switched from sharing a house with others to living alone in an apartment. In part, households used their growing affluence to separate themselves from lodging tenants and relatives. Basement and upstairs flats were walled off, private entrances constructed, and separate kitchens and washrooms added, thereby creating two or more households where previously there had been only one. Sometimes, such changes have important implications for living arrangements; at other times, the carpentry and its impacts on daily life may be minor.

In a similar way, looking at expenditures on rent, fuel, and power as an indicator of housing consumption can also be misleading. Economists like to think that a dwelling provides "services" that are consumed by the residents. A larger, better equipped, or higher quality dwelling is seen to provide more services, and hence greater consumption. The problem here is to clarify what constitutes housing consumption, how it is to be measured, and how its price is to be calculated.

Post-war Changes in Living Arrangements

In the early post-war years, the trend was to younger and more prevalent marriage, earlier first child births, and larger completed family size. The 1960s and 1970s marked a shift to later marriage, more bachelorhood, more divorces, fewer children, and postponed child birth. Changing family formation, in turn, translated into new patterns of living arrangements that shaped, and were in turn shaped by, the kinds of housing being built.

All of this occurred against a backdrop of sustained population growth. The pace was especially quick up to about 1961. Growth continued thereafter but at a reduced rate. Overall, changes in fertility, survivorship, and migration contributed to a doubling of Canada's population from 1945 to 1985.

- *Fertility*: From the onset of modern vital statistics record-keeping in the 1920s through 1939, the number of births hovered near 230,000 to 250,000 annually. After the start of World War II, the level of births rose steadily. Continuing to rise after the War, it reached a plateau in the late 1950s at just under 480,000 annually. The Period Total Fertility Rate (PTFR) rose from 2,654 in 1939 to a peak of 3,935 in 1959. The period roughly from 1946 to the early 1960s is commonly called the post-war "baby boom." By 1968, however, the birth rate had fallen to just 364,000. It continued to fall, stabilizing in the 1970s at about 340,000 to 360,000 births annually. By 1980 the PTFR had fallen to 1,746, well below replacement level. The period since the early 1960s has been dubbed the "baby bust."

- *Mortality*: The post-war period also saw increased longevity. From 1945 to 1980-2, life expectancy at birth increased by 7.2 years to 71.9 years for men and by 11.0 years to 79.0 years for women.[2] This increasing longevity itself was a source of population growth. The longer people live, the more likely they are to complete their child bearing years and to be alive still when their grandchildren or subsequent generations are born. Improved longevity could account for up to one tenth of Canadian post-war population growth (Miron 1988, Chapter 3). Improved longevity came about principally via two sources: reduced infant mortality and reduced mortality among the middle aged and elderly.[3]

- *Immigration and Migration*: The 1950s saw much immigration associated with European resettlement. A second wave, principally from Asia and the Caribbean, began about 1965 and subsided in 1974 with the introduction of more restrictive immigration controls. Since 1960 the annual volume has varied from 70,000 to 214,000. Because of this ebb and flow, the importance of immigration in Canada's population growth has varied. Immigration was an especially important source of growth for Canada's metropolitan areas. Internal migration was also important in this regard. Overall, migrants streamed out of rural areas and smaller towns into the major conurbations. In 1986, 31% of Canadians lived in the three largest metropolitan areas: up from just 19% in 1941. With the exception of the late 1970s energy boom that saw much migration into Alberta, the typical post-war patterns of migration were from the Atlantic provinces, Quebec, Saskatchewan, and Manitoba, into Ontario and British Columbia.

Taken together, these demographic forces created sustained population growth that accounts for part of the rapid post-war growth in number of households. At the same time, other factors also contributed to a reshaping of the size and composition of households.

One factor was the marriage rush from 1945 to about 1960, during which adults became more likely to ever marry and to marry young. Median age at first marriage dropped from 23 years for women in the pre-war period to just 21 during the marriage rush. In the subsequent marriage bust, the incidence of divorce and of bachelorhood rose, and the median age at first marriage for women, increased by about 0.5 years. A surge in the number of individuals living in families characterized the baby boom, fuelled both by a marriage boom and a high birth rate. In contrast, the post-1960 period saw relatively fewer children, fewer married, more single, separated or divorced adults, and more lone parents (Table 2.2). The number of persons living in families increased by 60% between 1941 and 1961, but by only 25% in the next two decades. In spite of the declining propensity to marry, the total number of persons living with a spouse continued to rise as the baby boom cohorts reached adulthood. At the same time, the number of lone mothers under 35 more than quadrupled from 1961 to 1986.

Also important were shifts in the number and spacing of children. Throughout the post-war period, childbirth among women over 35 years of age became less common, as did the incidence of fourth or higher-order births. During the baby boom, other patterns developed. Women tended to have their first and second child at younger ages; they became more likely to have a third child; and fewer women remained childless. These patterns were reversed in the subsequent baby bust. Overall, the baby bust period was characterized by fewer, more-closely spaced births. In the baby bust, couples spent more years together before the first birth, and again later after the children had left home. In earlier decades, when births were spread out over a longer period, couples spent more of their lives with at least one child at home.[4] Census counts of families by size reflect fertility and spacing decisions; they also reflect home-leaving among young adults. During the 1960s and 1970s, children became more likely to leave the parental home early in adulthood, a trend that was reversed in the 1980s.

Finally, another important factor was the growing gender differential in mortality. Women have traditionally outlived men in Canada. In conjunction with the marriage boom, increased longevity relative to their spouses meant that it became more common for women to experience widowhood at some point in their lives, and for greater lengths of time. Often, widowhood translated into living alone.

Coinciding with these shifts in family composition and size were important changes in living arrangement. For much of Canada's modern history, families have commonly maintained their own dwellings. What was surprising was the extent to which it became virtually universal. While the doubling-up of nuclear families (as in extended family households) was never commonplace, it became even rarer after 1945. Furthermore, it became even more common for

Table 2.2
Population and families by living arrangement:
Canada, 1941-1986

	1941	1951	1961	1971	1981	1986
	(thousands of persons)					
Total population						
(usual residents) [1]	11,490	13,984	18,238	21,568	24,203	25,207
In private dwellings,						
Family members						
Living with spouse [2]	4,432	5,923	7,600	9,184	11,222	11,763
Lone parent	309	326	347	479	714	854
Child	5,144	5,967	8,149	9,189	8,667	8,579
Non-family individuals [3]	1,237	1,384	1,659	2,323	3,195	3,578
In collective dwellings	368	384	484	393	406	434
	(thousands of families)					
All families [4]	2,525	3,287	4,147	5,076	6,325	6,735
Maintaining a dwelling [5]	2,333	2,967	3,912	4,915	6,133	6,534
Living alone	–	–	3,263	4,286	5,556	5,939
Others present	–	–	649	629	577	596
Not maintaining a dwelling [6]	192	321	235	161	192	201

SOURCE *Census of Canada*, various years. – Indicates data not available.
[1] Columns may not total due to rounding.
[2] Since 1981 common-law couples have been enumerated as marrieds. In earlier censuses, where such couples chose not to list themselves as married, they were counted as either non-family individuals (if no children present) or lone parent families (if children present). Thus, censuses since 1981 estimate more husband-wife families and fewer lone parents and non-family individuals than would have been the case previously.
[3] Includes individuals whose family status could not be ascertained.
[4] Census counts have excluded families in collective dwellings since 1981. Families in collective dwellings are included in earlier counts.
[5] Under the 1941 census definition of a household as a housekeeping unit, there could be two or more households per dwelling. Compared with the subsequently-used definition that assigns only one household per dwelling, the number of primary families was overstated, and the number of secondary families understated, in 1941.
[6] Since the 1981 census, maintainer status depended on whether the "household maintainer," (that is, the person chiefly responsible for financial maintenance of the dwelling) was a resident family member. In some cases, such as a family living alone that is financially supported from outside, there was no maintaining family. Prior to 1981 a family living alone was always enumerated as a maintainer. Thus, censuses since 1981 tend to overcount maintaining families relative to earlier censuses.

Table 2.3
Non-family individuals by age group: Canada, 1961 to 1986
('000s of persons)

	1961	1971	1981	1986
Under 35	673	937	1,429	1,501
35-44	205	220	266	399
45-54	240	267	264	294
55 or older	883	1,151	1,236	1,385
Total	2,002	2,575	3,195	3,578

SOURCE *Census of Canada,* various years. Columns may not total due to rounding. 1981 and 1986 data include only non-family individuals in private dwellings. Earlier data also includes non-family persons in collective dwellings.

nuclear families to live alone, that is, without any other persons present in the dwelling. This stripping away of non-family individuals constituted a second form of undoubling. The data presented at the bottom of Table 2.2 indicate that both forms of undoubling were important, although at different points in time. The percentage of families maintaining their own dwellings rose from 90% in 1951 to about 96% by the mid 1960s, thereafter remaining roughly constant. In 1986, 88% of all families lived alone, up from under 80% in 1961.

The growth of the one-person household has been a post-war phenomenon. As recently as 1951, individuals living alone were rare enough for the Census to make the following comment:

> The highest percentages of one-person households were found in rural nonfarm areas, and, as in the 1941 Census, one-person households were much more common in the provinces west of the Great Lakes than elsewhere in Canada. Judging from the geographical distribution of these households, it is probable that a fair percentage of them consisted of hunters, trappers, west coast fishermen, fire rangers, guides, and persons in similar occupations (*Census of Canada 1951*, 10: 368).

In the ensuing decades, living alone became predominantly an urban phenomenon, and it became commonplace. Where did these households come from? In part, they reflect the swelling number of non-family persons (Table 2.3). Overall, the number of non-family persons increased by almost 80 percent, and the number under 35 years of age more than doubled from 1961 to 1986. This reflected rising incidences of bachelorhood and divorce. Also, although the longevity gap between the sexes had begun to attenuate in the late 1970s, the post-war period, on the whole, was marked by a growing number of elderly widows. At the same time, the percentage of non-family persons who live alone rose sharply. As a result, the number of one-person households more than quadrupled

Table 2.4
Persons living alone by age group: Canada, 1961 to 1986
('ooos of persons)

	1961	1971	1981	1986
Under 25	17	70	201	154
25-34	40	96	347	395
35-44	45	73	159	241
45-54	64	99	162	185
55-64	86	154	247	280
65 or older	172	320	566	680
Total	425	811	1,681	1,935

SOURCE *Census of Canada,* various years. Columns may not total due to rounding.

overall and increased almost tenfold among the under 35 year-olds between 1961 and 1986 (Table 2.4).

Post-War Changes in Housing Consumption
Because every household occupies exactly one usual dwelling, the above description of changing living arrangements also largely describes the aggregate increase in number of housing units. However, this does not tell the entire story of post-war growth in the housing stock. The stock changed significantly in a qualitative sense – in terms of dwelling type, quality of construction, design, amenities, tenure, and state of repair. Also, although a lack of data prevents further examination of the issue, the consumption of secondary housing apparently has increased.

Chapter 1 identifies important changes. Since 1945 the amount of residential space per household and per capita has surged; while the number of rooms in a typical dwelling rose slightly, the number of persons per room fell sharply. The incidence of dwellings in need of major repair or without toilet facilities, central heating, or a refrigerator also dropped. In addition, the type of housing consumed by Canadians changed after 1945. As a percentage of the total private occupied housing stock, single detached units declined in importance after 1941. Particularly noteworthy was the great boom in apartment construction from the mid 1960s to mid 1970s and the ensuing boom in row housing (Figure 9.2).

In terms of the dollar value of housing consumption, post-war change is even more dramatic. The aggregate current dollar value of Canada's housing stock per household rose more than eighteenfold (Table 2.5). Even after adjusting for inflation in consumer prices generally, it rose fourfold. In current dollars, the value of the stock rose much faster than personal disposable income per household, especially during the late 1940s. In part, this reflected various post-war housing price booms. In part, it also reflects the volatility of new residential

Table 2.5
Current dollar value of housing stock, investment in housing,
housing consumption, and annual personal disposable income per household:
Canada, 1946-1986
($ per household)

	Housing stock	Residential investment	Personal disposable income	Housing consumption
1946	2,517	138	4,309	414
1951	4,412	235	4,585	441
1956	5,179	462	5,513	846
1961	6,022	391	5,950	1,043
1966	8,346	500	7,906	1,269
1971	11,930	928	10,132	1,823
1976	22,607	1,982	17,895	2,917
1981	34,291	2,487	28,699	4,902
1986	46,567	3,428	37,601	7,153

SOURCES Up to 1961, value of the housing stock is estimated from Statistics Canada (1984e, 284). After 1961, housing stock is estimated from total non-financial assets in residential structures. See Statistics Canada (1986a). The current dollar value of housing consumption (imputed rents, gross rents paid, fuel, and electricity), residential investment (gross fixed capital formation), and personal disposable income is taken from Statistics Canada (1988).

construction. During the 1970s, for example, residential investment outstripped the growth of income; in the recession of the early 1980s, investment tumbled, even though incomes rose nominally.

Why Did Living Arrangements Change?

Why did average household size decline so abruptly? The overall growth of population in Canada and important coincident demographic shifts have already been noted. The marriage bust was one. The decline of marriage and the rise of divorce meant more non-family individuals and lone parents. The closer spacing of child births (later first births and earlier last births), especially in combination with earlier home leaving among young adults and generally increasing longevity, increased the time that parents spent without children at home. Also important was the increasing gender differential in longevity that sharply increased the number of elderly widows.

Such arguments help explain why there were more non-family individuals and why families were typically smaller. Given a fixed propensity for families and non-family persons to live alone, in conjunction with the baby boom and with net immigration, these demographic changes might account for about

FIGURE 2.1   Real income of individuals and economic families:
Canada,1951-1986.

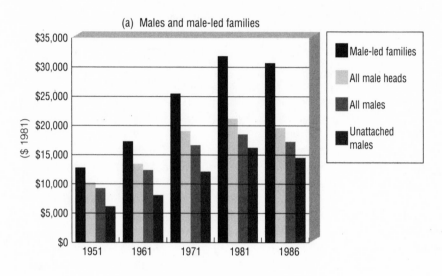

(a) Males and male-led families

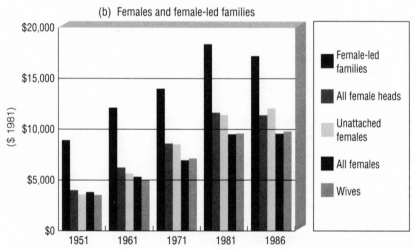

(b) Females and female-led families

SOURCE: Statistics Canada (1969). *Incomes of Families and Individuals in
Canada, Selected Years, 1951-65.* Statistics Canada (1973,1983, 1987).
Income Distributions by Size in Canada.

NOTE: Individuals without income are excluded. Family here includes any two
or more individuals related by blood, marriage, or adoption; all other
persons are "unattached". Where present, husband is head of family.
1951 and 1961 data exclude farm spending units. Incomes deflated by
CPI (1981=1.0). Data on " Wives" are for wives of spending unit heads.

two-thirds of net post-war household formation (Miron 1988, Chapter 5). They do not, however, explain the sharp rise in the propensity to live alone among both families and non-family individuals.

One possible explanation has to do with the subsiding of immigration into Canada. Immigrants from abroad have been considerably more likely to live in shared accommodation. The post-war decline in shared accommodation might partly be attributable to post-war changes in the importance or composition of immigration. What caused immigrants to share their housing? Was it a cultural norm carried from their homelands, or a rational strategy for coping with housing in a new, different, and costly housing market? If the latter, another explanation is suggested.

That explanation is post-war affluence. The average incomes of Canadians more than doubled (after discounting for inflation) from 1945 to 1981. In part, the affluence was augmented by the rise of the two-earner household that pushed up family incomes. In part, it reflected a relative improvement in the incomes of non-family individuals. (See Figure 2.1 which also show the effects of the recession of 1982 and the deteriorating real incomes of Canadians through the first half of the 1980s.) Important too were the redistributive effects of new, post-war income transfers and other social programs that were of particular benefit to low-income individuals and households. At the federal level, Unemployment Insurance began in 1941, Family Allowances in 1945, Old Age Security in 1952, and the Canada Pension Plan and Guaranteed Income Supplement in 1966. Government transfer payments to individuals rose from $1.1 billion (or 11% of total personal income) in 1946 to $62.0 billion (or 14% of total personal income) in 1986 (Statistics Canada 1988, 16-7).

Canadians used this growing affluence to purchase better clothing, food, health care, transportation, education, and household appliances. They also improved and upgraded their housing. In addition to purchasing dwellings that were larger, had more bathrooms, or were better appointed, they purchased a better quality of living arrangement. As noted earlier, this did not necessarily mean jettisoning those with whom one might otherwise have shared space; by walling them off and providing separate entrances, cooking and toilet facilities, these sharers may have been "externalized" into separate households and dwellings. Whatever the mechanism, rising affluence may have reduced the propensity for families and non-family individuals to share living quarters.

At the same time, the cost of shelter may have increased less rapidly than did other consumer prices. Since the mid 1950s rental housing has lagged behind other components of the overall Consumer Price Index (see Table 3.4). The home ownership price component increased more rapidly but, as it does not consider the capital gains benefit, it overstates the increase in user cost of owned homes over the post-war period. If housing became more affordable relative to other consumer goods, consumers may have substituted consumption of other, more expensive, goods in favour of now relatively cheaper housing.

Rising real incomes reshaped living arrangements. Evidence from cross section studies suggests that the propensity for non-family individuals to maintain a dwelling increases with income.[5] However, cross section studies using 1970s and 1980s data suggest that the propensity for families to maintain a dwelling was less sensitive to income. Put differently, high-income families were only marginally more likely to maintain a dwelling (or live alone) than were their poorer counterparts. This suggests a novel interpretation. The contemporary division of family and non-family individuals into separate households may have been instigated not by families seeking privacy but by non-family individuals purchasing their way out of shared accommodation.

This may not have always been true. In the 1970s and 1980s, "commercial" lodging was uncommon.[6] Where the decision to share space is an affair of the heart or of familial responsibility, it is not surprising that the family's income is unimportant. In earlier times, however, the taking in of unrelated lodgers may have been more widely used to supplement family incomes.

To understand better the decline of shared accommodation, it is helpful to think in terms of the costs and benefits of the arrangement to the person or family maintaining the dwelling (that is, the host) and to the lodger or partner.

THE HOST AND THE SUPPLY OF SHARED ACCOMMODATION

One precondition for a family or individual to share is sufficient space within the dwelling, usually at least a spare bedroom. Another is that the host be able to provide housekeeping as needed. The level of housekeeping can vary from a full room and board service where the host provides cooking, room cleaning, and laundry cleaning to the case where a single seeks a room-mate to share costs and housekeeping.

Between 1941 and 1961 the typical size (number of rooms) of an owned home remained constant. At the same time, families became larger with the onset of the baby boom. The resulting space squeeze made sharing less feasible for many families. This was reversed in the late 1960s and 1970s by a trend to larger dwellings in conjunction with the baby bust. At the same time, however, the rapid increase in paid work outside the home by married women meant that less labour was available in the home to undertake the housekeeping associated with many forms of sharing.

Regulation provided another disincentive to sharing. The effect of building and property maintenance codes, zoning by-laws, and restrictive covenants is often to limit housing to single families or lone occupants. Lodgers and households of unrelated individuals are often banned, whether explicitly or implicitly. Thus, even where the host so desires, regulation can constrain sharing.[7]

Sharing provides certain benefits to the host. In the case of commercial lodging, this includes a steady source of income that helps support the host in the event of illness, injury, or layoff. However, the attraction of this benefit declined with the introduction of Unemployment Insurance, Workman's Compensation, Old Age Security, and publicly-financed health care. Another benefit is the labour provided by lodgers in activities such as food preparation, laundry, house cleaning or maintenance, child care, or snow removal. In the case of ill or disabled relatives or friends, the benefit to the host may be primarily in terms of low cost, or easy access, in providing assistance or care.

Sharing requires trust among the co-residents. Because they occupy the same set of living quarters, it is easy to invade one another's privacy or misuse another's property. It is not surprising, therefore, to find that often the sharers are related. However, the post-war period was marked by extensive interregional migration. In many cases, these migrants left behind their parents and other close relatives. In part, the post-war decline in sharing may simply have reflected a paucity of relatives in the local community. [8]

The host may also consider congestion around the home as a cost of sharing. Bathrooms come specifically to mind. Most of us have experienced the morning congestion around family bathrooms. Over the post-war period, Canadians used their increasing wealth to purchase more, and better appointed, bathrooms. However, bathrooms are expensive to install or upgrade. For some families, getting rid of sharers may have been the least costly way of solving this congestion problem.

There is a related consideration here. Families may not have subsidized sharers simply because they were relatives. Rather, it may have had to do with transactions costs for home owners. Typically, it can be expensive for an owner to move if one considers both monetary and psychic costs. Furthermore, imperfect information in the real estate market adds an element of risk to the costs of moving. As a result, families move less frequently than might be expected given changes in their housing needs. A family that is planning to move now and expects to need a still larger house in a few years might well buy the larger house immediately and avoid the transaction cost of moving again in a few years. The marginal cost of making space available to a lodger, at least until the space is needed by the expanding family, can thus be small. Although adequate data are lacking, the level of residential mobility likely increased in post-war Canada. The higher mobility, in part, reflected a better organized real estate market (particularly in the larger urban areas); the flow of information improved (through means such as multiple listing services), and monetary transactions costs declined in relative terms. Where it became less expensive to move and adjust one's housing consumption, one's willingness to take in lodgers, and to discount their rents, declined. In part, the higher mobility arose for other reasons, such as job changes or transfers. Whatever the reason, it afforded some families (particularly those in metropolitan Canada) more opportunities to bring their housing consumption into line with their space needs, reducing the extent of short-term "overconsumption," and hence the space available to accommodate sharers.

THE LODGER OR PARTNER AND
THE DEMAND FOR SHARED ACCOMMODATION

The demand for shared accommodation by non-family individuals depends on the availability of alternatives (for example, living alone, living in a collective dwelling, or sharing with a family or with a partner or partners), the utility attached to the alternative, and its cost. Unfortunately, data on post-war changes in the cost, utility, or availability of sharing versus other forms of accommodation is lacking. However, alternatives to sharing became more widespread and relatively less expensive, especially with increasing urbanization. Cities, by virtue of their size, can sustain varied and profitable markets for rental housing – something not always possible in smaller towns and rural areas. Another reason pertains to changes in construction technology, of which post-war apartment buildings exemplify the breakthroughs. For the first time in history, multi-unit buildings could be constructed that provided many of the amenities of detached housing without the traditional drawbacks of tenement structures: noise, smells, lack of security, and fire hazard, for instance. At the same time, improvements in homemaking technology – including appliances that made it easier to do housekeeping – reduced the need for lodging and its associated housekeeping services. Effectively, these trends either raised the utility of living alone or reduced its cost, relative to lodging. Either way, they served to reduce the propensity to share.

In addition, social policy initiatives had a significant impact on the demand for shared accommodation. One was the concerted efforts by governments to reduce the size of Canada's institutionalized population that led to the development of group homes and other forms of shared "special needs" housing. Also important here were programs aimed at better housing the elderly (especially widows) and low-income families. From 1964 to 1980 NHA loans were provided for the construction of 142,345 new public housing units.[9] The construction of 126,158 new private dwellings and 42,034 hostel beds in new collective dwellings for the elderly were also assisted by NHA during this period.[10] This was subsidized housing, much of it on a rent geared to income basis. One impact of these policies was to drain off many potential sharers, that is, individuals and families whose low incomes would have prevented them from maintaining their own dwellings.

Why Did the Demand for Housing Change?

The preceding discussion of the causes of aggregate household formation is equivalent to analyzing the aggregate demand for usual residences since each household corresponds to an occupied private dwelling. It is also necessary to consider changes in demand for "housing services" (that is, in terms of the quality or quantity of housing consumed) by a typical household. Possible causes include changes in consumer preferences, incomes, or the price of housing.

CHANGES IN CONSUMER PREFERENCES

In light of the post-war changes in household formation already described, it is not surprising that the type, quality, and size of dwelling demanded also changed. After all, even ignoring the obvious impact of differences in income, the housing needs of elderly widows differ – in floor area, layout and use of

rooms, location, and access to education, health care, and other public services – from those of families with children. Different yet again are the needs of young singles and two-earner couples.

Why do different kinds of households have differing demands for housing? One set of reasons has to do with the size of the household. Although some rooms and facilities can be shared, larger households in general need more space. Related are reasons that have to do with the composition of the household. Households with children, for example, often need safe play areas and good access to schools. The presence of children may also make security of tenure more important to the household. As another example, two-earner households often need well-situated housing to balance career, home, and family responsibilities. Changes in overall housing consumption since 1945 thus reflect in part the changes in living arrangements already described

EFFECT OF RISING PROSPERITY AND INCOME DICHOTOMIZATION

In addition, housing consumption changed in response to rising incomes. Figure 2.1 suggests that the incomes of many categories of households increased apace from 1951 to 1981 before slowing down in the 1980s. Some did better than others. Because of the rising incidence of paid work among wives, husband-wife families did especially well on average. So too did the elderly living alone, many of whom benefitted from improvements to OAS and GIS, and the introduction of CPP/QPP. Overall, the trend among many household groups was to higher real incomes over time, at least until the 1980s.

In a sense, if one measures housing consumption using expenditure data, it is not surprising that total consumption kept pace with income. Throughout the post-war period, households of a given type tended to spend about the same proportion of income on housing. In other words, as aggregate income grew, so too did expenditure on housing. [11]

At the same time, Figure 2.1 obscures an important post-war dichotomization of households on the basis of income. The post-war explosion of non-traditional households (for example, persons living alone and lone-parent families) meant a rapid growth of households with low incomes compared to the more prosperous husband-wife family households. This trend is evidenced in the growing divergence between average per capita and per household income exhibited in Figure 1.2.

CHANGES IN THE PRICE OF HOUSING

It is relatively easy to observe how much people spend on their housing. It is something else to estimate how the price for a unit of housing services has changed since 1945 and to what extent consumption has changed as a consequence. For the most part, we must rely on the housing price indexes that form part of the CPI.

Available CPI indexes suggest that the price of housing just kept pace with other consumer prices (or even fell slightly in relative terms) and lagged

considerably behind the growth of income. Improvements in construction technology and plentiful factor supplies kept down the purchase price of new residential housing, although these were counterbalanced by sharply rising unit labour costs until at least the mid 1970s (see Table 5.3). Important also were changes in income tax provisions that reduced the after-tax cost of producing or consuming housing. [12] In addition, one should note government programs that subsidized the cost of constructing or renovating housing. Provincial governments also played an important role by easing the growth of property taxes, choosing instead to fund local governments from provincial income, sales, corporate, and other tax revenue.

Moreover, these changes in housing prices occurred against a backdrop of increasing urbanization. Most Canadians flocking to the major metropolitan areas were confronted with sharply higher land prices. Part of the increased post-war expenditure on housing is attributable to this. As well, households attempted to compensate for higher land prices by consuming less-extensive forms of housing, for example, apartments, duplex, semi-detached, and row housing, and smaller lots for detached housing.

Current and Emerging Issues

Canadian society has undergone a restructuring of living arrangements and housing demand. Much of the increased propensity to form separate households has been among individuals and families whose ranks include many with low incomes – for example, young singles and newlyweds, elderly widows and couples, the divorced, and lone parents. This trend has several important policy implications.

AFFORDABILITY

One implication is in terms of the impact on housing affordability. A commonly used goal in housing policy is the provision of adequate, affordable housing for everyone. In defining "affordable" housing, some policy analysts have used a threshold shelter-cost-to-income ratio of 25% or 30%. A household is said to have an "affordability problem" if it would have to spend in excess of the threshold to get adequate housing. Under this definition, many low-income consumers, especially persons living alone, have an affordability problem. However, to a certain extent, people choose their living arrangements. If consumers choose to pay more than 30% of their income to live alone, instead of lodging with others at a rent below the threshold, what is the nature of its "affordability problem"?

Some researchers have suggested using higher threshold percentages for small households: 40% of income for persons living alone, for instance. However, this is a makeshift solution. What we want to know is the extent to which consumers are unable to find adequate housing at a shelter cost at or below the threshold. Thus, we need to know about the alternatives available, whether these are "adequate" housing for the household concerned, and the associated shelter

costs. Unfortunately, we lack information on the alternative living arrangements (lodging versus maintaining a dwelling) open to individuals or families.
We do not know the extent to which consumers must lodge in inadequate housing because of the absence of something more suitable. We do not know whether
people who spend in excess of the threshold prefer to do so rather than occupy
an alternative that others (but perhaps not they) perceive to be adequate and
affordable. With the income dichotomization being brought about by ever
greater numbers of individuals and low-income families living alone, such data
are urgently needed to reassess the nature and extent of the housing affordability
problem.

CYCLICAL SENSITIVITY

Through much of Canada's modern history, husband-wife families have shown
a pronounced tendency to maintain their own dwellings. Except among
newlyweds and the elderly where low income is a constraint, almost all such
families live in separate accommodation. Couples may adjust the size of their
dwelling, or their tenure, in response to economic fluctuations; however, they
are remarkably persistent about maintaining a dwelling. If anything, this tendency has been reinforced by the post-war entry of married women into the
paid workforce. Effectively, this pushed most families over the income level necessary to sustain this living arrangement. As well, because of the presence of children, families are less likely to lodge with others. Children can create certain
externality costs (for example, noise and damage to property) that a host might
prefer to discriminate against. In addition, parents may prefer to maintain a
dwelling to control better the kinds of people with whom their children reside.

On the other hand, the living arrangements of non-family individuals are less
stable. Their incomes are typically low. Only a modest proportion can afford to
live alone. However, if and as incomes rise in the future, living alone will become
affordable to a larger group. Their demand for living alone is thus income elas-
tic. Not having children, they find fewer drawbacks to lodging. Because of this,
they are more are more willing to switch between lodging and living alone
depending on the relative cost and availability of these alternatives.

There is indirect evidence of this price and income sensitivity. During the
1970s households were being formed in Canada at a net rate of about 200,000
annually. During the recession of 1982-3, net household formation (as indicated
by net new dwelling completions) fell as low as about 120,000 units annually.
However, the underlying demographic patterns apparently were continuing
much as they had in the 1970s. I suspect that the difference was a lower rate of
household formation among non-family individuals and low-income families.

In the future, this cyclical sensitivity will become more important. As the
baby boom generation passes the prime marrying ages, the growth in husband-
wife families will attenuate. At the same time, little further increase can be
expected in their propensity to maintain a dwelling; at present, they almost all
do. However, if the marriage and baby busts persist and women continue to

outlive men, the number of non-family individuals will continue to rise quickly. Net household formation will come to depend increasingly on the propensity of this group to live alone. The larger price and income elasticities displayed by non-family individuals mean that future household formation and housing consumption will become increasingly dependent on economic conditions.

IMPACTS OF SUBSIDIZATION

A related argument concerns the impacts of housing policies that subsidize household formation. I am thinking here specifically of senior citizen and low-income public rental housing projects. [13] During the 1960s and 1970s, governments invested heavily in such projects. While the public housing stock has never exceeded 5% of all dwellings in Canada, it is largely targeted at two specific groups: the elderly and low-income families. Since much of this construction may simply have displaced existing housing or replaced the construction of unsubsidized units, the net impact of these programs is unknown. However, given the large number of subsidized units built, and the inability of many of the new occupants to afford accommodation in the private sector, the impact was potentially large.

Since 1945 in-home social services have proliferated. Programs like visiting nurses, "meals on wheels," and emergency hailer systems make it possible for some individuals to live alone. Other programs, notably subsidized daycare, help do the same for lone parent families. To the extent that these services were subsidized, they encouraged household formation much as did the subsidized housing programs.

Also important over the post-war period has been the expansion of income maintenance programs. The impacts of programs such as unemployment insurance and old age security on the decisions of hosts to take in lodgers have already been considered. These programs also bolstered the incomes of non-family individuals, making them better able to afford to live alone.

Many of these programs and subsidies were first put into place in the 1960s and 1970s. The 1980s have seen a stabilization in the growth of some programs and the retrenchment or elimination of others. As these programs and subsidies are so pervasive, it is difficult to assess their impact on net on household formation and housing demand. The propensity to live alone rose sharply at the same time as these programs were being introduced. It therefore is possible that, if stabilization, retrenchment, or elimination continue in the future, the growth in persons living alone might slow or even be reversed.

HAS NON-FAMILY FORMATION PEAKED?

In part, post-war household formation was driven by an undoubling of non-family individuals and low-income families from shared accommodation, associated with the formation of one-person households. With the aging of the baby boomers, and assuming a continuation of low fertility, differential mortality, and the marriage bust, the demographic conditions are ripe for a continued

explosion in the formation of non-family households over the next few decades. Some demographic forecasts suggest that the number of households in Canada might reach 11.4 to 12.1 million by the year 2001 (see Miron 1983, Table 2). Will it happen? Will the pace be similar to what we have witnessed over the last few decades?

In attempting to answer such questions, one should remember that what has happened so far was pushed and shaped by government policy. While it is not clear just how large a net effect any one policy has had in the last decade, the curtailment of growth in some programs and the scaling back of others makes it unlikely that the rate of growth of household formation that prevailed in the 1960s and early 1970s will return. If there is one lesson to be learned here, it is the danger of attempting to treat household formation as determined by demographics, and housing demand by economics. Household formation has increasingly come to be sensitive to price and income. Any attempt to forecast household formation in ignorance of this is foolhardy at best.

Notes

1 There is a paucity of data on secondary homes in Canada. A 1977 survey found that just under 9% of Canadian spending units with an above-median income owned a vacation home. Among units with a below-median income, the ownership rate was only about 3%. See Statistics Canada 1977, Table 16.

2 Calculated from Dominion Bureau of Statistics (1948, 5-10) and Statistics Canada (1984c, 16-19.

3 In 1931 the probability of death before one's first birthday was 8.7% for males compared with 1.5% in 1976. The probability that a one-year-old male dies before his 60th birthday fell from 28% to 19% over the same period. The improvements for women were even more dramatic.

4 Thus, the decline in Census-reported typical family size in the baby bust reflects both decreased fertility and closer spacing of births. Among young families, average family size peaked about 1961. In the 35-44 age group, the peak occurred closer to 1971, reflecting the changed spacing and the cumulative effects of fertility on family size. Among still-older families, average size remained constant or declined, even during the baby boom.

5 See Steele (1979), Smith (1984) and Harrison (1981). Miron (1988, 142) reports 1971 cross-sectional income elasticities of from 0.1 to 1.4 for the propensity for non-family individuals to head a household and much lower elasticities for the propensity of non-family household heads to live alone.

6 The 1981 Census reports that two-thirds of all non-family persons living in family households were related to the household head.

7 Of course, some of this housing was created surreptitiously and, as a result, may be more widespread than is reported in the Census.

8 However, this is not to say that post-war families provided less housing assistance than previously. Although we lack information about this, post-war families used part of their growing affluence to subsidize separate accommodation for needy relatives. This is

especially true of young adult children and grandparents. In other words, rather than taking relatives into their own dwelling, families increasingly subsidized the accommodation of kin elsewhere.

9 See *CHS 1980* (53). This consists of loans approved under NHA s. 43.

10 See *CHS 1980* (55). This consists of loans approved under NHA s. 6, 15, 15.1, 34.18, 40, and 43.

11 Another way of saying the same thing is that the income elasticity of housing consumption was about 1.0. There is a considerable debate about the magnitude of the income elasticity. Using cross-sectional data from the 1971 Census, Steele (1979) estimates elasticities of from 0.2 to 0.5 for different kinds of households. Miron (1988, Chapter 8) finds estimates of close to 0.3 using cross-sectional data for 1978. These tend to be lower than estimates found using longitudinal data. Reasons for such discrepancies are discussed in Miron (1988, Chapter 8).

12 The principal changes occurred with revisions to the federal Income Tax Act in 1972 that taxed some capital gains (though not on owned homes) for the first time and reduced the ability of small investors to claim losses on rental housing for tax purposes. See Miron (1988, Chapter 9) for more details.

13 Over the years federal housing policy has also provided subsidies, in the form of grants and forgiveable loans to home owners of modest income to make repairs and other improvements. These include the CHIP, CHRP, HIP, RNH, RRAP, the HIL Program, and Farm Improvement Loans. Arguably, these programs helped an unknown number of individuals and families to remain in separate households, who otherwise would have found it too costly.

===

CHAPTER THREE

Incomes, Prices, and Tenure Choice

Marion Steele

TO RENT or to own is a significant decision for Canadian households. While rural Canadians have always overwhelmingly chosen home ownership, before World War II urban Canadians were predominantly renters. Over the decade ending in 1951, however, many urban dwellers opted for home ownership and the costs and benefits it entails. A dramatic jump in the urban home ownership rate in that decade established home ownership as the urban norm and had a major impact on the overall rate for Canada. Inasmuch as single persons living alone and other non-traditional households tend to rent, the increase in their numbers since 1951 has reduced the overall home ownership rate. Otherwise, home ownership among traditional households has continued to rise since 1951. [1] This chapter analyses the demand forces underlying this change and those underlying the tenure choice pattern of different types of households.

Most housing analysts see tenure choice as a second-stage housing decision that follows the first-stage decision to form a separate household. In this view, the decision to form a new household is characterized as the rental of a separate dwelling, and the state of the rental housing market is more critical. Home ownership, if and when it occurs, is usually a separate and subsequent housing commitment, affected by investment considerations and having long-term implications. For this reason, it is attractive to analyze the tenure choice decision – whether to be a home owner or renter – as one in which home ownership is the active decision and renting is the default decision.

The Rationale for Home Ownership

Why might consumers prefer home ownership to renting? Owning gives the household more control over its housing than does renting. Tenants depend on a landlord to maintain the building and provide various services, and moving – the tenant's ultimate sanction against an unsatisfactory landlord – involves psychic and monetary costs. In contrast, home owners can adjust the level of maintenance to their preferences and can choose whether to buy maintenance services or use their own labour. The do-it-yourself option makes housing expense more flexible for a freehold home owner than for a renter. Elderly home owners

41

have the option, for example, of doing little maintenance and hence running down housing capital during their retirement; in this sense, housing depreciation acts as an annuity or a pension supplement.[2]

A second advantage of home ownership is its role in the accumulation of wealth and as an asset in the household's portfolio. Home purchase for the typical household is an investment that is easily managed, highly leveraged, and usually has predictable cash flow requirements,[3] a combination unmatched by other real investment options. As well, the standard level payment mortgage constitutes a forced saving scheme. Wealth held in the form of an owner-occupied home is also favourably treated by the tax system. Capital gains on a principal residence are not taxed;[4] neither are the implicit returns to home equity. Unlike the situation in the United States and Britain, mortgage interest is not tax deductible in Canada; nor are property taxes, unlike in the US. The tax advantages of home ownership are offset by the fact that rents tend to rise less than proportionally with inflation because landlords can fully deduct expenses (including the inflationary component of mortgage interest) and yet are taxed only at one-half income tax rates (three-quarters under post 1986 tax reform) on realized capital gains.[5] With this asymmetric treatment of capital gains and interest costs, investors are willing to invest in rental units even when rents are insufficient to provide a positive annual yield. Thus, rents will tend to be lower than they would be without this tax treatment.[6] The generous capital cost allowance provisions of the tax system (reduced under post 1986 tax reform), various subsidy schemes, and rent regulation, all tend to lower rents further.[7]

People wish to be home owners, also, for reasons having to do with incomplete markets.[8] Some kinds of housing may simply be unavailable for rent in some locales; other kinds may be available on a continuing basis only through home ownership.[9] The availability problem is acute for households with children because landlords perceive that such households are costly to serve and tend to discriminate against them.[10] Another advantage of home ownership is the obverse of this point. If landlords make accommodation available without discrimination – charging every tenant the same rent – and if some tenants are more costly to serve, then low-cost tenants overpay (see Henderson and Ioannides 1983). A household that does not cause damage will reap the rewards of its behaviour as a home owner, not as a tenant. Related to this is an environmental control point: if multiple-unit buildings have tenant mixes that are perceived to be undesirable, then home ownership may be a solution. For instance, an elderly couple not wishing to have a young family or partying singles living above them may avoid this by being a home owner either in a single-family house or in a condominium targeted at empty nesters.

Home ownership does have disadvantages relative to tenancy. For example, home owners have higher transactions costs than do tenants. The purchase and sale of a home involve out-of-pocket costs (real estate brokerage fees, land transfer tax, legal fees, moving fees) that are in the 10% range for households using a real estate broker. A risk in home ownership is that the owner may have to sell at a loss. Resale prices slumped nationwide, for example, at the beginning of the 1960s, and again more markedly during the recession of the early 1980s, and there have been marked regional and local variations in selling prices over the years. As a result, home ownership may not be cost-effective for a household that is likely to have to move again soon. Similarly, a household that is likely to become unemployed, and hence be unable to pay ownership costs, risks greater losses from home ownership than from renting.

Another disadvantage of home ownership is the burden of housing management and maintenance tasks. This is especially important for the single person household. Condominiums, however, greatly reduce this burden. Furthermore, condominium apartment buildings can provide protection and physical security arrangements that are uncommon in rental buildings.

PUBLIC POLICY GOALS AND HOME OWNERSHIP

The most frequently articulated goal (for example, Rose 1980, 7; CMHC 1983a, 34-5) of housing policy is the provision of physically adequate, uncrowded, affordable accommodation. Home ownership does not directly promote the first two aspects of this goal. The effect of home ownership on affordability is complex. In an inflationary environment, the cash flow cost of home ownership for a new purchaser is often greater than the rent previously paid. This point is emphasized by the fact that while the CMHC criterion of a housing affordability problem is a rent-to-income ratio of 30% or more (CMHC 1983a, 38), applicants may qualify for a NHA mortgage with a ratio of mortgage payment, property taxes, and heat expense, to income, of 32%. Indeed, when the expense of other utilities, maintenance, and insurance are added, the housing expense-to-income ratio may well exceed even this. With inflation, the nominal mortgage payment remains constant while nominal income typically rises. Ultimately, when the mortgage is paid off, the expense ratio will fall abruptly. Thus, the encouragement of home ownership among younger consumers can be viewed as a policy to ensure affordability for the middle-aged and elderly, or more broadly, as part of an income security system. The success of home ownership in this respect is illustrated by the experience of the Quebec housing allowance program. This program directly tackles the affordability problem of the low-income elderly (Steele 1985a), but few home owners have a sufficient affordability problem to be eligible; they account for a minuscule proportion of all recipients and receive only a small subsidy on average.

Home ownership directly and unambiguously provides security of tenure. So long as home owners pay their bills they need not fear having to move except in the highly unlikely event of expropriation. In contrast, a tenant may be legally evicted under provisions that vary from one province to the next. While the legal rights of the landlord to evict were increasingly circumscribed in the 1970s and 1980s, the economic forces that encourage landlords to seek eviction have also increased.[11] Public concern about the security of tenure of renters increased in the late 1970s and early 1980s because of low vacancy rates. Home owners who move know that so long as they are willing to pay the asking price, they will have little difficulty purchasing a home. A tenant forced to move when there are few vacant dwellings will not easily find alternative similar accommodation. Some renters may have difficulty finding accommodation because landlords regard them as costly. This problem tends to be acute in tight markets where landlords can afford to be more selective.

Home ownership may be seen as an end in itself, rather than as a means to other ends. Some regard widespread home ownership as the foundation for a stable democracy because of the belief that the ownership and care of property increases the responsibility and independence of the citizenry and increases the stake of citizens in their community. A second view is that home ownership merely represents an investment decision with no implications beyond economic ones. A third view – the Marxist view – is that home ownership undermines social progress because

> Pressures for better state pensions ... may ... be weakened by the existence of owner-occupation to the benefit of capital.... The nature of housing [provision] has widespread repercussions on personal life, acting as a severe restriction on attempts to break down the dominance of patriarchal nuclear family structures (Ball 1983, 365, 391).

Home ownership may make the achievement of some social goals more difficult. It may reduce national income by making the labour force less mobile because of the high transactions costs for home owners. However, there are also impediments to mobility for tenants both in the social housing system and in private markets where rent regulation depresses rents; such tenants will be uncertain of the availability of accommodation on the same terms at a new location. Membership in a non-equity housing cooperative can also be an impediment to mobility.

Home ownership may also make the achievement of income mix more difficult. This will be a particular problem if the home ownership rate increases, leaving few middle-income households in the private rental sector, and if the social housing sector increases without an explicit income mix component.

HOME OWNERSHIP ASPECTS OF CANADIAN POST-WAR HOUSING POLICY

One interpretation of CMHC activity for the first decade following World War II is that its main preoccupation was the encouragement of home ownership among middle-income families. A wiser view is that, confronted by thousands of returning servicemen and others needing housing in the early post-war period, and with an underlying assumption that home ownership was more desirable than tenancy, CMHC saw its main task as to ensure that mortgages were available and houses were built. It seemed natural that these new houses should be for home owners and thus home owner policies were the principal focus of CMHC's attention.[12]

In the 1940s and 1950s, CMHC transformed the residential mortgage market, helping to assist middle-income households anywhere in the country to obtain financing. CMHC did little, however, to help low-income families become home owners.[13] Indeed, it is often pointed out that CMHC policy militated against ownership for people of low income through the level-payment structure of an NHA mortgage, high construction standards (at the time) that put a floor on house price, unfavourable policies for lending on owner-occupied duplexes and triplexes, exclusion from "qualifying income" of the income of anyone other than the household head, and lending exclusively on new buildings, ignoring the cheapest acceptable houses – old ones (see for example, Dennis and Fish 1972; Rose 1980). In addition, large extended families as well as other household groups outside the traditional norm found it difficult to obtain NHA financing.

The exclusion of existing houses from NHA financing was in part attributable to the use of house construction for job creation. Reducing unemployment is a worthy objective in itself, and increasing construction is a particularly effective way to do it. Thus, a defence against the criticism that government policy did little directly to help the housing situation of low-income households is that housing policy was being used to achieve another important target – reducing unemployment.

Another defence was the "trickle down" or filtering theory which claimed that building new houses helped low-income families indirectly if not directly because, when middle-income families bought new houses, the older houses they vacated became available to poorer families. Increasing the overall supply of houses put downward pressure on house prices, helping to deliver older, lower quality houses to low-income families. In some circumstances, this contention proved to be correct. In recent decades, however, old houses in central locations became attractive to upper-income households. Gentrification stood the filtering process on its head as houses filtered up rather than down. The filtering down of high-rise rental apartments continues, but this kind of housing is often unsuitable for families with children and does not provide home ownership.[14]

While federal government policy in the 1940s and 1950s largely relied on filtering to help low-income households into home ownership, some provincial policies offered more direct help. For example, starting in 1948, Quebec's Family Housing Act provided a 3% interest subsidy to families. High-income families were ineligible (in contrast to NHA rules), as were expensive houses. Under a Nova Scotia program, low-income families were sold houses with unfinished interiors (shell houses). No downpayment was required; "sweat" equity, in the form of the labour required of the family to finish the house, replaced money equity.[15]

During this period, there is evidence that CMHC, unlike some provinces, regarded home ownership as unsuitable for low-income households despite the large percentage who had achieved this status without subsidy.[16] In 1949 CMHC resisted a proposal by Ontario for no-downpayment loans to low-income households.[17] In 1962, in reaction to a proposal for low interest, long amortization, loans to low-income families, CMHC made clear that it believed its assistance to low-income families should be confined to rental housing.[18]

Opposition to direct efforts to make home ownership accessible to low-income families gradually evaporated during the 1960s. Downpayment and qualification requirements were relaxed. By 1965, 18% of NHA borrowers fell in the bottom-third family income group, up from 6% in 1954 (*CHS* 1966, Table 70; 1968, Table 60). NHA coverage was extended in the late 1960s to include condominiums and existing houses, thus further opening home ownership to low-income families. Furthermore, in 1968 a decision was taken to target CMHC lending to low-income families (*CHS* 1968, x). At first, lending was directed at rental housing, but slowly CMHC moved towards the revolutionary step of a large-scale subsidy program for home ownership. The first step, in 1970, was the $200 million "innovative low-cost housing programme" (*CHS* 1970, x) that was aimed at low-income households and funded approximately 10,000 ownership units (see *CHS* 1970, x and Table 51; Dennis and Fish 1972).[19] This program did not, however, provide loans at below the CMHC direct lending rate. Its follow-up in 1971, the $100 million "assisted home ownership program" (*CHS* 1971, xii), did do this and also extended the amortization period.[20]

The giant step was the introduction of AHOP in 1973. The CMHC programs for home owners in the 1950s and 1960s – mortgage insurance and direct lending programs – were subsidy programs only in a limited sense and the $100 million program in 1971 represented only a small change. AHOP was a brave and path-breaking departure; it married large initial monthly subsidies to a mortgage design that differed radically from the standard, and it made new home ownership accessible to more families. Under AHOP, initial monthly payments on a new dwelling unit were cut by the use of a mortgage payment design under which payments gradually increased over time.[21] Underlying this design were the assumptions that the rate of inflation would not fall, that the inflation premium on interest rates would not change upon rollover, that incomes would rise with inflation (hence, the payment-to-income ratio under AHOP would remain affordable), and that house prices would rise (hence home owner's equity would not fall).

These assumptions were not realized. This bad luck along with imperfect design resulted in more defaults than had been foreseen. Of the 161,000 units funded by assisted home ownership programs over the period 1970-8, some 18,000 had defaulted by 1985, an 11% rate.[22] This is a high rate by the standards of the first two post-war decades, but it must be put in context. First, the defaults were largely an Ontario problem; 60% of all defaults occurred there, and the Ontario rate, at 20%, was about twice the next highest regional rate. Quebec and the Prairies, in contrast, had a rate of only 4%. Second, during this period the

Table 3.1
Home ownership rate by selected categories:
Canada, 1941-1986
(home owners as % of all households in category)

| | All households | | | Heads aged 35-44 | | |
| | All areas | Urban areas | Rural areas | Urban areas† | All areas | |
					Total	Male only
1941	57	41	76	23		
1951	66	56	82	65		
1961	66	59	83	63	68	70
1966	63	57	83		67	70
1971	60	54	82		67	71
1976	62	56	84		71	76
1981	62	56	84		72	77
1986‡	62	57	82		70	76

SOURCES 1941 *Census of Canada*, IX, Table 51; 1951, X, Table 91; 1961, II.2, Table 84;
1966, II, Table 3; 1971, II-3, Table 9; 1976, 3, Tables 5, 13; 1981, I, Table 4; 1986
Populations and Households Part I, Tables 5,11.
† In 1941 and 1951, only cities of 30,000 or more are included.
‡ In 1986, dwellings on reserves are excluded. If these are included and all are
assumed to be owner occupied, the rates for 1986 are 62, 57, 83, 70, and 76
respectively.

Table 3.2
Home ownership rates by age group:
Selected areas of Canada, 1931-1986
(home owners as % of all households in category)

| | 1931 urban | 1941 cities of 30,000+ | All areas, male heads | | |
			1961	1981	1986
Under 25	7	7	25	24	22
25-34	19	13	51	59	56
35-44	38	23	70	78	76
45-54	51	37	76	81	81
55 or older	61	50	79	77	78

SOURCE *Census of Canada*, 1931, Monograph No. 8, 98; 1941, IX, Table 50; 1961, II.2, Table
84; 1981, 1, National Series (92-933), Table 9; 1986, *Populations and Households Part I*,
Table 11.

default rate for regular NHA loans was also high,[23] presumably because of the greater house price volatility after 1970 compared to earlier post-war years.

AHOP was a popular, large-scale program that changed the nature of NHA borrowers. By 1975, 31% of NHA borrowers were from the bottom-third income group (CHS 1976, Table 103). Close to one-half of AHOP borrowers in 1973-5 fell into in this category – the proportion being even higher in the West (CHS 1975, Tables 97 and 99). A somewhat lower, although still high, proportion of AHOP borrowers in the next three years were also low-income families.

In 1978, as delinquencies grew, AHOP was terminated. CMHC programs for home owners of all incomes were henceforth confined to mortgage insurance and energy-saving programs, the Canadian Home Insulation Program (1977-86) and the Canadian Oil Substitution Program (1980-5). The only programs aimed explicitly at low-income families are the narrowly-targeted Residential Rehabilitation Assistance Program (RRAP) and the Rural and Native Housing Program.

Patterns of Home Ownership in Canada

The advantages of home ownership suggest that home ownership will be the tenure of choice for many consumers.[24] However, home ownership becomes accessible only when income is adequate. As well, the tax advantages increase with income. Because average real income has risen over time, one would expect home ownership rates also to have risen. As Table 3.1 shows, this indeed happened between 1941 when the rate was 57% and 1951 when the rate was 66%. Despite an increase in real income in most years since then, by 1966 the rate had fallen to 63% and has fluctuated around that level ever since. But increased urbanization and changing household composition play havoc with the meaning of these overall changes. Evidence indicates that, after controlling for the changing characteristics of households, the likelihood of home ownership was higher in 1986 than in 1961.

As illustrated in Table 3.1, the first part of that evidence is that home ownership is less common in urban compared to rural areas. In rural areas, almost all households, rich or poor, are home owners; from 1951 to 1986 the home ownership rate in rural Canada hovered around 82% to 84%, up from 76% in 1941. Thus, the increasingly urban nature of Canada by itself would push home ownership rates down nationally.

The changing composition of households is the second factor undermining the comparability of home ownership rates over time. Since 1945 the amount of household splitting has increased as has the number of non-traditional households: the separated and divorced, single-parent families, young singles living alone or with other singles, and widows and widowers living on their own. Non-traditional households are small, often childless, and typically less affluent. They are less likely to occupy single-detached or owner-occupied housing. As their numbers have increased over time, these households have depressed home ownership rates measured over all Canadian households.

To provide an indication of what the home ownership rate over time would have looked like if the nature of households had not changed, Table 3.1 gives the home ownership rate for households with a male head aged 35-44. This is a proxy for the traditional household of a wife, husband, and children. In 1971 the rate for this type of household was 70%. The rate did not fall in the 1960s, and indeed, it surged from 1971 to 1976. The latter rise was associated with a confluence of factors favourable to home ownership: rising real income, modest real interest rates, and encouraging CMHC policies. It is noteworthy that, despite the less favourable economic environment after that time, the home ownership rate did not fall.

FEMALE-HEADED HOUSEHOLDS AND CONDOMINIUMS

Most non-traditional households are female-headed; the latter more than doubled from 1971 to 1986 to reach 28% of all households. Lower home ownership rates among these households depressed the overall rate of home ownership in 1986. But modest as the female-headed home ownership rate was in 1986, it nonetheless had grown since 1971. This suggests that the depressing impact of these households on the overall rate will fade over time. Specifically, in 1971, 4% of households headed by females under 25 years of age were home owners, compared to 9% in 1986; the rate also more than doubled, to 28%, for those aged 25-34 years. [25] This is of course only one of many changes in the economic status of women over these years.

One factor contributing to increased female home ownership was reduced discrimination by lending institutions. Another factor was the increased availability of condominiums that had first appeared in Canada in the late 1960s. Earlier, a household wanting home ownership had to accept the attendant property and tenant management tasks. The condominium frees many home owners from these tasks. In 1986 the rate of condominium home ownership for women was more than twice that for men. Nonetheless, even in the age group where condominiums had the greatest penetration (under 25 years old), only 12% of female home owners were condominium owners in 1986. [26] While the rise in condominiums was important in encouraging home ownership among female-led households, it was not the only consideration.

THE RISE IN THE HOME OWNERSHIP RATIO WITH AGE

There is a pronounced association between home ownership rates and age. As can be seen from Table 3.2, only 22% of households with male heads under 25 years of age were home owners in 1986 compared to 81% among heads aged 55-64 years. Two underlying causes of this association are the rise with age in income and net worth (that make home ownership affordable and wealth portfolio considerations important) and the rise with age through the childbearing years in the likelihood that there will be children in the household (which makes single-detached housing attractive) (see Steele 1979, Table 6.4). [27] Even controlling for income, wealth, and the number of children, however, age is an important determinant of the home ownership ratio (Steele 1979, Table 6.4). The reasons for this are primarily the pension motive – the desire to provide for the housing needs of old age – and, secondarily, the reduced mobility associated with age.

The increasing ownership ratio in the last five decades was generated largely by earlier entry into home ownership. In 1931 only 19% of urban households with heads aged 25-34 years were home owners, as compared with 56% of male heads in this age group in 1986 (Table 3.2). The rise in real incomes and the availability of low-downpayment mortgages were important in reducing the age at which households first become home owners.

One curiosity in these data is the fall in the home ownership rate among the elderly during the last two decades, contrary to the trend for other age groups. Among male heads, only 74% were home owners in 1986, down from 81% in 1961. Most of this drop occurred in the early 1960s, before subsidized senior citizen housing was available. It seems likely that it largely is associated with the phenomenon of greatly increased household splitting. When adult children leave home to live on their own, there are fewer people to share expenses and maintenance duties, and of course, the need for space declines.

Prices, Rents, Interest Rates: Inflation and the Tax System

Relative prices affect tenure decisions, just as they affect other consumption and investment decisions. New house prices and rents, in real terms (that is, deflated by the consumer expenditure deflator, to remove the effects of inflation) moved in different ways over the post-war period. From 1945 to 1984 real house prices (Table 3.3) increased by 34%. Real rents, as measured by Statistics Canada's rent index (Table 3.3), on the other hand, fell by 46%.

Why the secular rise in real house prices? In part, productivity increases in the construction of single-detached houses may have been less than in the manufacture of other goods; although there have been remarkable advances in building materials and technology, and the amount of prefabricated components have increased steadily, it is still not economic to build dwellings in factories where assembly-line methods could be utilized. In addition, house lots have become more expensive in part because of a tightening of zoning and subdivision regulation; for example, in the 1940s and 1950s, septic tank subdivisions were more widely accepted than at present.[28]

While few find it surprising that real house prices have risen, most find it surprising that real rents have fallen. The explanation is as follows.[29] The return to rental residential property consists of three components: the net rental yield (that is, gross rents minus gross current expenses, including mortgage interest) on a before-tax basis), the capital gain, and what may be termed the tax loss plus subsidy yield. Developers will engage in rental residential property development if the return is greater than the return to other assets of the same risk class. Thus, the size of the sum of the three components of return is critical, not the size of any single component. The higher the capital gain and/or the higher the tax loss

Table 3.3
Housing prices: Canada, 1951-1984

	Real rent index	Real new house price index	% change new house price index	MLS real average price index	Five-year conventional mortgage rate	Financing cost indicator
1951	147	69	-0.1	n/a	5.5	378
1952	152	70	1.7	n/a	5.8	406
1953	158	70	0.7	n/a	6.0	423
1954	162	74	4.7	n/a	6.0	443
1955	166	76	2.8	n/a	5.9	447
1956	166	78	3.6	55	6.2	487
1957	164	79	0.5	57	6.9	544
1958	163	79	0.1	60	6.8	536
1959	163	79	0.6	61	7.0	556
1960	162	77	-2.7	60	7.2	557
1961	162	77	-0.3	60	7.0	539
1962	160	76	-1.1	60	7.0	533
1963	158	77	0.5	59	7.0	536
1964	156	78	2.1	61	7.0	547
1965	155	80	2.5	63	7.0	561
1966	152	84	4.4	67	7.7	644
1967	152	84	0.6	71	8.1	681
1968	152	85	1.0	76	9.1	773
1969	152	88	3.1	80	9.8	858
1970	152	89	1.4	79	10.4	922
1971	151	90	1.1	80	9.4	843
1972	147	95	5.6	82	9.2	871
1973	138	107	13.3	89	9.6	1,030
1974	108	125	16.1	102	11.2	1,396
1975	122	119	-4.3	100	11.4	1,359
1976	121	119	-0.2	104	11.8	1,403
1977	119	113	-4.8	101	10.4	1,178
1978	116	107	-5.5	101	10.6	1,135
1979	111	101	-5.6	101	12.0	1,213
1980	105	99	-2.5	97	14.3	1,419
1981	100	100	1.4	100	18.1	1,810
1982	98	91	-8.7	86	17.9	1,634
1983	100	86	-5.9	86	13.3	1,142
1984	99	85	-1.2	86	12.5	1,061

SOURCES Column 1: Rent component of Consumer Price Index divided by Consumer Expenditure Deflator. 1981=100. Column 2: Nominal new house price: Annual average of quarterly new house price index constructed (largely on the basis of Statistics Canada's new housing price price indexes for certain cities and on the basis of NHA cost per square foot data) in Steele (1987) linked at 1969 to the average cost per square foot of NHA singles (HSC, series S326) linked at 1952 to the residential building construction input index (HSC, series K136). Real new house price index: nominal house price index divided by Consumer Expenditure Deflator. 1981=100. Column 4: Nominal average MLS price index: annual average of quarterly MLS index constructed S319. Real average MLS price index: nominal average MLS price index divided by the Consumer Expenditure Deflator. 1981=100. Column 5: CANSIM B14024. Column 6: Column 5 times Column 2.

plus subsidy yield, the lower is the rental yield required for development to take place.

The tax loss plus subsidy yield tends to increase with inflation as follows. An increase in the rate of inflation tends to increase interest rates and nominal capital gain by the same amount, but while interest cost increases are fully deductible against current income, the nominal capital gain associated with inflation is effectively taxed at only one-half (three-quarters under post-1986 tax reform) income tax rates; indeed, before 1972 capital gain was entirely untaxed. As a consequence a typical landlord with a large mortgage-to-value ratio, by the late 1970s, could declare a loss on rental investment for tax purposes, thus reducing the taxes paid on other income (the tax loss benefit). When he sold the property, making a large capital gain, he was lightly taxed. With rapid inflation, this asymmetry in the tax system benefits principally investors with high marginal tax rates.[30] For many years, there was also the extra tax loss yield created by MURB provisions that allowed investors to deduct against other income rental losses created by capital cost allowances on new buildings. Added to this in the 1970s was ARP, a program delivering interest rate subsidies to new private buildings, and miscellaneous other federal and provincial subsidy programs.[31]

The pattern of the fall in real rents, as evidenced in Table 3. 3, supports this discussion. Immediately after World War II, inflation was widespread, and real house prices rose; but real rents fell at a rate of 3.0%. From 1951 to 1955 inflation was negligible, but real rents increased at 3.1% annually. From 1955 to 1964 inflation was modest (averaging 1.7%), but real rents now fell by 0.7% annually. With the heating up of inflation from 1964 to 1971, real rent fell slightly. Then, with the extraordinary inflation (averaging 9.2% annually) experienced from 1971 to 1982, the rate of fall of real rents at 3.8% exceeded even that experienced in the early post-war period.[32] There was thus a strong inverse association between the rate of inflation and the rate of change of real rents, despite a host of other factors that presumably also affected rents. ARP and other programs, along with rent regulation, may have helped reduce real rents in the late 1970s, but they do not explain why real rents dropped so sharply in the early 1970s.

The divergent movements of real house prices and real rents over the post-war period resulted in an enormous drop in the rent-to-house price ratio (Table 3.3). At first glance, this relative price shift suggests that there should have been a massive move towards tenancy. This did not occur because the cost of home ownership depends on more than house price.

House price is only one factor in the two more comprehensive measures of the cost of home ownership. One measure is cash flow cost, which indicates the affordability of home ownership under the assumption that it is financed out of current income. Its components are: the cost of utilities, property taxes, maintenance and insurance, and the mortgage payment (principal and interest).[33] The most important component of this total is typically the mortgage payment; this plus property taxes (PIT) is included in the mortgage service ratio that lenders

ordinarily use to determine whether an applicant is a good credit risk. As seen in Table 3.3, the real mortgage payment (as indicated by the "financing cost indicator") did not rise much until the big jump in mortgage interest rates in the late 1960s, rose steadily in the mid 1970s when house prices rose, and then climbed to extraordinary levels in the early 1980s. Interest rates then fell sharply; by 1984 they were back down to a level close to that in the early 1970s. In sum, the real cash flow cost of home ownership has moved in a more extreme way than has the real price of a house. It almost tripled between 1966 and 1981. Is it surprising that the home ownership rate for our proxy for traditional households rose over this period?

Many economists would answer "Notwithstanding these facts, no." For the cash flow cost of home ownership does not take into account capital gains. These are taken into account by the second comprehensive measure of the cost of home ownership, the economic cost (also known as "user cost"). This is a measure of the true cost of home ownership, under the assumption that markets are perfect and there are no cash constraints, so that a dollar of accrued capital gain is worth the same to a home owner as a dollar's reduction in the required mortgage payment. In particular, assuming for simplicity that a purchaser makes no downpayment and transactions costs are zero, the real user cost of home ownership is:

$$\frac{P_H(i + m + d - c)}{P}$$

where P_H is the price of the house, P is the overall price index, i is the nominal mortgage rate, m is property taxes plus maintenance, utility, and insurance expense as a proportion of house price, d is the depreciation rate, and c is the expected rate of capital gain of the house.

Minor differences between user cost and cash flow cost are the inclusion of depreciation and the exclusion of the principal portion of the mortgage payment. The critical difference is the deduction for capital gain, which recognizes that the cost of being a home owner is reduced by the existence of capital gain. If house prices rise enough in a year, capital gain may be more than enough to offset interest and other outgoings so that the cost of occupying the house in that year is actually negative. The distinction between user cost and cash flow cost is thus of critical importance when there is inflation, because inflation on average increases both nominal capital gains and nominal interest rates and only user cost takes into account both.

For a home owner with positive (rather than zero) equity, user cost must take into account the opportunity cost of funds the home owner has tied up in equity. This opportunity cost is the after-tax yield these funds would earn if invested in another asset;[34] in this case, the user cost expression is:

$$\frac{P_H[i(1-e) + i(1-t)e + m + d - c)]}{P}$$

where *e* is the ratio of equity to house price, *t* is the marginal tax rate of the home owner, and it is assumed that the best return the home owner could get (if the funds tied up in equity were invested in an asset of the same risk class) equals the mortgage interest rate.

Table 3.4 shows estimates in 1981 dollars of the annual cash flow cost and user cost of a standard new house for a home owner with an equity-to-value ratio of 10%. A marginal tax rate of 27% is assumed. Expected capital gains are estimated assuming that households base their expectations on past and current experience.[35] As can be seen, user cost is always well below cash flow cost. In 1984 it was $5,315 ($443 per month) while cash flow cost was $7,679 ($640 per month). User cost fell in 1972-3 despite a marked rise in cash flow cost; large expected capital gains helped make home ownership attractive. But expected capital gains fell in later years, and when rising interest rates pushed cash flow costs to $893 per month in 1981 – a 20% increase over 1980 – user costs went up even faster (35%).

Noteworthy here is that user cost rose more in percentage terms over the 1970-81 period than did cash flow cost. User cost fell in 1972-3, and this probably played a part in the rising home ownership rate among traditional households from 1971 to 1976. But the rise in user cost and cash flow cost have to be counted as depressing factors over the decade as a whole. Thus, strongly rising real incomes and encouraging federal home ownership policies, rather than changes in user cost (or its component, capital gain) were primarily responsible for the increase in the home ownership rate among traditional households over this period.

Home Ownership and Net Worth

It is plausible that the purchase of a home is usually more than a housing consumption decision; it may also be characterized as a saving and investment decision. Because of the nature of the standard mortgage plan, home ownership results in a build-up of net worth, especially in a period of inflation. The difference between home owners and others in net worth is dramatic. In 1977 the average net worth of home owners in Canada was over $71,000 while that of others was less than $9,000 (Statistics Canada 1977). Tenure status is thus a good indicator of whether net worth is substantial. The difference persists even when income is held constant; for those of about average income ($15,000 to $20,000), average net worth is more than five times as great for the 68% who are home owners as for others.

Regional contrasts are also great. Quebec, with its relatively lower home ownership rate, has the lowest average net worth in the country. Average income in Quebec is almost precisely the same as it is the Prairie provinces, but the market value of homes in Quebec is 28% less, the home ownership rate is 13 percentage points less, and net worth is 57% less. The modest net worth of Quebec families – lower even than the net worth of Atlantic families despite the lower average income there – is likely associated with the Quebec preference for renting. At the

same time, the gap between the net worth of Quebec and non-Quebec families will likely close rapidly in the future as the gap in home ownership rates continues to shrink.

Inflation and the Standard Mortgage Design

Inflation makes the standard mortgage design an important issue. Under inflation, this design has a "tilt" problem, that is, the burden of the mortgage payment on borrowers is higher at the beginning of the term than it is later. This can be seen as follows. The monthly payment over the term of the mortgage is constant; it blends interest and principal repayment so that, if the maturity term of the mortgage is the same as the amortization term, the mortgage is completely paid off by the end of the term. Now suppose that the income of a borrower rises at the rate of inflation. Suppose also that the initial mortgage-payment-to-income ratio is 28%. If there is no inflation this ratio remains the same over the life of the mortgage. But if the rate of inflation is, say, 8%, then income rises so that at the end of the first year the ratio has fallen to 26% and by the end of the fifth year to 19%. This is, in part, the tilt.

The effect of inflation on the borrower appears benign; indeed, under these initial assumptions, inflation confers a large benefit on the borrower as the mortgage ages, without imposing any cost. This would in fact be the case if inflation was not anticipated. However, if inflation is anticipated, lenders demand a higher interest rate to compensate for the decline in the real value of their principal. The inflationary premium they demand tends to equal the expected rate of inflation. (The nominal interest rate minus the inflationary premium is the real rate of interest.) Assume now that the real rate is 4%. Then, when the rate of inflation is zero, the nominal interest rate is the same as the real rate (4%); when the rate of inflation is 8%, the nominal interest rate is 12%. For a $50,000 mortgage the annual payment (assuming twenty-five-year amortization) is $6,375, or 28% of an income of $22,767, when the rate of inflation is 8%, but only 14% of income when the rate of inflation is zero. [36] Thus, while inflation results in a falling mortgage payment burden over the life of the mortgage, it also, if anticipated, increases the initial burden – in our example, from 14% of income to 28%.

An associated consequence of inflation is the faster build-up of equity. If house values increase at the rate of inflation, say 8%, then the increase in equity attributable to inflation – an amount equal to 8% of the value of property – dwarfs the increase in equity attributable to repayment of mortgage principal (if the mortgage is early in its maturity term).

A second problem associated with inflation is the increased variability in interest rates and house prices (compare the 1970s and 1980s with earlier decades in Table 3.4). This, together with the fact that in Canada, in contrast to the US, the 1970s saw the end of long-term mortgages, meant that Canadian borrowers bore increased mortgage rate risk. A home purchaser was no longer secure in the knowledge that mortgage payments were fixed for twenty-five or more years;

Table 3.4
New house user costs and cash flow:
Canada, 1965-1984 ($)

Year	Real user costs	Real cash flow costs
1965	–	5,057
1966	–	5,410
1967	–	5,583
1968	–	6,061
1969	–	6,522
1970	4,106	6,870
1971	3,508	6,548
1972	3,037	6,677
1973	2,842	7,479
1974	4,863	9,262
1975	5,348	8,969
1976	5,539	9,257
1977	4,827	8,270
1978	4,620	7,936
1979	4,973	8,132
1980	5,791	8,956
1981	7,793	10,711
1982	7,714	9,858
1983	5,071	7,621
1984	5,315	7,679

SOURCE Steele (1987, Table 11, 17). – indicates data are not available.

instead, payments could change upon renewal in five or fewer years. The magnitude of the possible change is indicated by the fact that some purchasers borrowing at 11% in 1976 faced a rate of 18% or more at renewal in 1981. A purchaser faced an increased possibility that a home that was affordable when purchased might later become unaffordable.

One implication of the tilt is reduced access to mortgage finance. The high initial mortgage payment under inflation means that some households cannot qualify for a mortgage, even though on average, over their working life, they can afford the payments. This reduction in accessibility especially hurts low-income households, an observation that motivated the design of AHOP and its successor, the GPM. The reduction in accessibility because of the tilt, however, is greater than it seems. Initially, the increase in the tilt during the 1970s was accompanied by a relaxation of lending rules that helped offset the reduction in accessibility caused by the tilt. The maximum ratio of PIT to income was increased; the percentage of spouse's income included for the purposes of this ratio was increased; and downpayment requirements were reduced. This relaxation allowed households to purchase even when their resources were tightly stretched. This probably makes sense only in the context of a tilt. A 32% ratio of PIT-plus-heat-to-income at the time of purchase would likely be a problem if it were not quickly eroded away by inflation.

Households can adopt strategies to produce a homemade flattening of the tilt. One strategy is to purchase a cheaper house than would otherwise be done, with the plan to move up from this so-called "starter," once inflation has sufficiently reduced the payment-to-income ratio and increased the equity-to-value ratio. The ready availability of condominium apartments and townhouses in the 1970s in many cities allowed purchasers to do just this. A second strategy is to buy a cheap unrenovated house with the plan to renovate it later. A third strategy is to rent out part of a house, initially, planning to occupy the whole house later.

All these strategies to deal with the tilt result in less consumption – equivalently, more saving, during the early years of home ownership than if there were no tilt. More saving before entering home ownership, to accumulate a large downpayment, is another strategy for gaining access, and this strategy was subsidized from 1974 to 1985 by the RHOSP tax shelter.

Some households with low net worth receive intergenerational transfers that enable them to make a large downpayment and reduce the burden of the tilt. Intergenerational transfers will be encouraged in times of inflation because of inflation's positive effects on the net worth of elderly home owners and the low after-tax real return to financial instruments.

An implication of interest rate variability is the increased possibility that home owners may have to sell or default on their mortgages because an initially affordable home becomes unaffordable upon mortgage renewal. This is relatively unlikely: in the previous example of the home owner renewing in 1981 at 18%, the monthly payment increased by about 40%. But the payment-to-income ratio was still less than in 1976, the date the mortgage was originally taken out, if the household's income increased by 53%, the average rate of increase over the period (Canada 1986, 82, 117). Thus, the tilt, even in this extreme example, rescued the home owner. Strong demand by households for short-term mortgages and weak demand for the Mortgage Rate Protection Plan suggests that households do not regard interest variability as an important risk problem, although high interest rates do negatively affect demand.

The great variability in house prices also has important implications. It means that great capital gains are possible. The lure of these tax free gains is enhanced by the availability of modest downpayment loans. The resultant leverage means that a purchaser with appropriate timing can earn a high return on equity. This fact receives attention in the popular press. What receives less attention is the fact that, for home owners with a large mortgage, the net return is usually much less than the gross return, because of high interest and other costs of home ownership, in a time of inflation. Furthermore, just as price variability means great capital gains are possible, it also means large losses are possible. This possibility became a reality for many home owners in Western Canada in the early 1980s. The losses suffered by home owners were reduced, however, by an asymmetry that does not exist in the case of business borrowers. Mortgage insurers sometimes accept a "quit claim" when a borrower is no longer able to make mortgage payments: the borrower's losses are limited to the downpayment plus the accumulated difference between periodic cash costs and gross imputed rent. The home owner "walks away" rather than bearing the difference between the mortgage principal and the market value of the house.

Issues

One issue raised by the preceding discussion is the appropriate policy response to the inflation and high variability of interest rates and house prices that was endemic in the 1970s and early 1980s. The mortgage design response is discussed in Chapter 6, but other responses should be considered as well. A consequence of great variability in house prices is large differences in households' net worth depending on when and where they happened to buy. Someone who bought a house in Vancouver in 1978, sold it on transferring to Toronto in 1980, and bought in that city in 1980 would be far better off in 1985 than someone who bought in 1981 in Vancouver and remained there.

Another issue is the appropriate response to the enormous increase in the number of non-traditional households. These households are on average small and tend to be headed by women. Their home ownership rate is lower than that of traditional households. Should there be programs to specifically encourage them to become or continue to remain home owners? Should there be a program, for instance, directed specifically at single parent mothers just as there is a program targeted at Aboriginal peoples?

Perhaps the best response would be non-categorical programs targeted at all low-income households. At present, there are no such policies, although low-income families make up a significant proportion of those benefitting from the extension in the late 1960s of NHA financing to cover condominiums and existing houses. In 1984, for instance, while borrowers with family income under $30,000 accounted for only 15% of all NHA borrowers for new single-detached houses, they accounted for 32% of borrowers for existing single-detached houses and 33% of borrowers for condominiums (CHS 1984, Tables 86, 87, and 88).

There are several rationales for special encouragement to low-income families. First, home owners are less likely than renters to need income supplements in old age. Home ownership, because of its associated forced saving, increases net worth and reduces this need. A small amount of assistance to low-income families with middle-aged heads will pay dividends in the future, in terms of reduced income supplement payments. Second, the tax breaks for home owners are less for low-income families (because of their lower marginal tax rate) than they are for high-income families. Third, low-income families with children are apt to be perceived by landlords as costly tenants, so they will find it difficult to obtain accommodation: home ownership is a solution to this problem.

Subsidizing home ownership for low-income households along with a housing allowance for private renters would mean that low-income households would no longer have to live in public housing, non-profit housing or non-equity cooperative housing to receive an explicit housing subsidy. Of course, for many non-traditional households – for example, many widows and single-parent households – home ownership of a non-traditional type may be the best choice. A condominium relieves the home owner of many of the management and maintenance tasks of a freehold home owner. Equity cooperatives – not just non-equity ones – allow the small household to share these tasks the way larger households in a single-detached house do. Co-operating by sharing a duplex with another household is a further example of the diversity of home ownership solutions for non-traditional households. Any home ownership subsidy policy should be sensitive to this diversity and should be comprehensive in the range of arrangements it supports.

Notes

1 A home owner is taken to include condominium owner-occupiers and owner-occupiers of equity cooperatives, but not occupiers of non-equity cooperatives. In the latter, which are funded by special programs under the NHA, occupiers are not owners.

2 The distinction between owners and non-owners is less sharp than this simplified analysis suggests. There is a continuum: a condominium or equity cooperative home owner clearly has less control over his environment than does a freehold owner, but more control than a renter. A non-equity cooperative member also has more control than a renter. Renters now have more control than they did in times past because of increased regulation of landlord-tenant relations.

3 Of course, in times of highly variable interest rates such as the late 1970s, the standard Canadian short-term rollover mortgage will have quite unpredictable cash flow requirements. Predictability is greater the longer the term, but at times terms as long as five years have been virtually unavailable. It is now possible to limit unpredictability by purchasing mortgage rate insurance. This and other aspects of mortgage risk are discussed in Capozza and Gau (1983).

4 Prior to 1972 capital gains on other assets were also tax-exempt; in 1985 these capital gains were partially sheltered with the phasing in of a capital gains exemption, now capped at $100,000 (and not available for real estate gains as of 1992).

5 Suppose, for instance, an apartment unit priced at $50,000 is financed by loans and a mortgage equal to this amount, and the interest rate on the financing is 11%. Suppose that the actual and expected rate of inflation is 5%. The full annual interest of $5,500 is deductible as an expense, including the $2,500 component attributable to inflation. Total expenses including interest will likely be greater than rents in these circumstances, so they will generate a net deduction for the investor. Now, at the end of five years, suppose that the investor sells the unit at a price, $63,800, which merely reflects the 5% inflation rate, compounded. The gain of $13,800 is taxed, in effect, at only 75 percent (50% prior to tax reform) of ordinary tax rates, even though the $2,500 per year of interest costs attributable to inflation have been fully deductible.

6 This point is fully discussed in Steele (1992).

7 See Clayton (1974) for a discussion of the position of landlords *versus* home owners and Clayton and Associates (1984) for a detailed accounting of the recent tax subsidies and explicit subsidies for landlords.

8 In the *New Palgrave Dictionary of Economics*, published by MacMillan in 1987, "markets are complete when every agent is able to exchange every good either directly or indirectly with every other agent." Incompleteness here is a consequence of fundamental demand and supply conditions.

9 Factors generating gaps in the housing market are analyzed in Bossons (1978).

10 Survey evidence of landlords' perceptions of the costs of low-income households with children is given in Steele (1985c, Chapter 2). Discrimination against children is sufficiently important that some jurisdictions have enacted laws prohibiting it (for example, Ontario in 1987, as part of human rights legislation). See Choko (1986) for comments on discrimination against children in Montreal.

11 Among these factors are gentrification and rent regulation. For data on the effects of gentrification and rent regulation in Toronto see Smith and Tomlinson (1981). For Montreal, the effect of gentrification and condominium conversion has been described as follows:

> These new phenomena produced a profound transformation of the old core areas, due especially to the forced departures (as a result of takeovers or as a consequence of rental increases) of the traditional rental households. The elderly and the inactive, especially households headed by women, were most affected. The old rental market in Montreal in the core area changed hands. The new residents have much higher income, many more academic credentials and are younger than the former residents: everything differentiates the new residents from the old (Choko 1986, 16).

Choko (1986, 20) cites studies of conversions finding that 90% of resident households were forced to move and the elderly are the most likely to be dislodged. Also see note 14.

12 At the same time, it should be noted that the NHA provided for joint loans (later replaced by insured loans) for construction of apartment buildings as early as 1944. Also, the federal Rental Insurance Plan operated from 1948 to 1950.

13 This is apparent in the income distribution data. *CHS* (1965, Table 61) indicates, for example, that 24% of all Canadian families had an income below $3,000 in 1959, compared to only 0.1% of families borrowing under the NHA; only 26% of all families had an income over $6,000, compared to 48% of NHA borrowers.

14 See Choko (1986) for the loss of low-rise low-income housing in Montreal. Also, see note 11. In the City of Toronto, according to Ward, Silzer and Singer (1986), about 1,000 units a year of low-rise stock have been lost; and about 2,000 moderate rental units per year have been lost in buildings containing six or more apartments because of demolition, conversion, and luxury renovation. In addition, according to Ward *et al.*, "Planning staff of other municipalities within Metropolitan Toronto and City of Ottawa planners indicate that their losses due to the same kinds of pressures have been substantial" (1986, 4). Ward *et al.* attribute a substantial part of the move up-market of these latter buildings to the renovation provisions in the pre-1987 rent review legislation. Perhaps the most notable

example of a high-rise complex which has filtered down from middle-income singles to low-income families and the elderly is St. James Town, a 6,000 to 7,000 unit development in the City of Toronto.

15 For further information on provincial programs see Dennis and Fish (1972, 276-7.).

16 At the same time, CMHC did focus its activities on modest-cost, rather than luxury, housing (for example, the Small Homes Loans Program introduced in 1957).

17 The president of CMHC in his reaction to this proposal, said: Indeed a rental purchase scheme with virtually nothing down is rental housing ..." (cited in Dennis and Fish 1972, 266). This odd statement ignores, among other things, the fact that house purchase provides a household with control over its environment, including security of tenure, and it generally results in a build-up of equity.

18 The proposal came from the builder, Robert Campeau. The president of CMHC, in his reaction to the proposal stated: ... "Mr. Campeau's proposal would undoubtedly enable families of lower income to achieve home ownership.... The National Housing Act recognizes that not all families are able to own their own home. The Act makes special provision for low rental housing projects...." Another official, in 1967, argued: "One of the objections to the principle of providing subsidies for home ownership has been reluctance to asking some people ... to pay for the acquisition of assets by other people." Both quotations are taken from Dennis and Fish (1972, 267-8). It may be inferred that CMHC felt ready to subsidize housing, but was not ready to let a low income family choose its tenure by offering a subsidy of the same present value, no matter what the tenure choice.

19 Dennis and Fish (1972) suggest that the typical recipient was a young man "on the way up" rather than someone with a low life-time income.

20 The "program permitted a below-market interest rate." See *CHS* (1971, xii). Over 20% of the units under this and the 1970 program were condominiums. The median income for the two programs was $6,112, about half the income of borrowers in the regular home ownership programs and only slightly higher than that of tenants in Section 15 non-profit rental units. The average age of borrowers was 31 years. See *CHS* (1971, xviii).

21 All AHOP participants (there was no income ceiling, although there was a house price ceiling) received an interest-free loan which increased by a decreasing amount each year for five years. At the end of five years, no further additions were made to this loan and the interest holiday ended; repayment started at the end of six years. Ontario and Nova Scotia piggy-backed a grant for lower-income families onto AHOP, so that in those two provinces the subsidy was particularly deep. For some further details see *CHS* (1973, xviii; 1974, xx) and Rose (1980).

22 Defaults are given in *CHS* (1985, Table 67) and units funded in *CHS* (1979, Tables 60 and 61). Defaults and units funded include those from the 1970 and 1971 programs discussed earlier, as well as from AHOP proper. (Information on coverage of defaults thanks to Paddy Fuller, CMHC).

23 The default rate for non-AHOP, non-ARP (that is, "regular") NHA new home ownership units plus NHA new and existing rental is estimated at 5%. This is computed by taking the ratio of defaults for regular new home ownership plus regular new and existing rental, to total units in three categories (new single-detached and multiples plus existing multiples) net of AHOP and ARP units. See *CHS* (1979, Tables 60 and 61; 1983, Table 60;

1985, Tables 66 and 67). This is an underestimate of the true rate for two reasons. First, some of the existing multiple activity would be condominium (that is, properly categorized as existing home owner), but no existing home owner defaults are included; this tends to make the denominator too large while not affecting the numerator. Second, both the default data and the activity data are for 1974 to 1985, which means that many of the defaults after 1985 are excluded. In contrast, the AHOP defaults are those occurring 1974 to 1985 for units built 1970 to 1978 (and few of these were built 1970 to 1973).

24 Two kinds of market imperfections reduce the force of this statement. First, just as rental markets are incomplete, so are home ownership markets. Thus, someone wishing to occupy an old, unrenovated, small apartment will find it difficult to become a home owner; few condominiums or equity cooperatives are offering this kind of accommodation. Indeed, in some locations, there are no condominiums of any kind. Second, a household which believes it can afford home ownership may find that it is impossible to borrow on optimal terms. For instance, a young person in a secure job with a highly certain, rising income profile might wish to purchase with a low downpayment and with payments rising over time to match his income profile. Despite a willingness to pay an interest rate with a risk premium which would cover the risk of such a mortgage plan, he is unlikely to get one. See Lessard and Modigliani (1975) for a discussion of this aspect of the imperfection of mortgage markets.

25 *Census of Canada 1971* (II.4, Table 35); *Census of Canada 1986 (The Nation; Dwellings and Households: Part 2*, Table 8, Catalogue 93-105). 1986 percentages exclude maintainers living on reserves from both the numerator and denominator.

26 *Census of Canada 1986 (The Nation; Dwellings and Households: Part 2*, Table 8, Catalogue 93-105).

27 Bossons (1978), using US data, also finds a strong positive relationship between age and demand for home ownership, other things being equal. He attributes this to the likely increase in leisure time with age and the complementarity of the attributes of owner-occupied housing and the consumption of leisure time. Jones (1984b, Table 1-25) finds a negative effect of age on the demand for home ownership, other things equal. Struyk (1976) regards age as such a fundamentally important variable that he stratifies his sample according to age.

28 As well, some of the observed rise in house prices may be a statistical artifact of the method of construction of the index. For example, although the index, in principle, measures the price of a house (plus lot) of constant quality, the lot price used in the early years of the index is simply the NHA average lot price. The average lot, however, changed over time. It tended to increase in value because of the increase in servicing, but it tended to decrease in value because of its decrease in size, a decrease that was made possible in some cases by the replacement of septic tank systems with municipal services. It is not clear whether the net result is a positive or negative bias.

29 Part of the decline in real rents is almost certainly a statistical artifact caused by downward bias in the rent index. See Loynes (1972) and Fallis (1980).

30 For expansion of this point, see Steele (1992). It should be noted that the highest corporation tax rate is not as high as the highest personal rate, so that the tax advantages outlined are not as potent for corporate landlords as for individual landlords.

31 For an indication of the quantitative effects of these, see Clayton Research Associates (1984) and CMHC (1983a, Annex 3). Both these studies use odd assumptions: the Clayton study assumes zero nominal capital gains; the CMHC study in effect assumes a real interest rate of about 10% – about double the standard assumption for real interest rates – and uses a discount rate equal to this, rather than to a nominal rate, to find the present values of nominal streams.

32 The inflation rate is measured as the rate of increase in the Consumer Price Index; this is available in *Historical Statistics of Canada* (Series K8) and Department of Finance (1986).

33 Omitted is transactions cost which, unlike the other cash flow costs given here, is not a regular periodic cost. Also, the transactions costs cash requirement on purchase is much less than that on sale because there is no real estate brokerage fee on purchase.

34 The opportunity cost is given in after-tax terms because housing expenses, like other consumption expenditure is paid out of after-tax income.

35 More specifically, expected capital gains are estimated as follows. Price change in quarter t is regressed on price changes in past quarters. This estimated equation is used to predict for each quarter the average annual compound rate of capital gain over the next five years. This is the expected capital gain used in the computation of user cost. Further details are given in Steele (1987).

36 The payment is calculated assuming that interest is credited annually; the formula used is

$$P \; = \; i \; \frac{50{,}000}{1+(\frac{1}{1+i})^{25}}$$

where P is the annual payment, and i is the rate of interest. Published tables are available giving monthly payments under the assumption that interest is credited more frequently than annually.

CHAPTER FOUR

New Forms of Owning and Renting

J. David Hulchanski

IN THE period immediately after World War II, tenure options were relatively uncomplicated. Accommodation was largely either owned fee simple or rented, with few variations of either. By the 1980s, however, two new forms of ownership had emerged. Condominiums, introduced to Canada in the late 1960s, allowed individual ownership of a dwelling in a multiple unit project together with shared responsibility for maintenance of common areas and facilities. Non-equity cooperatives, also introduced in the late 1960s, are a form of ownership in which members (that is, residents) jointly own the dwellings, land, and common facilities.

Immediately after World War II, owners were free to build or alter a house or apartment building and to lease a dwelling therein in whatever manner they chose, subject only to rudimentary building and land-use restrictions, common law, and market forces. Since then, the property rights of owners and tenants have changed. The property rights of all owners, whether home owners or landlords, or purchasers or vendors, became subject to increased regulation of the use and exchange of real property. The character of rental tenure has also changed dramatically for landlord and tenant. Over the past few decades, concern for security of tenure and due process in rental housing led to the adoption of two types of regulation. One is the landlord and tenant legislation that most provinces adopted during the early 1970s. The other is rent regulation that most provinces introduced in the mid 1970s to control changes in rents for much of the stock. Prior to this, the landlord-tenant relationship was a property one under common law. By the late 1960s, it came to be widely seen that leases and common law were not adequate to protect tenants. Since the 1970s, the relationship between the two parties has shifted from its feudal origins in common law to a statutory basis in modern contract law. By the mid 1970s, most provinces had adopted legislation which included the introduction of contract principles, the requirement for landlords to repair and maintain, the requirement that a tenant be given a copy of the tenancy agreement, and provisions affecting the manner in which a landlord can evict and regain possession of a dwelling. In short, as society's view of what was considered to be fair and reasonable in the landlord and tenant relationship changed in the 1960s and 1970s, provincial legislation was altered to conform with the change.

This chapter reviews these changes and examines how they have affected housing in post-war Canada. Basic to the issue of tenure is the question of property rights. The place and meaning of property, and changes in how society views the rights associated with property, shapes trends in housing tenure. Issues relating to the role of government in housing markets and debates over the various forms government intervention should take are all rooted in the concept of property and views about property rights. Though the demand for ownership and rental housing is influenced by the bundle of rights associated with each tenure option, the issue of property rights has generally been ignored in housing analysis.

New Forms of Owning Housing

The demand for home ownership after World War II was initially met by building detached houses on individual lots on the fringes of cities. The NHA emphasized this form of housing, providing subsidized mortgage funds and, beginning in 1954, mortgage insurance to make lower downpayments and longer-term mortgages possible. In spite of the subsidies available and the prosperity of much of the post-war era, it became increasingly difficult for many households to afford the traditional suburban house. Supply constraints on the availability of serviced land, and at times, mortgage funds, together with the post-war demographic pressures, contributed to housing supply problems and house price inflation. The rise of the two-earner family (60% of all families in 1981, up from 33% in 1951) is likely both a cause and effect of the post-war inflation in the cost of home ownership. It is a cause in that families with two earners could allocate more money to obtain the housing of their choice, thereby helping to bid up the price of housing, and it is an effect in that other families needed a second earner to afford home ownership.

The continuing demand for home ownership, combined with its ever rising cost, gave rise to two new forms of ownership that are now commonplace: condominiums and non-equity cooperatives. Each changed the notion of what is "owned" and how it is owned.

CONDOMINIUMS

The first condominium acts in Canada were adopted in 1966 by British Columbia and Alberta. By the end of 1970, all but one of the provinces had adopted condominium legislation; Prince Edward Island did not adopt legislation until 1977. Condominiums provide a package of property rights through a legal arrangement that makes it possible for an individual to own a dwelling without exclusive ownership of the land on which the structure is built. Together with the dwelling, each resident of a condominium jointly owns a proportionate share of the common elements, such as sidewalks, driveways, landscaped areas, recreational facilities, elevators, corridors, parking, and storage. A condominium is a form of ownership, not a housing form. Condominiums may be detached, semi-detached, row townhouse, stacked townhouse, or apartment structures. Although a condominium provides property rights similar to individual home ownership, its communal environment requires each resident to yield some individual rights for the sake of harmonious management of the project as a whole.

Condominiums became popular as a result of growing urbanization accompanied by high land values and a continuing demand for home ownership. Factors such as rapid population growth, demographic changes, decreasing household size, and rising household income, all contributed to a high demand for housing, particularly ownership housing. At the same time, other factors, including rapid increases in the price of housing and the cost of residential land as well as increased commuting time to the new suburbs gave rise to a demand for changes in the laws governing the ownership of housing. By separating ownership of the dwelling from ownership of the site, more Canadians could become home owners at a potentially lower cost as a result of savings achieved by collective ownership of the land, common elements, and shared maintenance.

Though traditional fee simple ownership is usually the preferred form of home ownership, condominiums became more widely accepted as the cost of and the demand for ownership increased in the 1970s. Tax changes introduced in 1971 further stimulated demand by exempting privately-owned houses from the newly-introduced capital gains tax and removing the tax shelter provisions which applied to rental housing. These Income Tax Act changes made home ownership more attractive and the ownership of rental housing less attractive.

All these demographic and tax considerations increased the demand for home ownership, causing many developers to shift their activities from the development of rental to ownership housing to satisfy the demand. The multiple unit condominium project was a legal innovation which provided home ownership at a potentially lower price. Though many factors contributed to the decline of the private rental market, an important one is the increase in condominium starts. Developers could obtain an immediate return on their investment, rather than the gradual return obtainable by rental housing investment. Purchasers of condominiums obtain the benefits of ownership, usually in better locations than otherwise possible within their budget constraints. Not all condominium units are owner-occupied. Many have been purchased as investments and are rented.

Local market conditions affect the demand for condominiums. Condominiums have become a significant part of the housing market in the metropolitan areas with the highest housing costs. Toronto and Vancouver accounted for one-half of all condominiums in Canada in 1981. Condominiums are uncommon in cities where land is more affordable.

In the early 1970s young couples from the post-war baby boom purchased condominiums as an alternative to either renting or single-detached ownership.

As the decade progressed, more empty nesters entered the market, and builders aimed new projects at this market segment. A 1984 study concluded that the residents of the more than 200,000 owner-occupied condominiums in Canada represented a broad spectrum of households and that the market has three basic components: "One caters to the under 40 year old apartment dweller without children who eventually plans to buy a single detached house; the second is the young family; the third is the empty nester seeking apartment condominiums" (Skaburskis 1984, 34-5).

The introduction of condominiums to Canada has broadened the range of available home ownership options and increased the supply of dwellings, thus benefiting both suppliers and consumers. However, this innovation has not been without its problems. The quality of construction in many condominiums may prove to become a serious problem as the buildings age. The nature of condominium development and ownership means that, other than for warranty or performance guarantees, the builder is free of long-term responsibility for the project once the units are sold. The quality of construction, materials used, and design considerations relating to life cycle maintenance costs may be compromised. The fact that the units are sold to individuals who rarely if ever know each other prior to purchasing their units means that the initial purchasers of a newly-built project have little or no opportunity to influence or supervise the quality of the design and construction process, as is sometimes the case in the construction of new houses. The quality and dedication of the management of a project is also extremely important, though it is easy for poor management to go unnoticed until serious maintenance or financial problems arise.

Condominiums have also played a role in the decline of the rental sector. Until prevented by regulation, landlords were free to convert existing apartment buildings to condominium ownership, contributing to the loss of the existing rental stock. In addition, many moderate-income households which would otherwise have demanded rental accommodation have opted for the condominium option. This means that the demand for rental housing in the 1980s falls increasingly into the category of social need rather than market demand. As the private sector responds only to market demand, not social need, unsubsidized private rental starts have been in decline since the early 1970s.

COOPERATIVES

Cooperative ownership of housing is not new to Canada. In the 1930s a number of "building cooperatives" were established in smaller communities, mainly in Nova Scotia and Quebec. In a building cooperative, a group of people join together to build each other's individual homes. Though this kind of housing cooperative was successful in small communities, it was a difficult model to apply to an increasingly urbanized nation and to large groups.

Advocates of housing reform then began to focus on the "continuing cooperative" model whereby the members jointly own the entire project on a continuing basis rather than take individual ownership of the units once they are built.

As more and more households realized that they could not afford the purchase of a home, housing advocates began to investigate non-profit continuing cooperatives. There are a few equity cooperatives in Canada, but the vast majority have been non-equity cooperatives, built mainly since the early 1970s with the assistance of federal and, from time to time, provincial subsidies.

In 1962, with the financial assistance of CMHC, the Cooperative Union of Canada undertook research into the feasibility of non-profit continuing cooperatives. In 1966 the first large non-equity continuing cooperative, the 200-unit Willow Park Housing Cooperative, was built in Winnipeg. Enough momentum had developed during the 1960s for a national organization, the Cooperative Housing Foundation of Canada (CHF), to be established in 1968. The Hellyer Task Force recommended that greater emphasis be placed on finding ways for moderate-income households to obtain home ownership and that the public housing program be discontinued in favour of socially-mixed assisted housing projects. Cooperative housing was one of the options recommended. When a special $200 million fund for social housing was established in 1970, the CHF succeeded in having eleven housing cooperatives financed to test the cooperative model further. When the NHA was revised in 1973, non-profit continuing housing cooperatives were recognized by policy makers as a desirable and feasible tenure option as well as a social housing program option. The number of housing cooperatives and sponsoring organizations increased, and when the program was further revised in 1978, the federal commitment grew to about 5,000 units per year.

Much controversy surrounded that program, focused mainly on program targeting. The social mix component of the cooperative program means that not all the financial assistance benefits low-income households. To some degree, this debate was settled when the cooperative program was further revised in 1985. The federal government not only decided to continue its commitment of about 5,000 units annually under a new funding formula but also clarified its objectives. The Minister Responsible for CMHC explained that the main objective is "to provide security of tenure for moderate and middle-income households as an alternative to home ownership." The program is to "provide a level of assistance intended to help that group of people with incomes above those in core need that are unable, through no fault of their own, to afford home ownership." A rent supplement subsidy will also be available to enable low-income households to access cooperative housing (Canada, *House of Commons Debates* 12 December 1985, 9433). The cooperative housing program can be viewed as not only an alternate to traditional home ownership, but an accessible version of condominium ownership for low and middle-income households.

In terms of the entire housing stock, the 40,000 cooperative dwellings completed as of 1986 represent a small share, less than 1%. Most of these units, however, have been built since the late 1970s. They represent a larger share of annual housing starts, and they provide the only means for many households to achieve the benefits of home ownership in the high cost metropolitan housing markets.

The 40,000 households who live in one of Canada's 1,000 cooperatives do not own the individual unit they live in, nor do they make a traditional downpayment. Like tenants, cooperative members move into and out of their housing without making an investment and without earning capital gain. Like home owners, however, they have security of tenure and the right to make all decisions about their housing environment. There is no outside owner or manager. Members of a cooperative jointly own the entire project and jointly share in its management. A board of directors is elected and members are appointed by the board to sit on a number of committees, usually a maintenance, finance, and membership committee. The monthly housing charge is set each year by the members at a level sufficient to cover the mortgage payment, operating costs, and replacement reserves. The entire process of ownership and management is a democratic one, each resident having one vote.

New Forms of Renting

What distinguishes rental from owned housing is that rental tenure separates ownership from occupancy. The owner of rental housing becomes an investor in a good (accommodation) that can be treated like any other typical investment, whereas the occupant is the user of the good, with little or no concern for its investment aspect. In contrast, home owners can decide what level of housing maintenance and renovation to sustain. The separation of ownership from occupancy creates the potential for conflict when investment and occupancy interests differ. In Canada, up until the early 1970s, the owner's interest dominated. There was no balance of rights, responsibilities, or power in the landlord-tenant contract. Common law treated residential tenancies the same as it treated commercial and industrial tenancies.

Since then, governments have sought, in housing as in other areas, to eliminate abuse of basic rights on the basis of such grounds as race and sex and to protect consumers from misleading and arbitrary actions. Intervention by regulation has focused especially on basic human needs, such as good health and physical safety as well as fundamental principles of justice and due process. Out of this, landlord and tenant legislation has emerged. In some jurisdictions, rent regulation has followed the adoption of landlord-tenant legislation because of the potential of economic eviction, that is, the possibility that landlords will circumvent security of tenure regulations by using higher rents to evict tenants. There is, therefore, societal recognition of the uniqueness of rental housing which separates it from the host of other goods we treat as normal market commodities.

SECURITY OF TENURE

The notion of security of tenure has found its way into residential landlord-tenant law over the past two decades just as "stability of employment" has permeated employment law over the course of this century (Glendon 1981, 176-7). Landlord-tenant law changed little before the 1960s. Changes in employment law began much earlier, largely because of labour organizations. Although some 40% of Canadians have been tenants since the 1940s, tenants were not well organized, at least not until recently, and then only in a few of the larger cities. Legislatures and courts, following public opinion, especially with regard to eviction, increasingly came to the view that the landlord, like the employer, controls a basic human necessity. Shelter is as crucial to subsistence as a job. "Increasing acceptance of this idea as an implicit premise has made legislative, as well as judicial, regulation of the residential rent contract as inevitable as it was of the employment contract" (Glendon 1981, 177).

As a result, both employment and rental relationships have passed over time from status to contract, from contract to regulation, and to some degree, from regulation to administration. "Such a view," according to Makuch and Weinrib, "supports the notion that freedom of contract is no longer the norm and that society can and should impose social values in landlord and tenant relationships as it does in consumer affairs and other contractual tortious relationships based on reasonable reliance and expectation" (1985, 8). Social attitudes are changing, bringing subtle but significant changes in our institutions. Security of tenure and rent control in Canada's residential rental sector are prime examples. Anglo-American lease law had developed the rule of "mutuality" that either party might terminate a tenancy at will or a month to month tenancy for any or no reason upon complying with statutory notice requirements. Over the twentieth century this rule of free terminability has been replaced by its opposite, security of tenure. Just as discharges of employees at-will became increasingly unlawful, so too legal termination of leases began to be forbidden if they contravened public policy.

If security of tenure is to be achieved, there is need also to deal with demolitions and conversions of rental units to other uses or to owner occupancy. As the first province to adopt legislation permitting condominium ownership, British Columbia had to deal with the problem of the conversion of rental units to condominiums in the early 1970s. With vacancy rates as low as 0.4% in Vancouver and Victoria in 1973, and conversions causing controversy because of the displacement of lower income renters, British Columbia's condominium legislation was amended in 1974 to allow municipalities to stop conversions (Hamilton 1978, 136-8). During the 1970s, many provinces and municipalities adopted legislation regulating and usually preventing such conversions. This represents a further change in the nature of the bundle of rights associated with owning rental accommodation.

RENT CONTROL

Rent control was first used in this century during war time. Britain imposed rent controls during World War I and virtually every major combatant, including Canada, imposed rent controls as part of more general price controls during World War II. Canada's Wartime Prices and Trade Board imposed rent freezes in fifteen cities in September 1940. A year later, rents in the rest of the country were frozen. This was "simple" rent control, that is, an absolute freeze. There was no complex formula for permissible rent increases or exceptions.

Beginning in 1947, a period of rent decontrol began in Canada, as it had in other western countries. Federal housing policy sought to establish a private rental housing development industry. During the war, Wartime Housing Ltd., a crown corporation, built rental housing for war industries and after the war for veterans. In the post-war period, in addition to decontrolling rents, the federal government introduced subsidies for the private sector, both directly and through the tax system. With these direct and indirect subsidies in place the federal government ended rent controls in 1951, along with the remaining wartime price controls. Only the province of Quebec maintained rent regulation beyond 1951. From the early 1950s to the mid 1970s, rent regulation was largely non-existent in Canada.

Rental market pressures were such that, in the early 1970s, demands for rent control became more common. Security of tenure legislation had been adopted by most provinces but, in the inflation of the 1970s, it meant little if economic eviction was possible. Several provinces had either adopted rent controls or were about to adopt them as of 1975, and the decision of the federal government to impose wage and price controls in October 1975 carried with it a request to the provinces to impose rent controls. By April 1976, all provinces had rent controls in place. Most of the legislation was retroactive, commencing on or before the October 1975 speech of the Prime Minister which announced the wage and price controls. Most of the legislation was comprehensive in terms of the types of residential premises covered. Each province established a system of rent tribunals separate from other price control mechanisms (Patterson and Watson 1976).

Though rent control was considered to be temporary by many, all but three provinces maintained rent controls into the mid 1980s. They have remained a permanent feature of the rental market in much of the country. The critics of rent control rightly point out that an attempt to use rent control to redistribute income would lead to deterioration in the housing stock and disinvestment in the rental housing sector. Moreover, it is a blunt and unsuitable instrument for the positive redistribution of income. However, a rationale for rent controls is that they can be effective in retarding undesirable redistribution in the opposite direction.

When housing supply fails to keep pace with demand, as it has since the early 1970s, the scarcity value of the existing rental housing stock becomes a powerful force reallocating income in favour of landlords. Rent stabilization or control, in the short run at least, frustrates this undesirable change in the distribution of income. In this sense of protecting the status quo and frustrating undesirable income redistribution, rent control is a conservative instrument. It cannot improve the situation, but it can at least prevent it from getting worse for tenants caught by the failure of the rental market supply mechanism. While not all tenants are disadvantaged, the majority do have low incomes and are unable to afford home ownership. They have no place to go.

The benefit of rent control – frustrating undesirable income redistribution – is hidden in the sense that it can be measured only against a hypothetical rent that would have come into existence if the control had not been there. It is a cost to the landlord also, but it is not one which involves a real loss, since expenses have not increased, but rather a hypothetical cost – the loss of what the landlord would have made if rent control had not been imposed (Patterson and Watson 1976).

Implications for Housing Policy Debate

There are two categories of significant implications of these trends in post-war tenure for future housing policy debate. The first relates to the broad institutional framework in which housing policy is situated, and the second relates to the high demand for ownership housing over the post-war period.

THE PROPERTY RIGHTS DEBATE

Over the past several decades, the growth in the role of government in matters affecting the development of land and housing has made it is easy for some to claim that there has been an "erosion of property rights." According to a study published by the Ontario Real Estate Association, for example, "property rights are being eroded at an ever increasing pace" as a result of an unabated "avalanche of legislation that affects the citizen's property rights" (Oosterhoff and Rayner 1979, v, ix.). There is indeed a great deal of legislation affecting all aspects of owning and renting housing as well as the site on which it is built (see, for example, Hamilton 1981). However, "erosion" assumes some defined set of property rights which is ideal and that any change away from that is a step backwards.

Property and the ownership of property is a social and juridical institution. Property "rights" are socially defined. Property rights are "the creation of positive law whatever social or political theory may presuppose about their metaphysical origins in the natural or supernatural order of things. The legislature can give and take, allocate and reallocate titles to them" (Denman 1978, 3). As a result, the meaning of property is continually in a state of flux. The "actual institution, and the way people see it, and hence the meaning they give the word, all change over time ... [and these] changes are related to changes in the purposes which society or the dominant classes in society expect the institution of property to serve" (Macpherson 1978, 1).

Such extensive intervention in the housing market seems to be based on two rationales: the need to correct the real or perceived failings of the market; and the desire to achieve certain social objectives. In technical terms, the first rationale refers to the market's failure to allocate efficiently in terms of the supply and demand aspects of rental housing as a market commodity. The second rationale refers to the political decision in a democracy to achieve certain social objectives even if this means making a trade-off with the efficient operation of the market.

At the broad institutional level, the meaning of property rights and tenure in

Table 4.1

Changes in home ownership rates within and between income quintiles:
Canada, 1967 to 1981
(% of households owning their unit)

	1967	1973	1977	1981	1967-81
Lowest quintile	62	50	47	43	-19
Second quintile	56	54	53	52	-3
Middle quintile	59	58	63	63	+4
Fourth quintile	64	70	73	75	+11
Highest quintile	73	81	82	84	+10
Total	63	62	64	63	+1

SOURCE Statistics Canada (1983).

Table 4.2

Renter households by income quintile: Canada, 1967 to 1981

Income quintile	1967	1973	1977	1981	% Change 1967-1981
Lowest quintile	20	27	29	31	+11
Second quintile	24	25	26	26	+2
Middle quintile	22	23	20	20	-2
Fourth quintile	19	16	15	14	-6
Highest quintile	14	10	10	9	-5
Total	100	100	100	100	

SOURCE Statistics Canada (1983).

Canada is continuously evolving. This type of change will continue to be controversial, forming the overall framework within which specific housing policies and programs will be developed. Various groups with differing interests and philosophies in the housing sector will be promoting and defending their "rights." Broadly-defined social and community rights will conflict with more narrow definitions of private property rights. The controversial nature of this broader institutional change process will help guarantee that comprehensive and long-term housing policies will not be politically viable. Too many conflicting demands will be placed on elected decision makers, and too little consensus exists.

C.B. Macpherson has defined this problem as the central difficulty of our liberal-democratic institutions:

The central problem of liberal-democratic theory may be stated as the difficulty of reconciling the liberal property right with that equal effective right of all individuals to use and develop their capacities which is the essential principle of liberal democracy. The difficulty is great (1978, 199).

The only resolution seems to be a broader conceptualization of the nature of property and of property rights. The problem is that "we have all been misled by accepting an unnecessarily narrow concept of property, a concept within which it is impossible to resolve the difficulties of liberal theory" (Macpherson 1978, 201). The difficulty disappears once we broaden our concept. Property, although it must always be an individual right, need not be confined, as liberal theory has confined it, to a right to exclude others from the use or benefit of some thing, but may equally be an individual right not to be excluded by others from the use or benefit of some thing. The right not to be excluded by others may provisionally be stated as the individual right to equal access to the means of labour and/or the means of life (Macpherson 1978, 201).

As a basic and essential "means of life," housing is and will continue to be at the forefront of this fundamental philosophical and political debate. Government intervention in the housing sector is controversial because, in a fundamental sense, it affects property rights. Unlike most other consumer durables, housing is intimately tied to the problem of property and property rights. The rights and obligations associated with owning and renting will continue to undergo significant changes.

POLARIZATION OF HOUSEHOLDS BY INCOME AND TENURE

The second implication of post-war trends in tenure relates to the demand for home ownership. Fuelled by many factors, not the least of which has been government housing and tax policy, this demand has led to a situation in which households have become polarized on the basis of income and tenure.

Trends in the distribution of home ownership among income groups are evidenced in Table 4.1. There were gains among the top two quintiles and declines in the bottom two from 1967 to 1985. Eighteen years is a relatively short time for such a marked change in tenure distribution patterns to take place. It indicates the significant impacts of changing macroeconomic conditions since the 1960s have had on the housing sector. The overall percentage of households owning their own units, about 63%, remained virtually the same during the period.

What was dramatic was the change in who were the home owners. During a period in which many subsidies were provided to the ownership sector and to first-time home buyers, the two highest income quintiles made gains in home ownership rates (about 10 percentage points each), whereas the households in the two lowest quintiles increasingly became tenants. The home ownership rate of the middle quintile remained about the same. In short, fewer households in the lower 60% of the income range are home owners today compared to 1967. The temporary home ownership programs introduced since that time have not kept pace with the tide of rising house prices and mortgage interest rates. This trend was of course not due solely to the regressive nature of housing program subsidies. Macroeconomic trends continued to work against low-income households.

The increasing rate of home ownership among the upper-income groups also indicates a significant and troubling trend for the rental housing sector. The rental sector became increasingly a residual one, restricted primarily to low-income Canadians. This has not always been the case, as Table 4.1 indicates. In 1967 tenants were found with similar frequency in each income quintile except the highest. By 1985, however, the incidence of tenants in the two highest quintiles declined, and in the two lowest quintiles increased – both by significant amounts. This means that those households able to take advantage of the home ownership option did so, leaving virtually all those who had no choice in the rental sector.

Is it any wonder, then, that private investors cannot supply new rental units and make a return on their investment? How can Canada have a viable private market in an expensive consumer good when its consumers are increasingly limited to low-income groups? The private rental supply "market mechanism" has not functioned since the early 1970s and most likely cannot function in the future, because of the upward cost pressures on supplying a rental unit and the downward trend in the income profile of renters. The vast majority of private rental starts over the past fifteen years have been subsidized. There are not enough tenants any longer with the incomes necessary to support the economic returns required to make most new rental projects viable. In addition, the costs of building a unit are so high that tenants who can afford to pay the required rent levels are usually able to afford the purchase of a condominium at about the same monthly cost.

It is unlikely that this trend in the polarization of Canadian households by income and tenure will be reversed. It is also unlikely that the decline in the private rental sector will be reversed. The post-war demand for home ownership and the creation of a new form of ownership, condominiums, which essentially permits higher income tenants to own what would otherwise be rental apartments and townhouses, has helped create a socio-tenurial polarization of the country. This has serious social, neighbourhood, and housing market consequences which the current and future generation of Canadians will have to face.

CHAPTER FIVE

The Suppliers of Housing

George Fallis

THIS CHAPTER sets out a supply perspective on housing markets. The supply perspective does not focus on the housing conditions of Canadians, on the number of houses built, or on the price of housing. These are outcomes of the housing market; they are the joint result of demand-side and supply-side forces and of government programs. Supply-side analysis considers the agents making supply decisions (the suppliers of housing), the production technology, and the prices of inputs used in production.

What constitutes progress in the context of the supply side of housing? Discussion of progress on the supply side considers whether there has been increased technological efficiency in the sense of being able to produce services or stock using fewer inputs and whether there has been a reduction in barriers to the free flow of inputs into the production of housing services and stock.[1] This approach to housing progress is helpful in thinking about housing affordability. Housing becomes "more affordable," for example, when technical progress permits housing services or stock to be produced more efficiently, when input prices fall, or when barriers to the flow of inputs are removed. In this sense, progress in improving efficiency is progress with respect to affordability. The government programs of most interest are those which influence technical change, the flow of inputs, and the profitability of being a supplier of housing services or stock: for example, income taxes, property taxes, rent regulation, and land-use and building regulations. Government programs aimed directly at improving housing conditions, such as public housing or non-profit housing, are dealt with in other chapters.

Perspectives on the Supply of Housing

In the analysis of housing supply, the distinction between consumption goods and capital goods is important. A consumption good is something people use (or consume) to increase their well-being. A capital good is something which is long-lived and used in the production of consumption goods. It is useful to distinguish between the market in which a consumption good is exchanged and the market in which a capital good is exchanged. These two markets each have a supply side. In housing analysis, the consumption good is called housing services. All households (except the homeless), whether renters or owners, consume housing services. The capital good used in producing housing services is called housing stock. Housing analysis must consider the suppliers of both housing services and housing stock.

On the supply side of the rental market, landlords are the suppliers of housing services. Housing services are produced using a capital good (a building and its land) as well as other inputs such as labour, heat, and light. There are three sorts of landlords: private sector (for profit) landlords; government sector landlords; and third-sector landlords such as non-profit groups and cooperatives. The vast majority of suppliers are in the private sector. Private sector landlords are a diverse group ranging from large public companies to owners of a small apartment building to individuals renting out a portion of their home. All of these private landlords may be thought of as "firms" with revenues from sales (that is, rents) and costs of production. Private sector landlords pay income taxes on their net earnings, and the income tax system is important in determining the profitability of these firms, and hence the rate at which new firms enter the market. Government and third-sector landlords must be cognizant of their revenues and costs, but do not pay income taxes and are motivated by factors other than profits.

While the supply side of the rental market is obvious, who are the suppliers of housing services to home owners? Home owners consume services each year just as do tenants. On the supply side, home owners are their own landlords; they produce services using a capital good (a house and its land) as well as other inputs such as labour, heat, and light. This fact is recognized in the calculation of Gross Domestic Product (GDP). A component of GDP is the sum of the value of consumption goods produced in the country – a summation that includes both the total rent paid by tenants to landlords and the estimated value of housing services produced by home owners for themselves, that is, imputed rents. Housing services to both owners and renters are about 11% of GDP. Home owners are simultaneously the demanders of housing services and the suppliers of services produced for themselves. This ownership market for housing services is not easily analyzed because no explicit market transactions take place between the household as tenant and as landlord. However, we need a conception of this market to analyze housing issues: for example, to understand tenure choice, given that home ownership is both a consumption decision (like that of a tenant) and an investment decision (like that of a landlord). An analysis of the supply side of the market must recognize the role of landlords and home owners as suppliers of housing.

At any particular moment, a supply of housing stock (the capital good), built in the past, is available to produce housing services. In assessing the amount of stock available, the tendency is to focus only on the margin, that is, on those who supply net additional stock. Housing stock can be created by new construction, for which the suppliers are builders and developers in the housing construction

Table 5.1
Components of housing investment: Canada, 1951-1986
(millions of dollars)

	Housing investment (HI)†	New construction		Alterations and improvements	
		Total	% of HI	Total	% of HI
1951	1,054	725	69	252	24
1956	2,219	1,635	74	447	20
1961	2,156	1,446	67	406	19
1966	3,166	2,148	68	578	18
1971	5,589	4,050	72	1,056	19
1976	14,140	9,452	67	3,193	23
1981	20,569	11,122	54	6,353	31
1986	30,669	15,348	50	10,167	33

SOURCE Statistics Canada (1986).
† The third component of housing investment is transfer costs relating to the purchase and sale of existing residential properties. Conversions are considered as new construction.

industry. Some small amount of additional stock comes from people who build their own houses. However, new housing stock can also be created by renovation of an existing building. The suppliers are renovation firms and residents who renovate their owned or rented units by themselves.

These suppliers use a production technology to combine inputs (labour, building materials, and land) to produce housing stock. With the exception of do-it-yourselfers, these suppliers have revenues from sales and face costs of production. Again, income taxes influence the profitability of being a supplier of housing stock and therefore influence how much stock is produced in one year.

Additions to housing stock are housing investment, that is, the production of capital goods. The calculation of GDP also includes the value of capital goods produced. Housing investment has hovered near 6% of GDP since 1951. Housing investment in GDP may be categorized into new construction, alterations and improvements, and transfer costs relating to purchase and sale of existing residential buildings. This decomposition is reported in Table 5.1 for selected years from 1951 to 1986. Until about 1970 almost 70% of housing investment in Canada was new construction, but since then its importance has declined. Renovation has emerged in the last fifteen years as a source of stock change.

Suppliers of Housing Services

The suppliers of housing services have been identified as landlords (private, government, and third-sector) and home owners. This section focuses on the decisions made by these agents. In the short run, suppliers take account of the revenue from each output level, the technology, and the prices of inputs.

Little is known about the technology for producing housing services, but casual observation suggests it is relatively simple. Once a stock level has been chosen, in the short run the possibility for input substitution is limited; the stock, labour, heat, and light are combined in fixed proportions. The technological choices available to the producer have not changed much since 1945 and are unlikely to change in the future. But the production activity of an existing supplier is not trivial; there are bills to be paid, including property taxes, minor repairs to be arranged, insurance to be arranged, and exterior grounds to be maintained. Private, government and nonprofit landlords produce services for their tenants.[2] In cooperatives, tenants are also the producers of services. Home owners produce for themselves, although they may contract others to do much of the work for them; in condominiums, work on the common areas is done by a condominium corporation. Thus, each form of tenure can mean a different involvement in the production of services. Tenure choice is arguably influenced by the differing abilities and desires of consumers to produce housing services.

THE HOUSING STOCK INPUT

The most important input is the capital good – housing stock – that is, the building and the land. Tables 9.1 and 9.2 summarize data on Canadian housing stock – measured as dwellings – by tenure, type, and region. Unfortunately, no data exist to measure the housing stock properly. Although counts of dwellings and data on their age, state of repair, and plumbing, electrical and heating systems do exist, two dwellings similar in all these characteristics can differ in size and quality.

What is known about the housing stock available to produce housing services for Canadians? Data on whether a dwelling had hot and cold running water, a private bath, and central heating were customarily used as proxy measures of the quality of the housing stock, but these are no longer useful proxies; almost all stock is now adequate by these proxies. The 1981 Census tried to establish whether dwellings were in need of regular maintenance only, minor repairs, or major repairs. About 76% of dwellings needed only regular maintenance; 17% needed minor repairs; and 7% needed major repairs. However, these data represent only the opinions of occupants, among whom perceptions and expectations vary greatly.

Nevertheless, one can confidently state that the vast majority of the Canadian housing stock is in good repair, with less than 10% needing significant repair. Surveys by building professionals suggest that almost all the stock is structurally sound (see Klein and Sears et al. 1983; Barnard Associates 1985). However, the important issue is not the current state of Canadian housing stock, but how it may depreciate over the next decades as it is used in producing housing services. Buildings, like all capital goods, depreciate through use. Although they can be improved or held at the same level through additional capital expenditures (renovation), what might happen in the absence of substantial capital expenditures? The future path of depreciation depends upon the age of the building, the quality of the original design, materials, and construction, and the maintenance and renovation expenditures over the life of the building to date. Although little is known about the rate at which buildings depreciate, it is likely to be slow over the first fifteen or twenty years (or longer depending on the quality of the original building) and then more rapid. In 1981 about 45% of the stock was built from 1945 to 1970, which suggests almost 45% of the stock could be reaching the age when major capital investment may be needed. But these data cannot form the basis of conclusion because we do not know the original quality or capital expenditures since construction.

Has Canada's rental housing stock begun to depreciate rapidly? Barnard Associates (1985) reported on the low-rise rental stock based on inspections by building professionals of over one hundred buildings in Toronto, Ottawa, and Hamilton. These buildings were initially of good quality and have lasted well, but more than one-third are over 40 years of age. Age is the factor explaining the coming need for major rehabilitation.[3] The study found that one-quarter of the units needed major repairs to bring the buildings to minimum property standards. Almost all the buildings were structurally sound. The main problems were exterior building components such as walls, stairways, windows, and doors where moisture penetration was evident and deterioration increasing, along with heating systems needing major repair or replacement. Klein and Sears et al. (1983) studied the high-rise rental stock in Ontario and also reported a coming need for major capital expenditure. The problem in these buildings was not age but the "inadequacy of initial design and construction." The buildings were structurally sound, but there were problems of keeping them weather-tight (roofs, walls, and windows were all problems); parking garages, slab balconies and railings had structural problems; and often galvanized pipes and heating boilers and pumps needed to be replaced in the coming years. A similar situation was reported in private, government, and third-sector housing stock.

It cannot be determined whether this situation is common across Canada; further data collection and analysis are necessary. However, the broad picture may well be the same because the age of the stock and the financial circumstances of landlords are roughly similar. The rental housing stock is presently in good repair, but it may depreciate rapidly over the next ten years because of age, inadequacies of original design and construction, and the accumulated effects of "patch-up" maintenance where practised. If so, investment in the existing stock on a scale not seen over the last forty years may be necessary in order simply to maintain the current level of rental housing services. This could represent a challenge for future housing policy because the situation is arising after a prolonged rent-cost squeeze (see subsequent section). For the landlord, forecast net rental revenue must be sufficient to generate a return on invested funds if the project is to be undertaken.

This issue is important in thinking about housing adequacy and affordability because many of these dwellings, especially the low-rise buildings, have low rents and are occupied by low-income households. Demolition would remove

Table 5.2
Indices of prices of inputs into the production of housing services:
Canada, 1949-1985

	Rent	Property taxes*	Mortgage interest*	Insurance*	Water, fuel, electricity	All items CPI
(a) 1949=100						
1949	100		100†		100‡	100
1956	135		128		117	118
1961	143		152		123	131
(b) 1961=100						
1961	100	100	100	100	100	100
1966	104	119	120	125	100	111
1971	123	159	199	198	117	133
(c) 1971=100						
1971	100	100	100	100	100	100
1976	120	126	199	266	173	149
1981	157	173	296	435	342	237
(d) 1981=100						
1981	100	100	100	100	100	100
1985	128	137	137	128	144	127

SOURCE CHS (1961, 1971, 1981, 1985).
* Component of CPI Ownership Index.
† Approximated as CPI Ownership Index.
‡ Approximated as CPI Household Operation Index.

units such households can afford, often to be replaced by new units which they cannot afford, while major capital repairs would require either significant rent increases or significant subsidies.

OTHER INPUTS

Other inputs into the production of housing services are labour, water, heat, and light. These are widely available now and likely to be available in the future. Since 1945 there have been dramatic changes as indoor plumbing, central heating, and modern electrical systems have been added; changes have been greatest in small towns and rural areas. Virtually all Canadian housing stock now has complete mechanical systems. It seems unlikely that there will be problems with adequate supplies of labour, water, heat, and electricity over the coming decade.

The relative prices of these inputs have changed relatively since 1945. A proxy for water, fuel, and electricity prices is that component of the Consumer Price Index (CPI) (see column five of Table 5.2). During the 1950s and 1960s, these prices rose more slowly than the CPI and the prices of other inputs into the production of housing services, but in the 1970s, this changed as the water, fuel, and electricity index rose faster than other price indices. Into the 1980s, the index continued to rise but not as quickly.

Three other costs paid by producers of housing services are property taxes, insurance, and mortgage interest. Almost all housing suppliers mortgage to purchase housing stock, so a cost of production is the payment of interest on the loan. The measure of the mortgage interest costs faced by suppliers over time is not the index of market interest rates in each year but the index of rates on mortgages outstanding. No index is available for landlords, so the index for home owners will have to serve for all suppliers. Table 5.2 reports price indices for these inputs. In the 1950s and 1960s these indices rose faster than the CPI. In the 1970s property taxes rose less rapidly than the CPI whereas interest costs and the insurance index rose more rapidly. In the 1980s these costs have paralleled general price increases.

THE FINANCIAL SITUATION OF SUPPLIERS

A composite picture of the rent-cost situation of housing suppliers, particularly private landlords, since 1945 (see Table 5.2), shows that in the 1950s, rents rose faster than the CPI and roughly kept pace with costs. In the 1960s rents rose less rapidly than the CPI, but costs rose more rapidly than rents. In the 1970s, this rent-cost squeeze became acute. Rents rose less than all prices, and costs rose faster then rents. In the 1980s prices, rents, and costs have moved together. Thus, private producers faced a squeeze in the 1960s, which considerably tightened in the 1970s, and no readjustment has occurred in the 1980s. [4] Interestingly, the values of rental properties did not fall in the 1970s in spite of this rent-cost squeeze. Into the 1970s investors in rental real estate did well although values began to fall in some areas in the 1980s. [5]

The fundamental picture is thus clear. The business of supplying housing services as a private landlord became less attractive as rent increases fell short of cost increases. Unless there is an upward adjustment in rents or new government subsidies, there will be little incentive to renovate the existing stock, increasing pressure to convert to condominiums or to demolish buildings, and fewer persons or firms who wish to become housing suppliers.

Government and third-sector landlords face the same squeeze, although it manifests itself differently. The government is a landlord via its ownership of public housing units. Rents are not market determined but geared to the income of tenants. At best, tenant incomes, and therefore rents, might be expected to rise over time with the general price level. Because public housing landlords have long-term mortgages, they have not faced interest increases; yet they still have to face the same increases in other costs as private landlords. The subsidy cost per public housing unit has risen faster than the general price level. Private landlords face a profit squeeze; public landlords face rising subsidies. Ironically, the results have been similar; private individuals and firms have become less willing to become landlords, and governments have become less willing to build new public housing.[6] Third-sector landlords charge rents to cover costs. Long-term mortgages protected old landlords from mortgage interest increases; but new landlords began with high interest rates (or large subsidies). Other costs have risen rapidly for third-sector landlords, just as for private landlords, and rents have risen to cover them.

Finally, of course, the production costs for home owners have risen in the same fashion. They have faced large increases in mortgage interest, insurance, and fuel costs. The costs have been offset by capital gains on their housing stock – gains which are sustained by future strong demand for owner-occupied housing. It should be noted that these gains for many owners are accrued but not realized; some first-time buyers and some elderly owners have faced cash-flow problems.

OTHER SUPPLIER DECISIONS

In the short run, existing suppliers act as property managers. Other decisions facing existing suppliers are whether to supply to the ownership or rental market and how many dwellings to supply. Usually with minor alterations to a building, suppliers have flexibility in this regard. Private landlords and home owners often make changes; government and third-sector landlords usually remain in the rental market supplying the same number of dwellings.

Private landlords with detached or semi-detached houses can reduce the number of dwellings and can transfer their buildings to the ownership market. Young families may purchase a house and live on one floor while renting out the basement and other floors; gradually, the family ceases renting floors of the house and occupies the space itself. This was common among immigrants, who would frequently rent to still more recent arrivals. The transfer of a house from multiple unit rental to single unit ownership can also occur when a building is sold, a process now associated with gentrification. Also, multiple unit rental buildings can be converted to condominiums. These transfers depend on the value of the building if used in the rental market, compared to its value to home owners. Given the rent-cost squeeze on private landlords and the strong demand for home ownership, this transfer to ownership is likely to continue.

Transfers from the ownership to the rental market can also occur, often resulting in an increased number of dwellings. The most important transfers are when home owners rent out a portion of their homes. Other transfers occur when owners rent their homes, usually temporarily, and when condominium owners rent their units. The renting of a portion of an owner-occupied house depends upon the value of the space in its two alternative uses.[7] Such transfers may represent important sources of future rental dwellings and affordable units. Older home owners may wish to remain in their homes but also get some income from their housing stock. This is emerging as a policy issue in some cities. Existing regulations often restrict these transfers and an important question for housing policy will be whether to ease the regulations.

IMPACTS OF GOVERNMENT PROGRAMS

A thrust of federal housing policy has been to ensure an adequate flow of mortgage financing for those who wish to buy apartment buildings or homes and become suppliers of housing services. For example, early initiatives were in the form of joint loans, mortgage insurance, and direct lending; in the late 1970s the initiative was to develop a graduated payment mortgage in response to the tilt problem.

Another group of programs dealt with input costs to suppliers, mainly property taxes and heating costs. The entire complex of provincial-municipal finance and the assignment of responsibilities to the two levels influences the level of property taxes. There are great differences among the provinces (see Higgins 1986). From 1945 municipal expenditure responsibilities grew rapidly and most provinces acted to check the growth of property taxes by increased grants and a transfer of responsibility to the provincial level. There have often been calls to give municipalities other tax sources or to establish formal revenue sharing with the provinces to reduce property tax increases further, but these have not been heeded and are unlikely to be in the future. [8]

Governments also acted in the 1970s to cushion housing suppliers (and all users of oil products) against increases in oil prices. The federal government held Canadian oil prices below world oil prices, subsidized the conversion from oil to other heat sources, subsidized home insulation, and underwrote research on retrofitting existing buildings. Many provinces offered additional assistance.

Not all government programs have been to the advantage of housing suppliers. The most obvious example is provincial rent regulation. At different times and in different provinces, regulation has taken various forms. Most schemes have established an allowable annual rent increase, with provisions for extra increases if the landlord's cost increases warranted. Sometimes, new units were exempted, as were government and third-sector buildings. Binding regulations keep rents low and so benefit existing tenants at the expense of existing landlords. When rents are held down, maintenance is reduced, new construction is reduced, vacancy rates fall, mobility is reduced, and non-price mechanisms ration the available units. That these are the effects of binding regulation cannot be debated, but the magnitudes of these effects is debated. The magnitudes depend on the particular scheme and the particular housing market, but there is always a fundamental trade-off. The more rents are depressed and the more existing tenants benefit, the more landlords are hurt and the greater the negative effects such as reduced maintenance, construction, and mobility.

Although not intended, changes in income tax laws since 1945 have reduced the attractiveness of being a private landlord. Early on, investing in residential rental real estate had several advantages. As a tax shelter, losses generated by capital cost allowance (CCA) could be applied against other income. Also, CCA tended to be larger than actual depreciation. And, capital gains realized on the sale of a building were exempt from tax. The thrust of tax reform since 1945 has been to try to remove special advantages and treat all forms of income and all types of economic activity similarly. While tax neutrality is laudable, the advantaged sector is hurt when tax advantages are removed. [9]

The 1971 tax reform removed the tax shelter advantages for individual investors and taxed 50% of realized capital gains. Interestingly, home owners kept all their tax advantages and their removal has never seriously been contemplated. In 1974 the tax shelter was restored with the MURB program. Although intended to last one year, the MURB program was renewed and not finally terminated until 1981. In 1985 individual taxpayers were given a lifetime capital gains exemption that was capped in 1987 at $100,000. Also in 1987 CCA was reduced for new investors in rental real estate. On balance, tax reform removed advantages to investing in rental buildings, and the 1987 changes continued the trend.

Finally, many federal and provincial housing programs have assisted suppliers of housing services and so encouraged new rental construction. Some have required that assistance be passed on to tenants as reduced rents, for example, the federal LD program; but others such as ARP or CRSP have not. Annual funding of such programs has varied greatly. There has been an on-again off-again pattern. Some of the funding increases in the middle to late 1970s were attempts to offset the negative effects of rent controls and tax changes. Little is known about the net effect of these programs on the level of housing stock in Canada. Certainly, there are more housing suppliers than there otherwise would have been; but also many who received assistance would have become suppliers in any event. Furthermore, borrowing to finance a government housing program pushes up interest rates and reduces unsubsidized construction. [10]

Running throughout the second half of the post-war period has been the policy question of whether to subsidize rental accommodation for households that are not poor. Different housing programs suggest different answers. For example, rent controls imply the answer is yes. Removing tax advantages implies no. ARP and CRSP imply yes. At present, the federal government seems to be answering no and arguing that housing assistance should be targeted at those in core housing need.

The Demand for New Stock

Housing stock changes are market outcomes, combining demand, supply, and government. Considering the demand for housing stock follows naturally from the analysis of the suppliers of housing services. The supply side of the housing services market becomes the demand side of the housing stock market because existing suppliers or potential new suppliers are the demanders of new stock. People or firms demand housing stock because they want to become housing suppliers. New stock can be added by renovation, new construction, or conversion. The demand for owned homes is considered in Chapter 3 and is not explored further in this chapter.

For private landlords, the decision to renovate or to acquire a new building is an investment decision. Existing or potential landlords look at the rate of return and the risks and compare these to the rate of return and risks in other available business opportunities. The rent-cost squeeze, rent controls, and tax changes have made the business of supplying housing services less attractive. Although these programs may have other desirable effects, their implications for the demand for new stock must be acknowledged.

Without higher rents, lower costs, changes in the tax and regulatory environment, or new government subsidies, expansion of the private rental housing sector will remain unattractive.[11] This is widely recognized when discussing new rental construction. It is now recognized that the same picture applies to renovations of the existing stock. Much of our rental stock could soon enter a phase in which it depreciates rapidly. To maintain the existing stock will require substantial capital expenditures, but as a business proposition, this has become less attractive.

When private landlords acquire new stock, either through renovation or a new building, the purchase is usually financed with a mortgage. The price and availability of mortgage credit help determine the demand for new stock (along with forecast future rents and other costs). Canadian capital markets are now well developed and mortgage credit at the market interest rate is likely to be available for new construction in the future.

Lending for renovation, however, may present problems. Renovation loans differ from new construction loans. They are generally smaller per dwelling but carry higher administration costs. The quality of the original dwelling as well as the proposed renovation must be appraised, and the system of inspecting work in progress is not as developed. New construction loans are usually advanced as work progresses, and the value of the property rises as work progresses. With renovation, the value of the property falls in the early stages; when things have been torn up and nothing has yet been replaced. Loan advances are therefore more risky. But these problems are characteristic of the renovation process, and it is unlikely that public sector loans without subsidy would be any different from private sector loans.

The demand for new stock from government and third-sector landlords does not depend on the rate of return in housing; but the relationship of revenues and costs is still relevant. The government's willingness to expand public housing is not independent of the subsidy per dwelling. Renovation of existing public housing units will require either increased rents or increased subsidies. Third-sector buildings will also require capital improvements over the next few decades, which will require either rent increases or increased subsidies. Klein and Sears *et al.* (1983) examined a "typical" third-sector building and estimated the rent increase necessary to pay for the capital investments. The absolute rent increase needed was not unlike that needed by private landlords; but the percentage increase was much higher, almost 45%, because current third-sector rents are lower.

In summary, the demand for new rental stock, either renovation or new construction, has been diminished by the rent-cost squeeze and the changes in taxes and regulations (and by such other factors as demographics and household tenure choice). Also, government programs to stimulate the creation of new stock have been cancelled or curtailed. The strength of future demand will depend upon government's willingness to let private rents rise or to subsidize private activity, to expand its role as a landlord, and to subsidize third-sector landlords or to let third-sector rents rise.

Suppliers of New Housing Stock

The demand side of the housing stock market has been shown to be linked to the suppliers of housing services. The suppliers of new housing stock are the renovation industry, the new construction industry, and do-it-yourselfers. [12]

THE RENOVATION INDUSTRY

The renovation industry uses available technology to combine an existing building, building materials, and labour in the production of housing stock. Little is known about the industry supplying renovation in Canada. Impressionistic evidence reveals many small firms and almost no large firms; which suggests low fixed costs to establishing a business and few economies of scale in operation. Each job is unique because it must take place in an existing building – buildings which, even if similar at time of construction, have depreciated at different rates and have had different patterns of maintenance and renovation. Often, the structural characteristics and quality of the existing building are not fully known until the renovation project has begun. Firms often leave the industry because they fail or because their operators find better work elsewhere. Like the construction of new buildings, the pattern is not of one firm doing all the work but of a general contractor undertaking the job and subcontracting work to other firms or individuals: for example, framers, electricians, plumbers, drywallers, tile-setters, and roofers. A problem for the general contractor is the scheduling of the various trades because they cannot all be on site at once. The trades move from job to job.

The price and availability of trades depends on demand in new residential construction and in non-residential construction. Table 5.3 provides price indices for building materials and construction labour since 1949. Building materials prices have moved closely with the general price level (Table 5.2); construction labour increased in price more rapidly than general prices in the 1950s and 1960s, but has matched general price change since. It seems likely that the prices of these inputs will match general price changes, but over any short period, their prices may rise rapidly, and problems of availability may arise depending on the demand for trades and materials elsewhere.

In some ways, the renovation industry is like the new construction industry. It uses building technology, has the same general contractor-subcontractor structure, and uses the same inputs. However, there are also important differences. The separate building projects are more heterogeneous. The price and

Table 5.3
Mortgage interest rates and indices of prices of inputs
into production of housing stock: Canada, 1949-1985

	Building materials	Construction labour	Land	Mortgage interest rate	% change in GNE implicit price index	Real mortgage interest rate*
(a) 1949=100						
1949	100	100	100†	5.7	4.3	1.4
1956	129	152	210	6.2	3.6	2.6
1961	128	200	258	7.2	0.5	6.7
(b) 1961=100						
1961	100	100	100	7.2	0.5	6.7
1966	121	128	133	7.6	4.4	3.2
1971	145	210	187	9.3	3.2	6.1
(c) 1971=100						
1971	100	100	100	9.3	3.2	6.1
1976	154	173	201	11.9	9.5	2.4
1981	236	259	318‡	18.5	10.6	7.9
(d) 1981=100						
1981	100	100		18.5	10.6	7.9
1985	122	129		11.3	4.1	7.2

SOURCE CHS (various years), Economic Review (1984).
* The real mortgage interest rate is calculated as the nominal mortgage interest rate minus the price change in that year.
† The 1949-61 index is estimated from data on the cost of land used in NHA-financed houses.
‡ Refers to 1980. The series has not been published since 1981.

availability of land are not central issues. The construction process is more labour intensive, must occur in a more confined space, and must consider much more the disruption of adjacent activities.

Perhaps the most important difference is that the two industries are at different stages in their evolution. The situation of the renovation industry in the mid 1980s resembles the building industry in the early post-war years. And, the public policy problems now are like the public policy problems then. A strong demand for renovation is emerging, and one must ask whether bottlenecks or problems in meeting that demand exist. CMHC can play a role in funding research into building technologies and in developing standards. Both CMHC and provincial governments should consider disseminating information to households on the economics of renovation, the approval process, and the process of dealing with a renovation contractor. These governments could assist in formulating some form of regulation or self-regulation for the industry which would provide for warranties and liabilities in the event of faulty workmanship. Provincial and local governments should examine their existing regulations to see how the balance between the interests of individual owners and their neighbours affects the level of renovation.

THE NEW CONSTRUCTION INDUSTRY

The new construction industry combines labour, building materials, and land to produce housing stock. Chapter 8 considers the production technology and the organization of the industry in more detail. There has been much technological progress since 1945, some of it supported by CMHC and the National Research Council. The industry once characterized by a proliferation of many small firms now includes both small and large firms; some of the latter are publicly traded companies that build all over the world.

Table 5.3 provides indices of the prices of inputs into new housing and information on nominal and real mortgage interest rates. The data reflect the periodic problems of the industry since 1949 with the prices of inputs. Building materials prices have risen at approximately the same rate as general prices; and their supply has never attracted much attention. Construction labour costs rose more rapidly than general prices in the 1950s and 1960s. This did cause concern; labour costs, labour productivity, and the role of labour unions were key issues in the debates about the cost and affordability of housing. Since the early 1970s labour costs have risen at about the rate of all prices, and so have receded in policy debates. It seems unlikely that either labour or materials costs will rise more rapidly than prices, as a long-term trend, although these costs will fluctuate considerably in the short run.

In the 1970s the problem was land cost. The Federal/Provincial Task Force on the Supply and Price of Serviced Residential Land (Greenspan 1978) concluded that the boom in land prices was not caused by monopoly developers or land owners or by local governments restricting new developments and demanding "gold-plated" services. The 1970s boom was caused by an extraordinary conjunction of factors: inflation accelerated; real income exploded; the stock market dropped; the baby boom entered the new home buying years; capital gains taxes were introduced but exempted owner-occupied housing; downpayment requirements fell; and real mortgage interest rates fell. An unanticipated surge in demand sent land prices soaring and speculation sent them higher still. The Task Force identified factors leading to a longer term upward creep in prices, including federal programs which stimulate demand, provincial and local policies which restrict development, and increasing servicing standards. These programs pursue desirable goals, but they also have the effect of raising land prices. Careful monitoring of this undesirable side effect is warranted over the coming decades. The Task Force found no evidence of monopoly among developers or land owners, but noted several trends which favour large companies over small and warned that monopoly power could become a significant problem in the future. Continued monitoring of this seems appropriate as well.

The effect of mortgage credit on supply should also be recalled. Mortgage credit influences demand, and therefore the volume of new construction. Changes in overall economic activity – and in government fiscal and monetary policies – have led to great fluctuations in the levels of housing starts. Since at least 1945, the unstable nature of mortgage lending and of the residential construction industry has been of concern. In the 1960s and 1970s, this concern led to some major studies: for example, *The Royal Commission on Banking and Finance* (Canada 1962), *The Residential Mortgage Market* (Poapst 1962), and *Toward More Stable Growth in Construction* (Economic Council of Canada 1974). Over time, the response in Canada has been to integrate the mortgage market into the national system of capital markets and remove regulations on bank lending, interest rates, and NHA loans. As a result, cyclical effects on new housing starts became more the result of shifts in mortgage interest rates than of the availability of credit. The desire to moderate the fluctuations in residential construction has waned as economists and politicians have become less sanguine about their ability to use fiscal and monetary policy to stabilize short-run fluctuations. Housing will likely remain, however, a sector to boost when general stimulus is required.

DO-IT-YOURSELFERS

Much renovation is undertaken by home owners and small landlords themselves and by trades hired by the owner, rather than by general contractors. Also, much home building, especially in rural areas and small towns, is done by the home owners themselves. In another variant, building cooperatives buy materials, pool labour, and occasionally hire trades (which the cooperative members cannot supply themselves) in order to build a home that is then owned by the member rather than by the cooperative. Reliable data to measure the extent of do-it-yourself activity is lacking; by its nature, it is outside normal data collection procedures. When national income is computed, some of this activity is captured because sales of building materials are assumed to be used in the construction of housing stock; however, the value of labour supplied is not included in the calculation of the value of housing stock created. Much of the activity occurs outside any regulatory framework, and therefore, no government records are available. Some of the activity is illegal. Many small renovations in urban areas occur without securing the necessary minor zoning variance or building permit, and many rural homes are built without building permits. Much of the home building is not financed by mortgages from the formal financial system, so this way of monitoring building activity is also unavailable.

Do-it-yourself building is part of a rapidly expanding informal economy. Many believe that the informal economy is important for community development in rural areas, small towns, and within large urban centres. It is hard to establish whether proposals to expand the informal economy are the dreams of utopians or a practicable alternative to the market economy. However, the experience of rural areas and small towns makes one ask whether it might offer lessons for larger cities. Much of the new housing stock in rural areas is supplied by do-it-yourselfers – outside the market economy, outside the formal mortgage market, and also largely outside the purview of federal housing policies (as the latter have focused on established markets, the established building industry, and the formal mortgage market). The quality of housing in rural areas and small towns has improved since 1945. Whether the same approach is applicable for large centres is not, however, immediately obvious. In rural areas, low-income persons can be home owners by building themselves, but this is less feasible in large cities where land prices are higher. Similarly, it is not obvious what governments can do to encourage the informal economy because the informal economy is inherently antithetical to government. But we must recognize that the very regulated housing markets of large cities – with official plans, building codes, zoning by-laws, and elaborate procedures for approval, permission for variance, and enforcement – suppress certain undesirable sorts of outcomes, but also suppress the informal economy and so restrict use of a do-it-yourself strategy to improve housing.

THE IMPACTS OF GOVERNMENT PROGRAMS

Most government housing programs focus on the demand side of the housing stock market and were discussed above in the context of the suppliers of housing services. In the early post-war period, CMHC helped develop the residential construction industry; it had much success and further efforts are not needed. CMHC also conducted research into building technology, and here a continuing role is appropriate. Few tax advantages are granted to the building and land development industry, the exceptions being the deductibility at the time incurred of soft costs and carrying costs on vacant land. However, these advantages gradually were restricted and, under 1987 tax reform, phased out by 1991. However equitable, this reform raises the after-tax cost of producing housing stock and reduces the quantity of construction. Perhaps the most important influence of government on suppliers of new stock is through regulation at the municipal level. Originally, subdivision approval was the most important process, but the regulation of redevelopment is of growing importance.

Conclusion

Throughout most of the post-war period, consideration of the supply side of housing included only the production of new houses or apartments. The focus on new residential construction is understandable and was probably appropriate as Canada emerged from the depression and World War II, then sought to deal with economic growth, rapid city growth, and the repercussions of the baby boom; and then finally in the 1960s and 1970s, Canada sought to provide public housing for the disadvantaged. However, this narrow focus is less and less appropriate. More attention must be paid to the existing suppliers of housing services, particularly rental housing services and their decisions about renovation. This shift may be more difficult than at first appears. CMHC has had less involvement with suppliers of existing private rental housing. And even if the Federal government wished to become involved, the constitution may constrain it. The shift will likely require greater provincial involvement.

Within a supply perspective the questions about housing progress become: Are the industries that supply services and stock developing? Are they able to respond to the demand side? Have any shortages of inputs or barriers to the inflow of inputs been removed? Have we become more efficient producers of housing services and housing stock? The answer to each surely must be: yes, we have made much progress since 1945 on the supply side.

However, certain issues do require attention in the future. Many of the existing suppliers of rental housing services – private, government, and third-sector landlords – own housing stock which is likely to depreciate more quickly than before. But replacing this stock with renovation or new construction is presently not attractive. The rental stock may deteriorate unless there is an increase in rents or an increase in government subsidies. The challenge of the next decades will be to find the right balance of rent increases, increased government subsidy, and regulatory reform to ensure an adequate supply of rental housing services.

Notes

1 Supply side analysis also considers whether there is non-competitive behaviour by suppliers. However, the markets for housing services and housing stock are thought to be quite competitive. See Muller (1978) and Greenspan (1978).

2 There is often debate about the degree to which tenants should be consulted and involved.

3 For some units, maintenance has been reduced because demolition is likely; for some others, it has been reduced because of rent controls.

4 The CPI rent index understates true rent increases; but with an appropriate correction, the basic picture remains. See Fallis (1985).

5 See Smith and Tomlinson (1981) for evidence that Ontario's rent regulation scheme reduced the value of apartment buildings.

6 Interestingly, while the decline in new public housing is primarily caused by other factors, governments have been unwilling to promote rent supplements as an alternative – a program with similar expected subsidies but without most of the problems of public housing.

7 The context of local government regulation, especially zoning by-laws, specifies whether a conversion is possible and influences the value of space in each use.

8 In many jurisdictions, property tax assessments are regarded as unfair because the ratio of assessed value to market value varies across residences. A change to market value assessment would likely reduce property taxes on rental dwellings, on average, and

increase taxes on owner-occupied homes, especially older inner-city homes (see Kitchen 1984).

9 Conversely, when advantages are granted, the sector expands; thus, there are always pressures to restore advantages, and this has certainly been the case in housing.

10 A temporary government lending program will have no effect on the long-run stock of housing. The gross impact of these housing programs is easy to measure: it is the number of investors who are assisted. However for program evaluation the important result is the net impact: the number of additional rental units created. See Smith (1974), Fallis (1985), and Miron (1988).

11 This, of course, is apart from other factors which influence demand – and therefore rents in an uncontrolled market – such as demographics and income. Some argue that the rent-cost squeeze is not a problem but merely reflects the fact that households are transferring to the ownership market. When demand increases, rents will rise, and new supply will be forthcoming.

12 For a more recent survey of knowledge about suppliers of housing stock, see Clayton Research Associates Limited (1988).

CHAPTER SIX

Financing of Post-war Housing

James V. Poapst

THE AVAILABILITY and terms of financing influence how much housing is demanded, by whom, and when. Given other conditions of demand for housing, the size and quality of the housing stock, access to and affordability of housing, and security of tenure will be higher, the easier the availability, terms, and conditions of housing finance.

Housing is financed almost entirely through mortgage loans and owner equity. The major mortgage lenders are financial institutions: banks, trust and mortgage loan companies, credit unions and caisses populaires, life insurance companies and pension funds, and several other smaller institutions. These lenders are commonly referred to as the "lending institutions." Other lenders include government agencies, individuals, and non-financial corporations who make loans to further their sales or to assist employees with their housing. Owner equity is the difference between the value of the property and the amount owing, if any, on any first and junior (that is, second or subsequent ranking) mortgages, or other claims against the property. Owners obtain equity funds from household savings (including their own labour – "sweat" equity), other loans, and share issues in the case of corporations.

Housing requires large amounts of financing. Newly-approved residential mortgage loans by private lending institutions in 1985 amounted to fully $30 billion; this does not include credit unions and caisses populaires; nor does it include loans by individuals, non-financial corporations, or government agencies. By comparison, gross new Canadian dollar issues of corporate bonds amounted to $4 billion. Similarly, residential mortgage debt outstanding with private institutional lenders was $115 billion in 1985 compared to $34 billion for corporate bonds outstanding (Bank of Canada, *Review*).

Milestones of Progress [1]

The supply of residential mortgages at the end of World War II was composed of three main segments: joint loans made by private lending institutions and CMHC, conventional loans made by private lending institutions, and other loans. NHA lending (joint loans) was confined to new housing – partly to focus limited funds on expanding the housing stock and partly to create employment (which helped justify federal activity in an area of provincial responsibility). The major lending institutions were the life insurance companies, which accounted for 90% of private NHA lending before 1954. Joint loans were limited in size, but they could be made for higher loan-to-value ratios than could conventional loans; and joint loans were fully amortized with a term to maturity of twenty years or more at a fixed and subsidized rate of interest. The government protected lending institutions against default losses free of charge. Such free protection in itself is a subsidy. Also, CMHC had authority to lend directly to qualified borrowers if private loans and NHA loans were not available on "reasonable" terms and conditions.

All other institutional lending was in the form of conventional loans. These were made on new and existing property, were not subject to regulatory limits in amount, term, or interest rate, but were restricted to 60% or less of the appraised value of the property. Trust and mortgage loan companies typically made five-year loans, which was the maximum period that they could legally control prepayments and the period up to which they typically issued their own obligations.

The third segment, other lenders, included individuals, corporations with some special interest in housing, and government agencies other than CMHC. In the early post-war years, individuals were important lenders on existing property, particularly in locations not well served by major lending institutions, such as old areas of cities and small and remote communities. Individuals often made first or second mortgages when selling their property (vendor mortgages). Second mortgages commonly were made at first mortgage rates or other rates that were low relative to risk. This reflected the lender's dual interest in the property. Cheap credit improved the sale price or otherwise facilitated the transaction.

Faced with large demands for mortgage funds after the war, the government had a choice between vastly expanding its own lending or fostering the growth of private supply. The former was not feasible, whereas powerful levers could be applied to private lending institutions. The federal government had authority to regulate the activities of most major financial institutions either exclusively (banks) or with the provinces (life insurance, mortgage loan, and trust companies).

RESTRUCTURING FINANCING UNDER NHA, 1954

The first milestone of progress in financing post-war housing was reached in 1954 when NHA-insured loans replaced joint loans, and banks were admitted to NHA lending. This simultaneously increased the government's control over its own mortgage lending, expanded the private supply of NHA funds, and removed two mortgage-based subsidies. The government's share of joint loans had been provided at a below-market rate of interest, and the private lender's share had been guaranteed by the government free of charge. The borrower now paid an insurance premium. With the worst years of housing shortage over, and the veterans' needs met, housing subsidies for middle-and high-income groups were hard to justify; and NHA loans provided few owned homes for low-income households.

Bringing the banks into NHA lending was a big, bold step: big because the banking system was head, shoulders, and torso larger than each of the other types of lending institutions; and bold because of a strong tradition against mortgage lending in Canadian banking. Bankers saw their primary role as providing short-term working capital loans to business. Mortgage lending was considered risky, a view influenced by US bank experience. Mortgage insurance would protect against default risk, but the loans still were less liquid than demand loans and vulnerable to the risk of rising interest rates. That banks eventually accepted twenty-five-year fixed-rate lending undoubtedly reflects the relative stability of interest rates over the two preceding decades.

Unlike other lending institutions, the banks had branches in many small and remote communities. They were situated to reduce the demands on CMHC for direct lending in these areas, and hence reduced the need for subsidized lending.

Once authorized to make NHA loans, the banks moved quickly into the market. Initially, CMHC performed property appraisals for the banks and augmented its own branch system to do so. It was a time of easy money. By 1955 banks drove the going NHA interest rate below the ceiling rate then imposed on NHA loans for the only time in the history of the ceiling rate.

This milestone saw borrower affordability and accessibility to housing maintained or improved directly through improved efficiency of Canada's financial system. Easing the demand for government funds to finance market housing helped conserve government resources for later use in social housing. Bankers and central bankers played their role, but from the perspective of developing the residential mortgage market, this was CMHC's biggest step forward.

INCREASES IN LOAN-TO-VALUE RATIOS

In conventional lending by federally regulated life insurance, mortgage loan and trust companies, the longstanding maximum loan-to-value ratio of 60% was raised to 66⅔% in 1961, and 75% in 1964, which in time became the standard regulated ceiling for conventional institutional loans. The trend to higher loan-to-value ratios continued in 1966 when home ownership loans on existing housing became eligible for NHA insurance.

High-ratio lending reduces the demand for junior mortgages which, apart from those of dual-interest lenders, necessarily carry a high rate of interest because of the risk. Lending institutions could make a 75% first mortgage at a rate of interest below the weighted average rate on a 60% loan and a 15% second mortgage. The difference could be 2 percentage points or more. Private lenders, encouraged by the success of NHA loans, wanted these changes and responded to them. Consequently, 1964-6 stands as a milestone marking another mortgage-based advance in the affordability of, and access to, housing, achieved by lender response to the easing of a constraint.

THE BANK ACT, 1967

The Bank Act, 1967 authorized conventional mortgage lending by banks, which enabled them to make loans on existing rental property and on other properties not covered by NHA. More important, the Act, in a provision proclaimed in 1969, removed the historic interest rate ceiling (6%) on all bank loans. After 1959, when market interest rates rose above 6%, the banks virtually withdrew from NHA lending. Lending at 6%, or purchasing loans originated by others at the going rate, did not fit their investment strategy. It complicated an already difficult problem of credit rationing. At this time, banks treated residential mortgage lending as a residual outlet for funds, lending actively when funds were plentiful and cutting back heavily when funds were scarce, to serve better their traditional loan demands. Not lending also added to pressure on government to remove the 6% ceiling. The new Act opened the door for wider and more continuous involvement of the banks in mortgage lending. It also opened the door to more active personal lending, thereby facilitating equity investment in housing.

INTEREST RATE CEILINGS REMOVED

The 1969 removal of the interest rate ceilings for bank loans and NHA loans brought the banks back into mortgage lending and also made the supply of NHA funds less unstable. As interest rates in general rose during the 1950s and 1960s, monetary restraint could be imposed on house building simply by "holding the rate." The private supply of NHA funds would diminish, and house builders and home buyers would have to turn to more costly combinations of conventional loans plus junior mortgage financing or defer their plans. The upwardly sticky rate disrupted house building, and as a (passive) selective credit control, it was inequitable compared to general monetary control. Research showed mortgage borrowing was interest-rate sensitive, and consequently, general monetary control would still have a significant effect on house building (Smith and Sparks 1970).

SHORTER NHA MINIMUM TERM

A longstanding objective of the Housing Acts was to replace the historic five-year renewable (rollover) mortgage with a longer, fully amortized mortgage. This reduced the borrower's risk from interest rate variations (interest rate risk) and risk from non-renewal. With rising and more variable interest rates, however, banks, trust, and loan companies needed loans with maturities closer to those of their liabilities, that is, from demand up to five years, the maximum term for which they could legally control prepayments, as noted above. Also, for borrowers concerned that interest rates might decline sometime in the not-too-distant future, the five-year rollover loan gives access to those lower rates without the cost of refinancing. Accordingly, in 1969 the minimum term for insured loans was reduced to five years, and it quickly became the predominant maturity in loans for home ownership. Interest rate risk was shifted back to the borrower. Subsequently, the minimum term was reduced to three years (1978) and to one year (1980). Shortening minimum term, however, was progress in a defensive sense; it did maintain the private supply of NHA funds. On the positive side, a wider range of options can be provided to borrowers should interest rate risk decline sufficiently to make long-term fixed-rate lending viable again.

PRIVATE MORTGAGE INSURERS

In 1970 federal legislation authorized private mortgage insurance, and from 1971 to 1973 most provinces passed parallel legislation. By the mid 1970s three private insurers – Sovereign Mortgage Insurance Company, Insmore Mortgage Insurance Company, and the Mortgage Insurance Company of Canada (MICC) – had a combined market share of about two-thirds of total NHA and conventional mortgage underwriting. Insmore and Sovereign merged in 1978, and Insmore and MICC in 1981; but by 1985 MICC had a market share of only 20% of insured lending. Nevertheless, the introduction of private insurance is a milestone of post-war progress in housing finance, partly because it created competition for CMHC and partly for the issue it raises about how the mortgage insurance industry should operate in future.

GROSS DEBT SERVICE RATIO

In screening applicants for loans on owner-occupied housing, lenders calculate a gross-debt-to-service (GDS) ratio – the ratio of monthly mortgage payments plus municipal property taxes to the borrower's income. The GDS ratio is a lender's proxy for affordability, but it also affects accessibility.

Since 1945 the maximum GDS ratio for NHA loans changed several times, from 23% to 27% (1957), to 30% (1972), to 32% including heating costs (1981). Also, many changes occurred in the variables affecting the ratio itself. Changes in loan sizes, interest rates, and amortization periods affected the numerator, and the rise in borrower incomes affected the denominator. Of all the changes, the most significant one occurred in 1972 when NHA lenders became authorized to consider any or all of a spouse's earnings rather than only 50% (introduced in 1968) in calculating the applicant's GDS ratio. This recognized the already significant contribution to family money income made by married women.

The GDS ratio was best suited to the traditional family in which only the husband earned money income. The wife's contribution to the family's economic well-being in the form of the production of income-in-kind was recognized implicitly in the standards for acceptable GDS ratios. The situation changed when wives entered the labour force in large numbers. How was the wife's earned income to be treated? Would she soon stop working to raise a family? On the other hand, if she was not earning money income at the time of the application, might she do so later, or in a financial emergency? As jobs for women increased, maternity leaves increasingly made the wife's earned income less tenuous and more predictable, and more of it could be included in the GDS ratio.

SCOPE OF NHA LENDING

In 1979 two significant extensions were made in the scope of NHA lending. Existing rental housing became insurable, thirteen years after existing owner-occupied housing. So ended the link between government market housing policy and employment policy symbolized by different terms of financing between new and existing housing. If NHA financing stimulated house building, it would be because it increased the attractiveness of housing, thereby inducing expansion in the stock, not because it induced house building when suitable housing was already available.

The second extension involved the removal of maximum size limits on NHA-insured loans, allowing even housing for upper-income groups to be eligible for NHA financing. Limiting loan size made sense when NHA financing for market housing was subsidized, lender protection was provided free of charge, and mortgage money and housing were scarce. Those days were gone. Removal of the size limits appears to have been prompted by CMHC's desire to have more insurance business. Nevertheless, it signified that market housing conditions were now such that it was appropriate for the principal public program for assisting housing finance to include the rich, a situation which opened the question of the respective roles of government and private insurers.

MORTGAGE INSURANCE AND CROSS-SUBSIDIES

From 1954 until 1982 CMHC maintained the same application fees for mortgage insurance and a flat insurance premium structure. Application fees neither varied with the location of the property nor rose with inflation. In 1982 and again in 1984, the level of application fees was raised to make application processing activities more self-supporting. The new fee schedule distinguishes between type of tenure, new and existing housing, and type of structure, but not between locations. Similarly from 1982-5, insurance premiums (as a percentage of loan amount) were raised, and distinctions were drawn between type of tenure, new and existing housing, and loan-to-value ratio. Again, no distinctions were made on the basis of location. To the extent processing costs and loan risks vary with location, cross-subsidies remain in the pricing structure, even if the higher prices make operations profitable.

MORTGAGE-BACKED SECURITIES

The government has long wanted a liquid, long-term mortgage financing instrument. Liquidity reduces the risk of not being able later to redeploy funds committed to mortgages on short notice, thereby reducing the interest rate necessary on new loans. For the same reason, a liquid instrument encourages lending for longer maturities, thereby reducing borrower interest rate and renewal risks.

The government has made several attempts to develop an active secondary market in mortgages. Early plans to develop a central mortgage bank were suspended because of World War II. In 1954, when NHA insurance replaced the previous guarantee, the insurance was linked to individual loans to make them more marketable. From 1961 to 1965 CMHC conducted thirteen auctions of NHA mortgages to familiarize investment dealers with the instrument, but the experiment ended before the dealers took up mortgage marketing on a continuing basis. In 1973 legislation was passed to enable the formation of a joint publicly and privately financed market maker, but private financing was not forthcoming. The costs of transacting were high; NHA loans became more liquid when their term shortened to five years; and lenders could increasingly rely on the developing money market to meet their liquidity needs. From 1981 to 1985 secondary transactions in NHA mortgages averaged only $1.8 billion per year or 5% of NHA loans held outside CMHC in 1984. Many of these sales were between affiliated organizations, that is, were part of the loan origination process.

In 1984 the government adopted a different approach in that it shifted to the liabilities side of the market. MBS are "pass through" claims issued against a specific pool of mortgages. As borrowers make monthly payments of principal and interest, the total amount (less administrative charges) is passed through to security holders on a pro rata basis. If the pool consists of NHA mortgages, the security holders are protected against default risk. Issuers of MBS may also guarantee timely repayment. The guarantor takes responsibility for bridging any delays or interruptions of scheduled payments. Together, these arrangements create an investment akin to a term annuity, at an interest rate that is mortgage-based, and with risk close to that of a government bond.

MBS for one pool can be made similar to those of another. A large amount of standardized "debt" can be created and distributed among many holders. Standardized and with negligible default and low cash flow risks, MBS transaction costs should be low. The conditions then would be established for the development of a secondary market.

MBS should appeal to individual investors and others who want assets with low default risk – in amounts greater and for terms longer than are protected by CDIC – and which are marketable. While investors may be willing to hold longer maturities, extending term also depends on the length of period for which borrowers are willing to borrow. In the case of home owner loans, a change in the Interest Act is needed to tighten up prepayment privileges after five years. This Act, in effect, precludes borrowers from binding themselves beyond a period greater than five years.

Altogether, the changes described in this section converted a segmented and somewhat isolated supply of mortgage funds into one that was more competitive internally and with other markets. The mortgage market thus became a better allocator of resources to the benefit of lenders, borrowers, and the economy as a whole.

Market Growth and Structures[2]

Institutional residential mortgage lending surged in the post-war period and underwent big changes in structure. Mortgage loan approvals averaged $400 million per year in the period 1949-53 compared to $18,636 million per year in the years 1981-5. This is an increase of over fortyfold, more than 90% of which occurred after 1969. Such growth is more than would be expected simply from rising population, income per capita, and housing turnover at rising prices. Undoubtedly, the institutional share of total residential mortgage lending rose at the expense of other lenders. At the same time, CMHC's lending declined – $16 million in 1981-5, compared to an average of $82 million per year in 1949-53, and a peak of nearly $700 million in 1975.

Within the institutional sector, market shares shifted from about half-and-half to one-third NHA and two-thirds conventional. Loans for new construction declined from about three-quarters to one-quarter of total institutional approvals. This switch reflects an increase in housing standards, a decline in the rate of economic growth, and the extension of NHA loans to existing ownership housing in 1964 and existing rental housing in 1979. These conditions suggest a decrease in the ratio of investment in new housing to the value of the housing stock. It is also possible that more loans were made for non-housing purposes in recent years.

The market shares of lenders have changed since the early post-war years. In 1949-53 life insurance companies accounted for 95% of NHA and 53% of conventional loan approvals by lending institutions compared to shares of 6% and 8% in 1981-5. This decline in importance was caused by the slow rate of growth of their assets, the entry of the banks into mortgage lending, and a shift in mortgage demand to short-term loans. The rate of growth of life insurance company assets suffered from loss of annuity market share to the trusteed pension funds, and from the response of individual investors to inflation that reduced the sale of traditional life insurance policies with savings features. Meanwhile, high rates of monetary expansion favoured growth of banks, loan and trust companies, and credit unions and caisses populaires.

The structure of institutional mortgage lending varies among provinces (Figure 6.1). Shares of mortgage holdings vary provincially for all institutions, but mostly for credit unions and caisses populaires. These financial cooperatives are strongest in Quebec, Saskatchewan, and British Columbia. This strength is reflected in their shares of institutional mortgage holdings, 42%, 38%, and 20% respectively. Until recently, credit unions were noted for making loans that were completely open for prepayment without penalty at any time. This practice has declined in the face of increased interest rate risk.

FIGURE 6.1 Distribution of residential mortgage loans outstanding by type of lending institution: Canada and provinces, 1984.

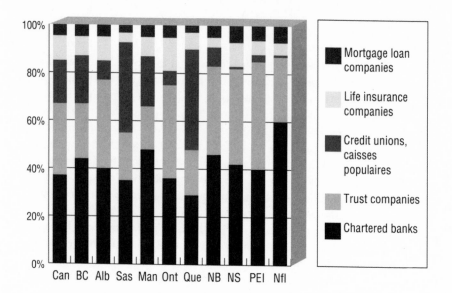

NOTE: Bank data include holdings of residential mortgage loan subsidiaries; national total includes Head Office and/or International mortgages. Data for mortgage loan companies and credit unions/caisses populaires were estimated to isolate provincial residential holdings from large aggregates.

Unresolved Challenges to Housing Finance

Three unresolved challenges to housing finance emerged in the post-war period: interest rate risk, default risk, and the preemption by employer pension plans of savings among younger employees.

INTEREST RATE RISK

The inflation-driven rising trend and increasing variability of interest rates in the 1970s and early 1980s reduced access to housing and its affordability. Access fell insofar as inflation altered only the time pattern of the ratio of mortgage payments to income of the borrower. A change in the expected long-term rate of inflation is quickly incorporated in the interest rate, whereas it affects borrower income year by year. Consequently, for an equal payment mortgage, the ratio of payments to income starts high because of the high rate of interest but declines year by year as income rises (the "tilt" problem). Affordability fell insofar as inflation raised real (inflation adjusted) interest rates rather than just nominal rates and shortened the term of available mortgages which made loans of a given size more risky.

Government responses to these problems included graduated payment mortgages (GPMs), the Mortgage Rate Protection Plan (MRPP), mortgage-backed securities and a direct attack on inflation itself. The last response was by far the most important, attacking the problem itself rather than simply its mortgage market consequences.

GRADUATED PAYMENT MORTGAGES, 1978

GPMs, with their low initial monthly payments, raise access to housing especially under conditions of inflation. But apart from their graduated payments, they do not help the borrower. They do not address the problem of interest rate risk, and they raise default risk compared to equal payment mortgages because the borrower's income may not rise as fast as payments. Also GPMs increase the risk that the loan balance will exceed the future value of the property. This can be offset by reducing the initial loan-to-value ratio, but this defeats the purpose. Lenders opposed GPMs (Clayton Research Associates Limited 1980). CMHC's loss experience appears to justify their concerns.

A better instrument for lending in inflationary conditions is the Price Level Adjusted Mortgage (PLAM). The PLAM features a price index applied *ex post* to the outstanding loan balance. The balance is adjusted on the basis of realized movements in the index and through amortization payments. Indexation of the principal is in lieu of the inflation premium in the interest rate. Consequently, the interest rate is low, which helps to address the tilt problem. Widespread private use of PLAMs would require large start-up costs, and borrower demand for them is not yet established. Nor is the market for indexed obligations which lenders would need to issue to undertake extensive indexed lending. Finally, interest rate risk would not be completely eliminated because housing prices do not closely track the rate of inflation (Pesando and Turnbull 1983). PLAMs were tried in NHA lending to non-profit housing cooperatives.

MORTGAGE RATE PROTECTION PLAN, 1984

The MRPP was introduced to offer partial protection to borrowers against large increases in interest rates on loan renewal. The MRPP issued only twenty-six policies in 1985 (CMHC 1986a). One problem with the program is its pricing. The same premium is charged for policies with protection periods that differ in term and starting date. Also, the premium is fixed, while expectations about future interest rates change often. The low level of sales is consistent with expectations that interest rates will not move adversely by enough to justify the purchase of protection. Pesando and Turnbull (1983) have suggested a market-based approach to premium setting. The MRPP is functionally equivalent to making longer loans callable by the borrower, for example, a ten-year loan callable after five years. If lenders issued ten-year non-callable obligations and ten-year obligations callable after five years, the difference in yields would provide a basis for establishing the "insurance" premium. Premiums for other renewal arrangements could be established in the same way. A market-based approach to premium setting merits examination.

INTEREST RATE FUTURES [3]

Recently, the government has become interested in another device for addressing the problem of interest rate risk: financial futures contracts. These are relatively new members of a broad class of bilateral contracts between purchasers and sellers that provide for deferred delivery and settlement at pre-specified dates and prices. They are known as forward contracts.

The *raison d'être* for such contracts is the shifting of risk-bearing they allow. A lender who agrees to make a mortgage loan at a specific rate of interest on a designated future date faces the risk that interest rates might rise in the intervening period. As a result, the lender would either have to pay more to raise money to finance the loan, or invest existing funds at less than the then-prevailing mortgage rate. In either case, he would be worse off than if he had not committed to lend in advance. But borrowers need such commitments. The lender can reduce his risk by entering a forward contract in which he agrees to deliver long-term federal government bonds at about the same date for a price close to today's price. Then, if interest rates rise by the time he must advance the loan funds, the price of the bonds will fall, and he can meet his forward contract at a profitable price. The gain on the forward contract helps offset his loss on the mortgage commitment. Alternatively, if interest rates fall, the margin between his lending rate and his borrowing rate will rise, but so will the price of the bonds he agreed to provide under the forward contract. Again, gain and loss are offset. Such "hedging" can also be used on the purchase side of transactions.

Forward contract transactions are effected through dealer markets. Futures contracts incorporate standardized terms for the subject securities. The terms cover quantity, delivery date, location and method of delivery, and penalties for substitution by the seller in respect to any specified term. The Chicago Board of Trade is the North American leader in the development of financial futures and related products. Canadian exchanges have been slow to add financial futures to their list of traded products. The Toronto Futures Exchange (a subsidiary of the Toronto Stock Exchange) began trading in futures contracts on Government of Canada Treasury Bills and long-term bonds in 1980. Contracts on foreign currency and the TSE 300 composite index were added in 1984. Because of the newness of these markets, trading is thin, and the maximum maturity on contracts is one year or less.

The example cited above actually constitutes cross-hedging. The original commitment was in mortgages, whereas the hedge was in long-term federal government bonds. Cross-hedging involves a risk that prices (or yields) for the two instruments are not perfectly correlated. The more alike are the two instruments, the closer the yields will be matched. Hence, a futures contract for an instrument closer to mortgages would be preferable. Some day, MBS might fill this role. MBS futures would be an improvement, although hedging imperfections from quantity and maturity mismatching would remain.

MORTGAGE DEFAULT RISK INSURANCE

Default risk differs from interest rate risk in an important way. Mortgage interest rate movements across Canada are closely tied, and consequently have a broad systematic effect upon a national portfolio of mortgages. Default risk, on the other hand, is affected by numerous variables, some of which are independent from one loan to another (for example, family divorce), or from one local market to another (for example, personal incomes in Windsor and Saskatoon), or from one collection of local markets to another (for example, provincial political developments in Quebec and British Columbia). While some variables are national in scope (for example, mortgage interest rates as noted above), the nationwide variations are considerably smaller for other variables. The greater number and diversity of variables that affect default risk make it more insurable than interest rate risk.

CMHC has provided mortgage insurance since 1954, first alone and then in competition with private insurers. Operations were generally successful until 1979, although in retrospect, by that time an actuarial deficit had accumulated in the Mortgage Insurance Fund (MIF). But then, the crunch came. In a period of stagflation, interest rates rose to unprecedented heights. Regional variables affected property values, notably political developments in Quebec and the National Energy Policy in Alberta. Loans under two heavily subsidized programs designed to help low-income groups, the Assisted Home Ownership Program (AHOP) and the Assisted Rental Program (ARP), proved vulnerable. By the end of 1984, the MIF had an actuarial deficit of $787 million of which AHOP (31%) and ARP (26%) loans accounted for 57% (CMHC 1986a). Of total claims on the MIF from 1954-84, 84% were made in the last six years. Faced with accumulating losses, from 1982-5 CMHC reduced its level of risk taking, raised insurance application fees to make them more cost related, raised premium structures, and altered them to better reflect differences in risk. The MIF's accumulated deficit fell about $50 million in 1985.

CMHC's underwriting practices can be improved (CMHC 1986a). The problem with the Mortgage Insurance Program (MIP), however, lies in the nature of its objective and the constraints on its pursuit. The stated objective is:

> To insure borrowers in all parts of Canada have access to high-ratio mortgages under the same borrowing terms and conditions, subject to the following three constraints;
>
> (i) The funds being provided by private lenders;
>
> (ii) The program being operated at no cost to the government; and
>
> (iii) That NHA insurance be provided in a competitive environment with private insurance (CMHC (1986a, 9).

The objective and constraints (i) and (ii) date back to the start of the program; constraint (iii) to the introduction of private mortgage insurance.

This objective is for loans that finance market housing. It is not clear why the "same borrowing terms and conditions" should be available for market housing "in all parts of Canada" when insurance application processing costs and default risks vary markedly from one locality to another. The cost of handling an application in a remote community may be five or ten times that of a large centre; risk in a single-industry community, for example, may be three times that in Toronto. CMHC's new fee and premium schedules are intended to be self-supporting and more cost and risk related than before. But they do not vary by location, so the possibility of cross-subsidization remains.

In principle, borrowers in low-cost/low-risk lending locations should not have to subsidize those in high cost/high risk lending locations any more than, say, home owners in communities with low land costs should subsidize those in communities with high land costs. Important as they are, borrowing costs are only part of the cost of housing, and application fees and premiums are only a small part of borrowing costs. The locational pattern of other housing costs may differ from that of financing subsidies. A preoccupation with making "the same borrowing terms and conditions" available "in all parts of Canada" may augment differences in the pattern of housing access, affordability, and security of tenure caused by other housing costs.

Cross subsidization also has undesirable competitive effects. Private insurers cannot compete in subsidized markets. They are forced to concentrate on low-cost/low-risk markets, a practice termed "cream-skimming." Cream-skimming is encouraged when CMHC runs a separator. Subsidizing borrowers in communities with high lending costs or risks has a long tradition. It became established when housing policy was dominated by employment policy, when monetary policy was less flexible and fiscal policy less developed, when banks were not in the mortgage market and CMHC was a more active lender, and when loan size and interest rate ceilings were in force. Under current conditions, the policy is an anachronism.

Another issue concerning the MIP is the optimum market configuration. What role should CMHC play? What role should be played by private insurers? The answer turns partly on the "contestability" of the market, that is, whether the threat of entry by new competitors keeps prices down. If so, then CMHC could withdraw from the market. Private insurers, however, are unable to cope with large, pervasive, systematic forces affecting default risk, for example, severe, widespread economic depression. Only the federal government can. Even if private insurers were to have the lion's share of the market, they would require a government backup agency as reinsurer. This is a role for CMHC.

CMHC has reviewed a number of options for expanding the role of private insurers (CMHC 1986a). An attractive option from a social standpoint is provision of a basic policy by a private insurer which gives less protection than currently provided under NHA, with an optional additional policy offered by

CMHC to provide protection up to the current NHA scale. This option ensures that all borrowers who are willing to pay the premium(s) required by insurance costs and risks have access to high-ratio mortgage loans, and it relies on market forces to determine where government insurance should be targeted. Government insurance complements private insurance which would restore investor confidence in the mortgage insurance business. The risk is whether competition would be sufficient to keep prices down in the long run.

EMPLOYER PENSION PLANS

Pension plans as currently designed require the employee to divide his or her long-term savings year by year over the household life cycle. Unlike other investments, the allocation is not made on the basis of expected returns relative to risk, but by preemption. An alternative approach might give households greater scope to apply annual long-term savings to the most productive uses at each stage of the household life cycle. Illustrative cost/benefit analysis suggests that for many households the optimum sequence of allocations is first, accumulate down payments and buy household "plant and equipment," then clear debts, and finally concentrate on financial assets, including pension claims. For many of the households for whom this approach is appropriate, this sequence of allocations would increase real income, and hence the capacity to save for long-term purposes including retirement. They could have more housing and more retirement financing.

The current approach to retirement financing reduces the accessibility and affordability of home ownership. For example, if the loan-to-value ratio is 90%, $10,000 preempted in pension savings (including the so-called employer contribution, and accumulated interest) reduces financing capacity as determined by this constraint by $100,000. If the GDS ratio is 32%, for every $100 of monthly income preempted, the amount of loan for which the borrower can qualify is reduced by $11,173 – at an interest rate of 10% and an amortization period of twenty-five years. [4] Alternatively, amortization periods could be cut roughly in half with access to pension savings.

This conflict needs to be examined. One possibility might be a revived and extended form of Registered Home Ownership Savings Plan (RHOSP) integrated into pension plans.

Lessons and Issues

One lesson evident from the record of post-war housing finance is that insured high-ratio residential mortgage lending is feasible for lending institutions, including banks. The loans must be made at interest rates and maturities consistent with the risks borne by the lender, including default risk and interest rate risk. As the failures of the Canadian Commercial Bank and the Northland Bank emphasize, bankers must be careful in risk selection and ensure adequate multiregional diversification of loan portfolios, and sufficient resources must be provided for regulatory supervision.

The lesson for the mortgage insurance business is that the government agency's operating objective must be appropriate, and consistent with the constraints upon it. Maintaining the same terms and conditions of borrowing in all parts of Canada is questionable. Reducing inequality is reasonable; pursuing equality is extreme. An extreme objective is likely to be inconsistent with the constraint that it be pursued "at no cost to the government." In the case at hand, it was also inconsistent with maintaining "a competitive environment with private insurance."

A corollary is that premium setting needs the consideration that the complexity of the underwriting situation demands. The accumulated deficit in the MIF was still large as of 1985 despite three decades of underwriting during which property prices surged throughout the nation. Not until 1982 was the MIF evaluated on an actuarial basis. Meanwhile, CMHC was required to turn "profits" over to the Consolidated Revenue Fund.[5]

If mortgage supply is to be improved by increasing the liquidity of mortgage financing, an indirect approach is better than a direct approach. Mortgage loans themselves are not readily amenable to trading. MBS offer a little more hope and would link the mortgage market closer to other long-term debt markets.

Inflation taught two important lessons. First, the regulatory framework needed to be less constraining. NHA ceilings on loan amounts and interest rates, and the minimum limit on term to maturity, proved to be disruptive to the continuity of supply of NHA funds. Second, when inflation has driven up the level of interest rates and interest rate risk is high, it is difficult to prevent a large increase in the interest rate risk faced by mortgage borrowers. The lending institutions' own sources of funds shorten, and they must shorten the maturities of their loans. The MRPP, MBS, and PLAMs have potential to help but require development. And borrowers themselves often prefer to shorten term, speculating on a future decline in interest rates. They become more vulnerable in the process. This is another argument for fighting the causes of inflation rather than the effects. It is also an argument for relying less on monetary restraint and more on fiscal restraint for the purpose.

Finally, private lenders demonstrated that in the absence of undue constraints on their activities, and with growing assets, they were capable of vastly expanding mortgage supply.

In the light of the post-war experience and its lessons, two main policy issues emerge. First, on what basis should mortgage loan insurance be provided? Is a public agency politically free enough to be able to operate on a sound financial basis? Is it forced into excessive risk-taking (for example, AHOP and ARP) in the pursuit of social housing and non-housing objectives? Should it concentrate on reinsuring private underwriters, direct lending, and research?

Second, given post-war progress in housing and its financing, should the focus of public interest be shifted from housing finance to household finance? The growing conflict between housing finance and retirement financing illustrates the need for this broader perspective. Improving the management of household finances creates better borrowers. It also improves their housing affordability. In the opposite direction, residential mortgage borrowing can be used advantageously to finance non-housing activities. Survey data indicate the ratio of mortgage debt to the market value of owner-occupied housing was only 20% in 1984 (Statistics Canada 1984a). With an aging population, should we be extending our thinking to the role of housing in retirement financing, for example, to the potential role of reverse annuity mortgages? More broadly, should we think of housing as security for loans for any legal purpose? Should governments now distinguish between residential mortgages for housing and non-housing purposes?

This brings us to a question that was purposely avoided at the start of the chapter. What is housing finance? We know housing transactions are not conducted in isolation. For example, a house buyer may take a larger mortgage loan than is necessary to buy the house to conserve funds to, say, buy a car, other durables, or to finance post-secondary education. What we commonly call housing finance then is only partly housing finance. Using data from Statistics Canada's 1977 Survey of Consumer Finances, and defining mortgage finance as housing finance only insofar as a household's entire net worth is absorbed in housing equity, Jones estimated "nearly one half of outstanding home mortgage debt is supporting non-housing activities" (1984a, 22).

In the light of post-war progress, has the time come to be less concerned with access, affordability, and security of tenure of housing *per se* so we can be more concerned with access, affordability, and security of consumption and saving in general?

Notes

1 Principal sources for this section are Poapst (1962, 1975) and CMHC (1986a).

2 The data in this section represent market outcomes. They reflect the interaction of mortgage demand and supply rather than supply alone. Nevertheless, they do convey information about the supply side of the market.

3 David Novak, Ph.D. candidate in finance at the Faculty of Management, University of Toronto, assisted with the work on this section. His help is gratefully acknowledged.

4 Pension contributions by employee and employer, plus interest earnings thereon are not taxed. These taxes would reduce the amount available for debt service, and hence this impact of preemption upon borrowing capacity. Alternatively, if arrangements were made by which pension savings could be borrowed to finance housing, the effect would be as shown. For an illustration of this arrangement see Poapst (1984).

5 CMHC recently completed an impressive review of the Mortgage Insurance Program.

CHAPTER SEVEN

Regulation and the Cost of Housing

John Bossons[1]

REGULATORY INTERVENTION in markets is a fact of modern economic life. Few markets, if any, are completely unregulated; at the same time, few are subject to the complex web of regulation found in the housing market. Other markets are regulated, but in few are there as many interacting reasons for regulation as in housing.

Government intervention in markets is ever changing, reflecting a kaleidoscope of conflicting political pressures. Regulation in the housing market has been no exception to this trend. Numerous federal and provincial task forces have been set up since 1945 to examine how regulation of the housing market might be made more efficient, and many of their recommendations have been implemented. The growth in regulatory intervention has not been uninterrupted; there have been periods of consolidation and simplification. Nevertheless, every decade of the post-war period has ended with more regulation of the housing market than it began.

Regulation is an economically valuable service provided by politicians that reflects an underlying demand among voters and supporters. Relevant to the analysis of regulation, therefore, is not the narrow issue of its cost, but rather whether alternative forms of government intervention can provide less costly means of satisfying the underlying demand. By its nature, regulation redistributes risks and associated costs; much of the cost of regulation is typically borne by individuals other than those who benefit from the regulations. As a result, the issues raised by regulation of the housing market have as much to do with redistributive equity as with economics. A second important issue arises from difficulty of regulating risks. It is not easy to design mechanisms to redistribute risks without creating unintended side effects, which in turn often generate demands for more regulation.

Analysis of the effects of regulation requires precise specification of the housing markets wherein regulatory intervention occurs. Because of the durability of housing capital, distinctions must be drawn between the markets for new and old housing, and between new housing built on vacant lots and housing renovation. Because of the varying importance of land costs and environmental impacts, distinctions must be drawn between inner-city and suburban housing in any one urban area, and between metropolitan areas and smaller cities. Finally, in the case of rental housing, distinctions must be drawn between the market for rental services and the market for the corresponding capital asset. There is substantial regulatory intervention in each of these markets.

The Growing Demand for Land-Use Regulation

The key element of the economic case for land-use regulation is the potential existence of externalities. Externalities are effects of private decisions that directly impact on individuals other than the decision makers. An unregulated market economy with perfect competition, in the absence of externalities, can yield an efficient allocation of resources.[2] Central to this assertion is that decision makers bear the entire cost and derive all of the benefits resulting from decisions that they make. Externality problems arise when this is not the case.

For example, the source of the market failure in the case of a lead refinery's pollution is that the effect of air pollution on nearby residents and owners is not reflected in the market prices of the output of the lead refinery nor in the costs of inputs used in production. Put differently, the neighbours do not own clearly defined property rights in clean air, and a refinery can (in the absence of government intervention) avoid compensating its neighbours for the effects of its pollution.[3] Zoning by-laws represent ways in which individuals attempt through the political system to obtain property rights not otherwise provided or to enforce them more cheaply than in the absence of such regulation.

Governments have other tools to ensure that land-use decisions take externalities into account. One tool is the taxation of sources of negative externalities and compensation of affected neighbours. However, the taxation approach has its own problems. One is the general inability of regulators to determine the optimal tax rate on the production of a negative externality. The second is the practical difficulty and cost of monitoring the production of the negative externality so that the tax can be imposed. It is often easier to come closer to the desired solution by regulation, since such regulation (at least in the case of land uses) generally consists of defining permissible locations for noxious activities or imposing limits on total development, thus directly controlling the production of negative externalities.

Another tool is to require land developers to obtain the agreement of nearby land owners to the specifics of a development scheme. A legal requirement for developers to obtain the agreement of neighbours would generally be deemed to be excessively restrictive. Nevertheless, one way of viewing the land-use regulatory process is as a means of introducing incentives for developers to negotiate with neighbours and to find ways of minimizing negative externalities.

In examining the sources of demand for land-use regulation, it is useful to distinguish among three classes of externalities: local, global, and fiscal. Local externalities arise from a specific, identifiable land use; the externalities affect nearby properties and presumably decline in importance with distance from the specific source of the externality. Global externalities are associated with congestion and other spillover effects of urban growth that impact directly on the welfare of individuals. Finally, fiscal externalities from new development are causes of higher taxes that reduce private consumption.

LOCAL EXTERNALITIES

Controlling the location of land uses that have negative external effects is the oldest function of zoning by-laws, so-called because of restriction of these uses to certain areas or "zones." Municipalities have been empowered to pass zoning by-laws since before the first world war in Alberta, British Columbia, and Ontario, and since the 1920s in most other provinces. [4] Land owners benefit from zoning by-laws principally by the reduction in uncertainty whether a producer of negative externalities can locate next door (and this may well translate into an attendant increase in property values). Political demands for more restrictive zoning have historically arisen when examples of the actual intrusion of undesired uses increase public awareness of the importance of these risks.

The practical definition of a local externality is a political matter, even though an economic rationale may be provided. How nuisances are defined has changed with tastes, incomes, and political sophistication. Economic growth has affected the demand for local environmental controls aimed at preserving neighbourhood character as well as simply prohibiting incompatible land uses. The perceived importance of the quality of urban life almost certainly is a function of rising household affluence. Moreover, the political sophistication of urban households has increased along with the awareness that government intervention can protect or enlarge an individual's range of consumption choices.

Land-use regulations dealing with sources of local externalities are chiefly concerned with increasing certainty as to what nearby property owners may do with their land. Success creates its own problems of unnecessary rigidities, that in turn have led to the establishment of complex adjudicative processes to permit acceptable relaxations of standards. Often, these also ensure that neighbouring owners have an input into the site approval process. Nevertheless, the conservative nature of local land-use regulation must be emphasized. Avoidance of uncertainty is inherently conservative.

GLOBAL EXTERNALITIES

Over time, the political demand for land-use regulation has come to include more than simply the regulation of local externalities. As urban populations have increased, political pressures have mounted to counter perceived negative consequences of growth, such as congestion and environmental degradation. During the rapid urbanization that has occurred after 1945, these pressures led to political responses ranging from major public investments in transportation and services to support growth to attempts to channel or control growth. In addition, the growth in environmental concerns – largely since the mid 1960s – has augmented political pressure for regulation.

These broader concerns reflect negative externalities that are global in that they arise from the collective actions of all decision makers rather than from the decisions of one landowner. The social costs of growth borne by existing residents and voters potentially include greater congestion and pollution, increased costs of travel between home and work (including the value of time spent in travel), and spillover effects on the quality of residential neighbourhoods. They also include costs arising from the impact of urbanization on surrounding areas: decreased access to recreation facilities and higher costs of recreation services – for instance, longer travel time, environmental degradation, and higher prices of vacation property. Increased awareness of these costs heightened the demand for environmental regulation.

This demand has also been fuelled by a fear among some that calculations of the impact of negative externalities give insufficient weight to the effect on future generations of environmental changes such as loss of farmland or of natural flora and fauna. In economic terms, this fear reflects a perception that the social rate of discount which should be applied in evaluating the cost of environmental externalities may be appreciably lower than market interest rates. The appropriate weight to give the interests of future generations in evaluating important environmental issues is, of course, contentious. A case can be made for a zero rate of discount with respect to environmental changes that in the aggregate may have severe and irreversible impacts on the quality of life for future generations (see, for example, Solow 1974).[5] Whether one agrees with such a position or not, such concerns have come to be politically significant and increase demand for land-use regulation.

FISCAL EXTERNALITIES

Many of the perceived social costs of urbanization and growth can be lessened through government expenditures on parks, transportation, and other services. But these expenditures may merely translate negative externalities into tax increases for existing voters. Fiscal externalities – tax increases for existing voters attributable to new development – may be as powerful a source of demand for regulation and for exclusionary zoning as are congestion and environmental effects.

Except in those Maritime provinces where education is fully financed by provincial governments, the current method of financing education is a source of fiscal externalities. A large fraction of education costs is paid for out of property tax revenues – in some municipalities, over 80% – and this creates a disincentive for municipalities to permit new residential development which would likely be occupied by low-income families with children. A strong argument for provincial assumption of responsibility for financing all education costs (and paying for them out of general taxes such as the income tax), as was done in

Nova Scotia in the mid 1970s, is that doing so would remove the largest negative fiscal externality currently associated with low-income housing.

One way of dealing with fiscal externalities is to impose special taxes on new development which are large enough to offset the costs and taxes which otherwise would be borne by existing taxpayers as a result of new development. Such taxes, currently imposed in the form of lot levies by municipalities in all provinces other than Quebec, provide an efficient means of offsetting negative fiscal externalities. The use of lot levies for this purpose is opposed by developers. Lot levies increase the price of serviced land relative to what it would otherwise be, and so are claimed to increase the cost of new housing. If municipal lot levies were held down by senior governments to levels which do not offset the negative fiscal externalities perceived to be associated with new development, municipalities might reduce processing of the supply of available land which could drive up the cost of new housing (and particularly the cost of new low-income housing).

POLITICAL TRADEOFFS AFFECTING LAND-USE REGULATION

It is not possible to prevent all potential negative local externalities or to eliminate the social costs of population growth. Moreover, welfare losses for existing residents may be offset by welfare gains for newcomers. The political pressure for regulation arises from the fact that the costs of new development are not all borne by those who gain. Regulation is the easiest way of diminishing the potential for undesirable redistributive transfers from losers to gainers. As long as voters see themselves as potential losers in the absence of regulation, the political demand for land-use regulation will continue.

The countervailing political force arises from the costs created by regulation, particularly as they are seen to arise from unpredictable red tape and delays. Nevertheless, while this creates a continued demand to make regulation more efficient, the continued existence of bureaucratic delay is a fact to which decision makers adapt. In addition, as developers and their advisers become expert with existing regulations, they have an advantage over new entrants and a vested interest in the status quo. This makes the forces countering regulation weaker than they otherwise would be.

Land-Use Controls in Already Developed Areas

Around 1945 municipal governments in Canada used zoning largely to maintain existing land uses. Most zoning by-laws separated residential from industrial and commercial uses and regulated the density and type of development permitted in neighbourhoods. These latter were thus typically divided into different zones, ranging from exclusively single-family residential through zones in which houses could contain apartments to zones in which apartment buildings could be constructed. A 1949 report prepared for CMHC described the existing system as "a concept of zoning which concentrates mainly on the fixity of land values by preventing changes in the established usages within an area"

(Spence-Sales 1949, 78). Zoning for existing neighbourhoods has largely continued to adhere to this concept.

Over the four decades since 1945, land-use controls in already developed neighbourhoods became more complex and sophisticated. In terms of complexity, regulation was extended over more and more details of the design of new development. In terms of sophistication, control techniques have been fine-tuned to permit a more graduated hierarchy of controls and to provide for area-specific differentiation in their rigidity. On one hand, this differentiation has resulted in increased rigidity of regulation in existing residential neighbourhoods. On the other hand, where redevelopment is permitted, regulation has been made more flexible, though at the price of increasing the discretionary authority of municipal officials and politicians.

The increased sophistication of land-use controls has increased their effectiveness. It has also increased the impacts of regulation. Increases in the complexity of regulation have been accompanied by and (some would argue) have been exceeded by improvement in the efficiency of the regulatory process. Increasing complexity has also progressively increased the importance of the role played by lawyers and professional planners in land-use regulatory decisions.

ZONING IN INNER-CITY LOW-DENSITY RESIDENTIAL NEIGHBOURHOODS

Perhaps the strongest evidence for the economic significance of land-use controls is the persistence of low-density residential neighbourhoods in the central areas of Toronto and Vancouver. In both cities, the unanticipated intrusion of high-density residential developments in middle and upper-income low-density neighbourhoods in the 1960s led to political pressures for more protective zoning. The result in both cities was a virtual cessation of high-rise development in existing low-density neighbourhoods[6] where, moreover, citizen pressure on politicians has led to the progressive enactment of increasingly detailed regulation of low-scale redevelopment.[7]

The Vancouver and Toronto experiences provide illuminating evidence of the nature of the demand for certainty which underlies the pressure for zoning controls in existing neighbourhoods. On the margin, the shadow price associated with the density constraints built into zoning regulations would appear to be high.[8] Nevertheless, the demand for inner-city land used for single-family housing has been increased by the perceived stability of land uses in such areas, and this has encouraged substantial investment in the renovation of inner-city housing along with an increase in its market value.

The value of single-family housing in inner-city low-density neighbourhoods has been increased by restrictive zoning, whereas the value of land assemblies has fallen. Because of the risks inherent in land assembly, much of the return from redevelopment to higher densities accrues to speculators who assemble land. Without restrictive zoning, the difference between the value of land in its current use (that is, fragmented ownership) and its potential value if redeveloped would be greater, hence increasing the profitability of redevelopment to higher densities. Without restrictive zoning, uncertainty regarding the timing and location of such redevelopment reduces the value of properties which are not redeveloped and discourages renovation and maintenance of existing low-density housing.

The trend to increased protection of existing low-density inner-city neighbourhoods has occurred at the same time as a change in public perceptions concerning viability. In the 1950s and early 1960s, it was common to regard inner-city residential neighbourhoods as blighted and to assume that they would be replaced by new developments. This assumption was explicitly reflected in planning for these areas, many of which were targeted for redevelopment. The resurgence of demand for single-family housing in these areas by middle and high-income households over the past two decades has caused planners' assumptions of blight to be discarded. The increase in demand for owner-occupied inner-city housing has reinforced the demand for restrictive zoning in these areas. [9]

CONTROLS IN RECENTLY BUILT SUBURBS

While economic pressures for redevelopment are greatest in inner-city neighbourhoods, there is substantial regulation of redevelopment in suburban residential neighbourhoods. The relative rigidity of this regulation has become an important policy concern in rapidly growing metropolitan areas. In part because of the greater homogeneity of post-war suburbs, it is easier to enforce rigid regulations in these areas than in inner-city neighbourhoods. Moreover, the political demand for protection of property values has been more uniformly high in suburban areas than in the more socially mixed inner-city neighbourhoods. For both of these reasons, land-use controls implemented in suburban areas have been more exclusionary and have made land-use intensification difficult in post-war suburbs.

The issues that arise in the analysis of political pressures underlying suburban land-use controls are not much different from those associated with the inner-city. In both cases, the strong demand from existing residents for protection from the risks of possible negative externalities of new development makes it politically difficult to reduce the rigidity and complexity of current rules. In both cases, intensification would reduce the cost of new housing.

EXCLUSIONARY ZONING IN BUILT-UP AREAS

The use of land-use regulation to reduce uncertainty can take many forms. There is, however, only a subtle distinction between preventing noxious uses of nearby land and maintaining the social homogeneity of a neighbourhood. Zoning has been widely used to exclude low-income households from suburban and exurban municipalities in the US, primarily through legislated minimum lot sizes or minimum housing standards. [10] In part because local government is less fragmented in most Canadian metropolitan areas, there has been less use of zoning for blatantly exclusionary purposes in Canada. However, the difference has been slight.

Exclusionary objectives are an important element in Canadian zoning practices, particularly in suburban municipalities. Maintaining the exclusivity of upper or middle-income residential neighbourhoods is accomplished through the exercise of discretionary powers in subdivision approval and through a variety of zoning practices (such as minimum lot sizes and setback requirements) which put a floor under the land cost component of new housing. Other regulation (such as zoning requirements for minimum apartment sizes or prohibition of basement apartments) may in effect put a floor under the capital cost of new housing. In addition, particular classes of individuals may be excluded by special-purpose prohibitions (such as exclusion of group homes).

An important difference between Canada and the US is that provincial governments generally exercise more control over local governments and have in some cases put pressure on municipal councils to change exclusionary zoning. Nevertheless, while intervention by higher-level governments has resulted in somewhat more uniformity in zoning practices than would otherwise be the case, provincial politicians are also subject to pressures from their constituents which reflect the demand for exclusionary zoning. Provincial intervention has not significantly reduced the occurrence of exclusionary practices.

Exclusionary zoning practices raise difficult conflicts between objectives. On one hand, they provide a means by which fiscal externalities may be reduced and local property values increased. A study in the US published in 1974 provides some evidence on the external effect of investment in one property on nearby house prices, concluding that the return to nearby property owners is from 10% to 15% of the cost of housing investment in structures (Peterson 1974). This evidence is a specific instance of the well-known effect of neighbourhood quality on land prices. The return to property owners from investments in nearby properties provides an economic rationale for zoning practices which increase the average investment in nearby houses; the economic or private-market optimum (from an efficiency viewpoint) is an allocation of housing to districts that maximizes homogeneity.[11]

On the other hand, exclusionary zoning reduces the supply of land available for low-income housing, relative to what would occur in a completely unregulated market (either by zoning rules or private agreements). It can thus increase the cost of land used for low-income housing, particularly if local governments compete among themselves for high and middle-income residents who pay more taxes per household. At a minimum, such cost increases generate a need for greater subsidies for low-income housing supply.

The difficulty in dealing with the effect of exclusionary zoning practices and minimum housing standards on the supply of low-income housing is that there is an identifiable potential cost to locating new cheap housing next to existing housing that is more expensive. This cost arises from the potential reduction in the value of nearby properties, and so is borne by those properties. While this cost may be reduced by sensitive design, nearby property owners have no way of ensuring design quality and understandably fear the worst.

The demand for minimum quality standards for neighbouring properties is a predictable political phenomenon which cannot be ignored. Though no prospective solution is ever easy, it is perhaps easier from a policy viewpoint to raise broad-based taxes to pay for more social housing subsidies to offset the cost increases incurred, than to attempt to reduce or eliminate minimum quality standards.

This conclusion is strengthened by the likelihood that private exclusionary practices would tend to arise in the absence of exclusionary zoning, and indeed, they have done so in jurisdictions where zoning laws are weak. In Houston, Texas, widely cited as an example of a metropolitan area without zoning, private subdivision agreements are widespread. These typically include restrictive covenants registered on land titles, mostly limiting further subdivision of lots and restricting lots to single-family usage. Many also control uses in commercial areas adjoining residential subdivisions. The economic significance of exclusionary zoning or restrictive covenants as a means of reducing uncertainty is illustrated by the fact that the use of such covenants has been a condition of loan approvals for most new residential subdivision developments in Houston (see Siegan 1970, 94-5). [12]

One needs to be careful in analyzing the impact of exclusionary zoning. While zoning can be used to reinforce covert racial and social discrimination, there is a substantive economic rationale for exclusionary zoning that justifies its existence. The issue is how it is used.

ALTERNATIVE FORMS OF CONTROL IN REDEVELOPABLE AREAS

The City of Vancouver has been a pioneer within Canada in implementing a flexible land-use control procedure for areas in which redevelopment is anticipated and encouraged. Provincial legislation passed in 1953 has empowered Vancouver to use a development permit system rather than zoning in designated areas; this system has been used in Vancouver's West End since 1956 and subsequently for new neighbourhoods created in False Creek. Vancouver's West End contains one of the greatest concentrations of high-density accommodation in Canada.

The development permit system used in Vancouver is one in which density, use and design are negotiated by the city and the developer. The process provides discretionary authority to city officials who are responsible for the negotiations. In large measure, the initial use of this discretionary zoning procedure was similar to the use of site-specific rezoning in other jurisdictions. [13] Subsequently, however, its use has been extended to include comprehensive development plans for new neighbourhoods in which discretionary authority is delegated to municipal officials in implementing the plans. [14]

The discretionary zoning system has been implemented in other jurisdictions. The City of Winnipeg Act has since 1971 provided for the establishment of development control areas within which control is by negotiated site-specific development permits rather than through pre-defined zoning, subject only to the provisions of a previously-adopted official plan for the district. Similar powers have been extended to municipalities by planning legislation in Alberta and Nova Scotia.

Conditional zoning and other innovative zoning techniques have been introduced to achieve more finely articulated means of control. In Toronto, a special form of zoning (so-called mixed-use districts) was adopted for areas of permissible redevelopment within the Central Area in 1976; the zoning provides flexibility in range of uses and has in practice provided room for negotiation of density, while at the same time putting limits on the exercise of discretion.[15] Transfers of development rights between locations have also been increasingly used as a device to permit the scope of negotiations with a developer to be widened.

Other elements of discretionary land-use controls have been introduced through the widespread adoption of supplementary development controls to regulate development details not controlled by zoning ordinances. Legislation empowering municipalities to introduce such controls was passed in most provinces in the 1970s. At a minimum, these supplementary powers allow municipalities to control siting, setbacks, and access through development agreements negotiated with the developer. In Ontario, planning legislation specifically precludes municipalities from using such controls to reduce permissible heights and density. However, in other provinces, supplementary municipal powers of development control are more broadly defined. Several provinces expressly empower municipalities to regulate matters of design.[16]

Supplementary development control powers should be distinguished from discretionary zoning, since (at least in concept) such powers merely supplement the more general specifications of permissible uses and density set out in a zoning by-law. Nevertheless, in practice the existence of any powers of discretion provides a basis for negotiations between municipalities and developers over any aspect of the development. The potential for arbitrary delay introduced by any discretionary system of development control provides municipal officials with substantial negotiating power, particularly in dealing with developers who expect to have to work with the same officials on subsequent proposals. Legislative limits on the matters subject to municipal review are consequently seldom effective.

INCREASES IN PROCEDURAL COMPLEXITY

Since 1945 the process by which regulations are changed and administered has become more complex. As the rights of citizens have become better articulated, the costs of this process have risen. In virtually all jurisdictions, there is provision for a local board to which individuals can appeal for minor variances. Where notice is provided to adjacent property owners and there is a public hearing of appeals, this process typically provides a reasonable forum for their adjudication. Important procedural changes have also occurred in the amendment of zoning regulations. In all provinces, individual citizens have gained increased rights to intervene in rezoning or other modifications of land-use restrictions. Public notice of a proposed zoning by-law is mandatory in every province; public hearings by municipal councils are now mandatory in six provinces. [17] In Quebec, local public hearings are not mandatory, but zoning by-laws must be submitted to a local referendum if sufficient citizens object.

In several provinces, decisions at the municipal level may be appealed to a provincial tribunal either by the developer or by citizens. [18] Because it is normally necessary to have legal counsel and to hire expert witnesses in order to participate effectively in hearings before such tribunals, the costs associated with this process can be substantial, and the difficulty faced by citizen groups in raising the necessary funds biases the hearing process in favour of developers. However, this is partly offset by the fact that the time taken up can impose significant costs on developers waiting for a decision and increase the pressure on them to negotiate a compromise with objecting citizens.

While the growth in procedural complexity is often condemned by developers as undue red tape, it has been a response to issues of individual rights which arise in the administration of discretionary authority by regulators. Canadian concepts of regulation have been heavily influenced by the US concept of rule-of-law, which Makuch summarizes as follows:

> Since we are dealing with property rights, rule-of-law values become paramount and the rules for controlling physical development must be set out in advance. Those rules should be clear, concise, predictable and understandable, and should be decided upon and applied by impartial arbiters (1986, 168).

This concept is the central feature of the US approach to zoning, in which a distinction is made between legislative and judicial functions, and in which (at least in theory) political choices are legislative and confined to the determination of general rules. The US rule-of-law approach is different from the concept of regulation in most European countries, where government discretionary authority is more the norm. In England, for example, developments are approved on a case-by-case basis by the central government.

Canadian regulatory practice is a compromise between US and European approaches. The post-war history of planning legislation and jurisprudence in Canada has reflected a continuing tension between a desire to adhere to the rule of law and the practical advantages of flexibility provided by reliance on discretionary authority. In many cases, it is unfeasible to attempt to control details of prospective development through pre-specified ordinances; the nature of the regulation required is too dependent on the nature of the development proposed. Moreover, if control on a site-by-site basis is ruled out, the next-best regulatory response may consist of the imposition of excessive standards and regulations. Indeed, most of the legislative changes introduced in the 1970s to permit supplementary development controls were largely motivated by a desire to reduce the extent to which municipalities achieved control through the otherwise unnecessary use of site-specific rezoning.

At the present time, municipalities exercise substantial discretionary power in regulating land use. Case-by-case regulation is widespread. How is the arbitrary exercise of discretionary power limited? Legislation controlling its exercise has generally relied on four methods: the first is to prevent conflicts of interest and to penalize corrupt practices; the second is to require that decisions affecting individual rights be made through a process which ensures that the affected individuals are given a fair hearing in accordance with generally-accepted notions of "natural justice";[19] the third is to provide an appeal process; and the fourth is to require decisions to be made on a basis that either adheres to prescribed criteria or is consistent with applicable precedents. In land-use regulation, the courts have distinguished between the adoption of area-wide zoning by-laws and site-specific rezoning. In the former, the courts have ruled that a municipal council is acting legislatively, and that rules of natural justice do not apply in such cases. However, in cases involving site-specific rezoning or other narrowly-defined issues where a municipal council is adjudicating between neighbouring property owners, judicial decisions have established that affected parties have a right to a hearing that is subject to such rules.[20]

PLANNING AND THE REGULATORY PROCESS

To encourage consistency in municipal zoning decisions over time, there has been a growing tendency for the provinces to require municipal zoning decisions to conform to an already-adopted official plan. A municipal official plan may be adopted for an entire municipality or for an area within it; and generally such a plan sets out policies and criteria for which the municipal council shall have regard in making subsequent zoning decisions. The use of both discretionary zoning and supplementary development controls have generally been required to be preceded by the adoption of official plans, with subsequent site-specific decisions required to conform to the policies set out in such plans.[21]

The preparation of official plans has been made a mandatory prerequisite to the use of normal (that is, non-discretionary) zoning powers only in Alberta and Quebec.[22] Nevertheless, most large municipalities have adopted official plans to guide development in areas where redevelopment proposals are frequent. Once an official plan has been adopted, subsequent municipal zoning by-laws are generally required to conform to the provisions of the plan.[23]

The use of official plans as a device to provide more certainty works only if such plans constrain subsequent site-specific regulatory decisions. While generally the case, there have been frequent exceptions. Moreover, where site-specific official plan amendments have been adopted to relax constraints on municipal zoning, such amendments typically are subject to added approval requirements. In most provinces, the adoption and amendment of official plans is subject to provincial ratification. In British Columbia, official plan amendments require approval by a two-thirds majority of the municipal council.

Regulating New Suburbs

Unlike land-use regulation in already developed areas, where the norm has been for regulation to be implemented through previously-enacted zoning by-laws, the control of new suburban development is almost entirely through the exercise of discretionary powers in the process of subdivision approval. Virtually all new suburban development requires the subdivision of existing lots into smaller lots. To ensure valid land titles for the owners of the new lots, developers must have their plans of subdivision approved by a government agency.

Such approvals, together with applicable zoning by-laws, are an additional instrument of regulatory control. While originally such approval was concerned with little more than the accuracy of surveys, governments had used the requirement for approval as a means of imposing land-use controls on new suburban development even before World War I.[24] Early concerns reflected the effect of smaller lot sizes on the adequacy of septic tank waste disposal systems as well as other considerations. By 1945 subdivision control had become an important component of land-use regulation in all provinces. However, the approval process focused on the adequacy of roads and municipal services. The use by municipalities of subdivision approval powers to control the location of new development emerged as a response to the pressures on municipal services created by the rapid growth that occurred in the years following 1945.

Although the nature of subdivision control varies from province to province, the basic features of such control are the same. An applicant must submit a plan showing the location and boundaries of lots, roads, parks, school sites, water mains, sewers, and other services. The plan must also show the uses to which lots shall be put. The importance of subdivision plans is not in what they contain but in the requirement for approval. Unlike zoning regulation, which defines pre-established rights of use for existing lots, there is no pre-defined right to subdivision approval. The approving agency, whether municipal or provincial, has discretion to approve or reject a plan.

Where provincial legislation constrains such discretion, it is to define circumstances in which a plan must be rejected.[25] Discretionary powers are generally used to delay approval of proposed subdivisions to stage new development in accordance with municipal servicing plans. Indeed, the Ontario Planning Act explicitly requires approving agencies to consider whether proposed developments are premature or in the public interest. In all provinces but Quebec, subdivision approval may be withheld unless adequate municipal services have been installed up to the site to be subdivided.

While the principal control tool for new suburban development is the subdivision approval process, zoning by-laws also apply to such development. In effect, subdivision approval plays the same role in the process (though with greater discretionary authority for the approval agency) that supplementary development review does for redevelopment in existing built-up areas. Zoning by-laws normally impose pre-specified limits on density, as is the case in already-developed areas. Moreover, they are of course required to control subsequent land-use changes once a subdivision is built.

The different process required for the enactment of subdivision zoning by-laws (often done at the same time as subdivision approval) has resulted in a seemingly complex system of development control. Some observers have proposed an integrated system of development permits in place of the existing dual control system.[26] It is not obvious that an integrated system would be preferable. While an integrated system could conceivably result in a system in which fewer approvals were required, it would be necessary to strengthen the role of municipal plans to maintain the protection of existing interests now provided by the zoning approval process.

In many cases, municipal plans and zoning are not set out for undeveloped areas prior to the submission of a development application. In part, this reflects a desire to avoid prematurely committing a municipality to a particular form of development. In addition, the approval of secondary official plans and zoning is often delayed for particular areas so as to direct growth to areas in which the municipality intends to provide servicing infrastructure (see, for example, Proudfoot (1980, 45-7).[27] The length of time required for approval of new subdivisions generally is considerably longer for land for which secondary plans have not been prepared.

Analysis of the lag between submission of a proposed subdivision plan and its approval indicates both that the average time required for a decision increased during the 1970s and that there is uncertainty as to the length of approval time.[28] The increase in the average length of time required for approvals seems primarily to have reflected two closely related factors: an increase in the number of provincial and municipal agencies which review subdivision applications; and an increase in the number of points of detail subject to negotiation with these agencies. In addition, requirements imposed in Ontario, Alberta, and other provinces for the preparation of regional plans temporarily slowed down the processing of subdivision applications affected by such plans.[29]

While the length of time required for subdivision approval may have some marginal effect on the concentration of ownership of developable land, it is likely to have little long-run effect on the supply and cost of new housing. The primary potential effect of an increase in approval delays is a one-time reduction in the value of non-approved developable land. Put differently, the long-term incidence of the cost of approval delays is borne primarily by land speculators rather than by prospective home buyers.[30] However, increases in average approval times can have a significant impact on short-run market responses to unanticipated increases in demand for new housing or to changes in the composition of such demand.[31] Whether temporary price increases occur depends on the size of the buffer stock of previously approved unbuilt subdivisions. The inventory of approved land on which building permits have not yet been taken out is normally sufficient to absorb most fluctuations in demand.[32]

The planning of major public infrastructure investments in expressways, regional public transit, and regional water and sewer mains has a greater impact on lot prices than does the subdivision approval process. For example, in the late 1970s Calgary froze the development of over twelve square kilometres of land on the south side of the city because of inadequate transportation facilities. The Ontario provincial government similarly froze most development north of Toronto for a period of fifteen years until a major water and sewer trunk servicing scheme was implemented in the early 1980s. In British Columbia, a freeze on conversion of agricultural land to housing in the Lower Fraser Valley was implemented through the establishment of the Agricultural Land Commission in 1972. Such freezes potentially have both temporary and long-run effects on lot prices, as have plans for green belts and regional environmental plans in other areas. The extent of the impact on lot prices depends on investor and developer expectations as to the durability of the freeze as well as on the supply of developable land not affected by the freeze.

DEVELOPER CONTRIBUTIONS TO MUNICIPAL COSTS

By the end of the 1960s, it had become normal practice in all provinces other than Quebec for subdivision approvals to be made conditional on the provision by the developer of required on-site services and on the donation of land for roads, parks, and other public purposes such as school sites. In addition, developers are frequently required to pay lot levies to offset off-site municipal costs.[33] Such conditions are normally detailed in supplementary agreements between the municipality and the developer.

Requiring developers to pay lot levies and to pay for the installation of required on-site services has shifted most of the public costs associated with new residential developments to the purchasers of the new housing. This is almost certainly the most efficient way of distributing the cost of servicing new developments. If municipalities were required to increase taxes on existing residents to subsidize new developments, the consequent voter resistance to new development could shrink the amount of new residential development approved, leading to higher land prices for approved land.

The costs imposed on developers have been the subject of much controversy. Since municipalities are responsible for the maintenance of services once installed, there is an incentive for municipalities to require a high standard of installed services. The resultant proliferation of "gold plated" standards for developer-financed services has been widely criticized by the development industry. The practice may be efficient from a social viewpoint.[34] Nevertheless, little analysis has been done of the real long-run marginal costs associated with alternative servicing standards. As Hamilton (1981, 63) notes, "It's surprising that more provinces have not required that some careful [cost benefit] analysis be undertaken to justify current subdivision standards."[35]

The normal requirement (outside Quebec) that developers pay at least the direct costs of roads and services within a development has the merit of internalizing the effects of varying development patterns on servicing costs. The added servicing costs associated with larger lot sizes and frontages are thus borne by the purchaser of the lots, as they should be. To the extent that higher servicing standards yield an efficient substitution of capital costs for subsequent maintenance costs, as Goldberg (1980) argues, then this results in further internalization of subsequent direct cost differentials.

A second advantage of requiring that developers pay all direct costs is that, for institutional reasons, it is easier to finance them through the new home purchases than by municipal borrowing. Municipalities tend, to a greater extent than federal or provincial governments, to balance annual budgets on a cash-flow basis. A requirement that direct servicing costs be paid for by municipalities out of current revenues would force taxes on existing residents to be increased, generating greater taxpayer resistance to new development.

Even in British Columbia, where legislative changes have limited developer payments to costs directly attributable to new subdivisions, the presumption that developers should pay for the direct costs of services installed within a new development is generally accepted. The question from a policy viewpoint is the extent to which developers should also be required to make payments (lot levies) as a contribution to the indirect costs of servicing a new development. Such contributions have been normal practice in all provinces but Quebec.

Little analysis has been done of the marginal costs to existing residents of new developments.[36] Analytically, the case for lot levies must be that they roughly correspond to the difference between the present value of added costs borne by existing residents of the municipality and the present value of incremental tax revenues attributable to the new development. The costs borne by existing residents include privately borne costs associated with growth (the negative global externalities described above) as well as the tax-financed costs of services required by residents of the new development. The effect of new development on existing residents in a municipality depends on many factors, including the aggregate rate of development in a region and the extent to which previous infrastructure investments can accommodate more growth. The social costs associated with growth will vary from region to region, and also among municipalities within a region. It is consequently difficult to generalize about the appropriateness of current lot levies without detailed analysis of the circumstances of specific municipalities.

One general point should, however, be emphasized. From an efficiency viewpoint, the social costs of growth should be internalized in the prices paid for new housing. While this has the undesirable side effect of increasing the cost of new low-income housing, it is better that the true costs of new development be faced explicitly (and reflected in private choices of alternative development forms). The subsidies required to make new social housing viable should be financed by all taxpayers. Attempts to reduce such subsidies by making neighbours bear some of the costs will merely engender more political opposition to the construction of social housing.

EXCLUSIONARY ZONING IN NEW SUBURBS

The most important impact of suburban land-use controls on housing markets has occurred through the imposition of minimum quality standards on new suburban housing developments. Minimum lot sizes and other land-use standards have become a significant barrier to the construction of low-cost housing.

As noted earlier, there is a justification for imposing minimum quality standards on new construction in existing neighbourhoods to protect the value of existing housing and so maintain incentives for residential improvements. However, the potential for negative local externalities is reduced in the case of new suburban developments in which all of the housing is constructed at the same time. In addition, the discretionary powers built into the subdivision approval process provide ample room for the imposition of site plan controls through which a municipality can ensure that local externalities are minimized for properties adjacent to the new subdivision.

A better justification for exclusionary zoning in new suburbs is the existence of fiscal externalities arising from the use of property taxes to finance education costs. Because low-cost housing is likely to be occupied by younger families with children, education costs may be increased by new low-cost housing developments to a greater extent than is the property tax base. This can result in property tax increases for existing residents, particularly in wealthier suburban municipalities in which an above-average fraction of education expenses are financed from local property taxes. This source of fiscal externality could be eliminated by a reform of education financing to eliminate the current degree of reliance on local property tax revenues.

In the absence of such reform, fiscal externalities could of course be eliminated by increasing lot levies sufficiently to cover the present value of the added taxes which would otherwise have to be raised from existing residents. Alternatively, provincial governments could provide larger grants to subsidize the costs of educating children in new low-income housing developments. In either case, the cost to government of building new low-cost housing is increased.

Even if fiscal externalities were eliminated, it is unlikely that the political pressures for exclusionary zoning would ease. Such pressures arise from many sources and motivations, only a few of which are economic. These pressures are reflected in the actions of municipal politicians. Any attempt by a higher-level government to force municipalities to become less exclusive will generally be politically costly.

The prevalence of minimum lot size restrictions and other minimum housing standards in new suburban housing developments imposes a serious constraint on the supply of low-cost housing. This constraint is particularly important in rapidly growing metropolitan areas. While finding ways of over-coming this constraint is important, a frontal attack on existing minimum qual-ity standards will require an investment of political capital.

Other Supply-Side Intervention

Government intervention in the housing market is not confined to land-use reg-ulation. It also includes regulation of new construction and of renovation through building codes, the licensing of tradesmen who work in the building trades, and a variety of tax incentives and direct subsidies. Changes in subsidies and tax preferences extended to the housing industry have had a particularly important effect on the supply of new housing since 1945.

REGULATION OF BUILDING QUALITY

Governments regulate construction primarily to protect the purchasers of new homes from unobservable quality defects. This is unlike land-use regulation which is most often concerned with avoiding negative externalities. Economists characterize the problem confronting consumers as primarily one of moral haz-ard. Where buyers cannot distinguish easily between construction of differing quality, market pressures may force quality standards to a lowest common denominator.[37]

Liability laws provide, in theory, an alternative to government-enforced min-imum quality standards. However, for these to be effective, it is necessary to remove the protection provided to shareholders by the limited liability of corpo-rations. Doing so is neither practical nor desirable. The effects that governments seek by regulation could be achieved in other ways, for example by requiring performance bonds from builders or by compulsory liability insurance. The lat-ter would result in the imposition of quality and inspection standards by private insurers as a condition of insurance.[38]

Some provincial governments have introduced warranty schemes. These schemes also generate problems of moral hazard and cross-subsidization. The moral hazard problem arises from the potential incentive effects of such schemes for high-risk producers. Even without such incentives, it is difficult to design an industry-financed warranty scheme which does not force low-risk, high-quality producers to subsidize high-risk, low-quality producers. Regula-tion limits the scope of such problems.

Since 1945 there has been some rationalization of building codes through replacement of local by provincial codes. In addition, more relaxed standards have been promulgated for renovations of older housing which does not con-form to the codes applied to new housing. Though criticisms of the rigidity of building codes and their bias against the introduction of new building-systems and other innovations in construction technique are commonplace, it is inevit-able that such rigidity and biases exist. Similar biases would exist in standards set by private insurers in an unregulated market with compulsory liability insur-

128   *John Bossons*

ance. Presuming that the risks associated with innovation are not to be borne in ignorance by the consumer, it is necessary for innovators to bear the costs of persuading regulators to develop standards that permit the proposed innovation.

TAXES AND SUBSIDIES

Governments intervene in the market for newly constructed housing primarily through tax preferences and direct subsidies. These tools have been applied on both the supply side and the demand side of the market. On the supply side, the most important have been federal tax incentives and direct subsidies for the construction of rental market housing (MURBs and the ARP program and its successors) and social housing. On the demand side, the most important have been the federal tax preferences for owner-occupied housing (the tax exemption of imputed rental income and capital gains on owner-occupied houses) and a variety of special-purpose schemes (for example, RHOSP). The most important effect of these tax and direct subsidy programs has been a distortion of the taxation of savings and investment, causing aggregate investment in housing to be greater than it otherwise would have been. This distortion in favour of housing has more than offset any aggregate effects of land-use regulation on the size of the housing stock.

Regulation of the Rent Market

In the 1970s two related forms of provincial regulatory intervention in rental markets were introduced in all ten provinces. These were rent controls and a substantial expansion in tenants' rights.[39] Some of the regulation was temporary, notably in most western provinces, where rent controls were removed in the 1980s. Rent regulation, in various forms, continues in Saskatchewan, Manitoba, Ontario, Quebec, Nova Scotia, Prince Edward Island, and Newfoundland.

The reasons for rent regulation are different from the reasons for land-use regulation. In land-use regulation or the establishment of construction quality standard, the substantive economic rationale for regulatory intervention is allocative failure in an unregulated market. It is difficult to argue that there is an allocative failure which justified either rent controls or tenant protection.

THE POLITICAL DEMAND FOR RENT CONTROLS

As discussed in Chapter 1, the average real rental price of apartments has declined over most of the post-war period. Indeed, during the 1970-5 period leading up to the introduction of rent controls, average rent rents fell by almost 20%. It is difficult to make a case for rent controls on the average behaviour of rental prices prior to the introduction of rent controls.

The political demand leading to the introduction of rent controls was largely a result of the high dispersion of changes in rents in the early 1970s. Neither the magnitude nor the duration of the inflationary upsurge in the early 1970s was anticipated by investors.[40] The consequence was a substantial variance in price increases, particularly for rent. The sudden, large increases in rents that occurred for a number of tenants was interpreted politically as "rent gouging." Fear of being subject to further such increases led to strong demands by tenants for protection from the perceived potential for abuse.

Had rent controls not been established, it is likely that average real rents would have risen over the 1975-85 decade. Important supply-side investment incentives were withdrawn, most notably the cancellation of MURBs in 1979. More important, interest rates rose through the 1970s and early 1980s, resulting in a sharp increase in the supply price of new rental units. [41]

The joint effects of rent controls and the increase in the real supply price of new rental units had created a serious disequilibrium in some metropolitan rental markets by the early 1980s. Such disequilibrium became particularly severe in Ontario, which experienced substantial population growth in 1980s. The major exceptions were Calgary, Edmonton, and Vancouver, where severe regional economic contractions reduced the demand for rental units in the early 1980s.

It is unlikely that rent controls can be eliminated in central Canada unless and until the current market disequilibrium is eliminated through the creation of a (temporary) excess supply of rental units. The conditions under which it is politically possible to eliminate rent controls can only arise through reductions in real interest rates, through subsidies of new rental construction, or through recession-induced reductions in aggregate demand for housing.

RENT REVIEW AND COST PASS-THROUGH

The predominant form of rent control, outside Quebec, has been "rent review." In effect, there is a two-stage process of rent control: rent increases up to a specified limit can occur by right, but increases above this limit may be permitted after review. The flexibility introduced by discretionary authority has permitted rent controls to be looser than otherwise would have been the case.

One of the more important elements of looseness in rent controls has been the allowance of "cost pass-through." [42] In particular, increases in financing costs arising from an involuntary refinancing have generally been deemed to be a basis for approving rent increases. Unlimited cost pass-through provides an incentive for transfers of ownership of old rental buildings that cause financing charges to be based on the current market value of the building. Since the benefits to the new owner from cost pass-through are likely to be capitalized in the price, rent increases attributable to such "voluntary" changes in financing costs have been limited. [43]

It is clearly desirable to permit landlords to pass increases in operating and maintenance costs through to tenants along with amortized capital expenditures. Without such flexibility, a rigid system of rent control would create serious disincentives for landlord investments in maintenance and renovation. Most existing rent review schemes provide limited opportunities for such recovery.

THE EFFECTS OF RENT REVIEW

The most important effect of rent controls has been to reduce the supply of new rental buildings, particularly as rent controls have been made tighter in recent years (in provinces where they are effective). This has shifted part of the welfare losses from rent controls to newly-formed households and to residents who move to rapidly growing metropolitan areas.

The side effects of rent control have magnified this effect. There has been a substantial incentive to renovate old buildings and to convert them to condominiums, thus reducing the supply of rental housing. (This has led to further regulation of condominium conversions in some jurisdictions.) In addition, since rent review is less effective in its application to new tenants, there has been an incentive for landlords to evict tenants in order to raise rents. [44]

An important by-product of rent control has been an increase in the demand for tenant protection. Until the early 1970s landlord and tenant regulation was primarily concerned with the enforcement of contracts against a tenant. However, in the 1970s there was a substantial increase in tenant protection, in part predating the introduction of rent controls and in part resulting from their introduction. In some jurisdictions (for example, Ontario), residential tenant protection now virtually accords tenants a right to indefinite occupancy of an apartment (at rents in accordance with rent guidelines) following the expiration of a lease. The consequent difficulty of removing an undesirable tenant has resulted in further decline in the supply of rental units, particularly of rooms previously rented to lodgers in owner-occupied housing. The net return from renting has become more uncertain, causing further increases in the required rate of return on new rental units.

Conclusion

Regulatory intervention is inevitable. The political demands by voters for regulation is deeply rooted in perceptions and buttressed by desired allocative effects which increase the aggregate welfare of the community. Nevertheless, there are opportunities for improvements in the efficiency of regulation. In land-use regulation, the most fruitful avenue of reform is likely to be through the use of official plans as quasi-constitutional instruments through which to reduce uncertainty. Such instruments can, if properly used, also serve as the basis for making implementing decisions more expeditiously. In construction standards, reform is likely to come through development of a better process for incorporating innovations in building technology standards.

It is difficult to foresee complete elimination of the significant tax distortions favouring home ownership. Indeed, the relevant political issue is to attempt to ensure that these distortions are not exacerbated by periodic pressure to allow the deductibility of mortgage interest and other expenses of home owners. Nevertheless, the magnitude of the tax distortions are currently enhanced by the effect of inflation on the tax system, and it is possible to eliminate this source of distortion by adjusting all income from capital for the effects of inflation.

In all of these endeavours, policy makers need to know more about the effects of regulatory intervention. In the case of land-use controls, important research issues concern the evaluation of indirect costs and benefits to existing residents from new residential development, both in new suburbs and in inner-city redevelopment. In the case of construction standards, there is a need for research on the effect of such standards on the cost of innovation and on the distribution of risks. Other key research questions include the evaluation of the interaction between the distribution among governments of fiscal instruments and the consequent incidence of fiscal externalities.

The impact of tax preferences and direct regulation on the housing market has important effects on society. There are potentially large returns to research that increases our collective understanding of the complex ways in which government interventions interact with one another and affect individuals in the economy.

Notes

1 I am indebted to John Hitchcock for his contributions to this chapter and to George Fallis, Jim Lemon, John Miron, John Todd, and an anonymous referee for their comments on an earlier draft.

2 "Efficiency" is here defined as the maximization of individual satisfactions given the initial distribution of human capital and other resources. The assumption that no externalities exist is but one element of a more general assumption made in deriving this neo-classical welfare proposition. The more general assumption is that the consumption alternatives available to each individual (and their utility to each individual) is independent of choices made by other producers and consumers. See, for example, Koopmans (1957, Sections 1.3 and 2.2).

3 In some cases, it may be possible for a property-owner to prove damages and so enforce a right to the "untrammelled enjoyment" of his or her property. However, the high costs and uncertainty of legal action rule this out in most instances.

4 A useful summary of early land-use legislation in Canada is provided in Hamilton (1981, Appendix V). Comprehensive zoning was introduced about the same time in the US. The first major use was in New York City in 1916, though widespread use of zoning ordinances did not occur in the US until after a 1926 Supreme Court ruling (in Village of Euclid v. Amber Realty Co.) that zoning was a legitimate use of police power not requiring compensation of injured property owners.

5 The case is essentially a negative one and is due to an influential paper by Ramsey (1928): Can one find an ethical ground for giving less weight to the utility of future generations' consumption than to that of the current generation? As do most ethical questions, this question raises complex issues.

6 In the City of Toronto, no privately-owned assembly of already-developed land in an existing low-density residence area has been rezoned to permit high-rise development since the early 1970s. Restrictions in Vancouver have not been quite so severe.

7 For example, in the City of Toronto the additional matters regulated in low-density

residence areas since the early 1970s include height, the length of buildings, and restrictions on the number of group homes (through enactment of a minimum distance between group homes). In addition, permitted densities have been reduced and minimum lot frontages increased in a number of low-density residence areas. Infill projects are prevented by a general prohibition of residential units behind other residential units; ways in which this prohibition could previously be avoided have been ruled out. Finally, additional regulation of design details of new developments has occurred through the extension of development review procedures to many inner-city neighbourhoods. These regulatory changes have been supported by the adoption of secondary official plans in the areas potentially most subject to redevelopment pressures. The effect of the plan adoption in the context of Ontario planning legislation is to make the regulations more difficult to change.

8 For example, in upscale inner-city residential neighbourhoods in Toronto's Central Area, the potential net return from a zoning change which would permit high-rise residential development at the residential densities permitted elsewhere in the Central Area is currently well in excess of $100 per square foot of developable floor space. The current values of land assemblies in these areas are less than half of what they would be if land uses were unregulated. Indeed, since the value of assembled land is now little different from the value of that land in fragmented ownership, land assemblies in Central Area low-density residence areas have virtually disappeared.

9 Inner-city regulation has permitted an expanding variety of uses within low-density neighbourhoods, including group homes and halfway houses. In addition, medium density redevelopment has been permitted on the fringes. Inner-city political decisions have generally reflected more compromises between protection of property values and other social concerns than has been the case in newer suburbs.

10 Limits on the use of zoning for blatantly exclusionary purposes have been imposed in a 1975 decision of the New Jersey State Supreme Court (Burlington NAACP v. Mt. Laurel Township), in which the Court ordered the community to change its zoning practices to enable a fair share of the region's poor to live in the community. Exactly what practices constitute "unfair "exclusion is a difficult question. The Mt. Laurel zoning law declared invalid by the Court included particularly offensive features such as a minimum floor area for single-family homes, severe limits on the number of apartments with more than one bedroom, and quality standards such as mandatory air conditioning.

11 This efficiency argument is in fact increased in an environment of fragmented local government, since it reinforces the well-known conclusions of the Tiebout model regarding the efficient provision of public goods; see Tiebout (1956). The Tiebout model is an application of the theory of clubs (non-profit cooperatives); this interpretation is set out in detail in Henderson (1979), in which it is shown that it is efficient for a suburb to be homogeneous. The essence of the Tiebout-Henderson model is that aggregate welfare losses associated with the provision of local public goods are an increasing function of the average absolute deviation of voter preferences for tax-financed public goods from those of the median voter. Municipal fragmentation and exclusionary zoning are both devices through which such welfare losses can be reduced.

12 Even though zoning ordinances have twice been rejected in city-wide referendums in

Houston, a number of other land-use regulations have been implemented. These include setback and parking requirements for new apartment and commercial buildings along with subdivision controls that include minimum lot frontages and setback requirements. Moreover, all land in a 2,000 square mile region surrounding the city is subject to Houston's subdivision controls (Siegan 1970, 76-7, 99, 116-7).

13 Spot rezoning (site-specific zoning by-laws) was declared legally valid in a 1959 Supreme Court decision (Scarborough Township v. Bondi). Since then, site plan by-laws have become the predominant land control tool in areas subject to extensive redevelopment.

14 Corke (1983) provides a detailed description.

15 Permissible density is regulated by height as well as by a prescribed relationship between the maximum number of residential units and commercial floor space. In addition, density bonuses may be awarded for the preservation of buildings designated as historic by the municipal council and for the provision of agreed-upon community services. Density transfers between sites are also permitted.

16 Hamilton (1981) cites Alberta, British Columbia, New Brunswick, Prince Edward Island, and Nova Scotia legislation authorizing municipal design control.

17 The six provinces are New Brunswick, Nova Scotia, Ontario, Saskatchewan, Alberta, and British Columbia. Requirements for public hearings by municipal councils on zoning or official plan amendments were enacted in Ontario in 1983 and in British Columbia and Alberta in the 1970s.

18 Appeals to provincial hearing agencies are provided for in Nova Scotia, Ontario, and Manitoba; in Alberta and British Columbia, there are provisions for appeal to the provincial Minister. (In the case of British Columbia, the provisions for appeal apply only to municipalities other than the City of Vancouver).

19 Such rules generally are interpreted to include rights to adequate notice, to information concerning a prospective decision, and to a hearing before all members of a decision-making body. In such a hearing, "natural justice" is normally presumed to imply that an affected individual should be able to present evidence, cross-examine opposing claims, and be represented by counsel.

20 See, for example, Re McMartin *et al.* v. City of Vancouver (1968), 70 D.L.R. (2d) 38 and Wiswell v. Metropolitan Corporation of Greater Winnipeg (1965), 51 W.W.R. 513. The courts provide a remedy only for extreme violations of natural justice. Generally, the courts are reluctant to intervene in municipal political decisions.

21 See for example the British Columbia Municipal Act, the City of Winnipeg Act, and the Alberta and Nova Scotia Planning Acts. Ontario is an exception, in that prior adoption of criteria in official plans is not required to use supplementary development control powers. However, the supplementary powers are more restricted than in other provinces, and the empowering legislation provides a right of appeal to the Ontario Municipal Board.

22 The preparation of municipal official plans was mandated by Alberta in 1977 and by Quebec in 1980.

23 The requirement that by-laws conform to official plans has been successfully used as the basis for appeal to the courts to have a municipal by-law disallowed. See, for example, Holmes *et al.* v. Regional Municipality of Halton (1977), 2 MLPR 149.

24 In Ontario, for example, the Ontario City and Suburb Plans Act of 1912 required that any plan of subdivision for land within five miles of a city having a population of more than 50,000 be submitted to the predecessor of the Ontario Municipal Board for approval prior to registration.

25 For example, the Alberta Planning Act of 1980 specifies that a plan of subdivision must be rejected unless the land is suitable for the intended uses and the proposed subdivision conforms to municipal and regional official plans.

26 The Ontario Planning Act Review Committee (1977, 101) proposed that a separate study be undertaken of the implications of instituting such a system. While it advocated consideration of an integrated development permit/zoning approval system as a means of reducing the "substantial inefficiencies" of the current system, it was not itself able to recommend a substitute.

27 In many provinces, a two-stage planning process is followed. The first stage is the preparation of a regional plan, which identifies the areas of potential development and is used as the basis for planning major regional infrastructure investments. The second stage is the preparation of a so-called "secondary" or "local" plan, which sets out local roads and zoning.

28 See McFadyen and Johnson (1981) and Proudfoot (1980). For an earlier analysis, see Greenspan *et al.* (1977, 125-30).

29 Proudfoot (1980, 46) notes evidence from Waterloo that secondary plans there took an average three years to process.

30 The approval process normally occurs in two stages (formal approval in principle of a draft plan, followed by subsequent final approval after all associated agreements have been entered into, lot levies paid, and performance bonds posted for developer undertakings. Consequently, the inventory most likely to be affected is developments which have received draft approval.

31 Greenspan *et al.* (1977, 128-9) suggest that this may have been an important factor affecting the rapidity of the rise in lot prices that occurred in the early 1970s. However, other factors were probably more important, notably the upsurge in expected inflation and the increased incentive for home ownership provided by the 1971 tax reforms.

32 For example, in Mississauga in 1978, the inventory of undeveloped lands which had received at least draft plan approval amounted to approximately 60,000 units, more than four times the number of units for which building permits had been issued. See Proudfoot (1980, 47).

33 The provision of services is entirely the responsibility of municipalities in Quebec, though the fiscal consequences of this are ameliorated by provincial grants to municipalities. More generally, there has been a lower rate of investment in servicing infrastructure in Quebec. As the long-term consequences of past under-investment become increasingly a political concern, political pressures may also mount to shift the financing of services for new development to developers.

34 Goldberg (1980) uses City of Vancouver engineering data to suggest that the present value of total social costs may be minimized by high servicing standards.

35 Hamilton (1981) refers to British Columbia as the one province in which such a requirement has been legislated. However, the 1977 amendment which constrained municipal

lot levies and servicing standards merely restricted them to "direct capital costs" (British Columbia Municipal Amendment Act, 1977, Section 702C).

36 Volume 2 of the Greenspan Report summarizes the results of several studies; see Greenspan *et al.* (1977, 135-8). The majority of these studies concluded that new residential developments generate more tax-financed costs than they provide in additional revenues.

37 This process of adverse selection, if not limited by other factors, can lead to continuous decline in product quality and, in the limit, to disappearance of markets. See for example Akerlof (1970) and Hirschleifer and Riley (1979, Section 1.2.2).

38 This is indeed the case in France, where there are no comprehensive building codes but designers and contractors are liable for ten-year period for major defects. Liability insurance is almost universal, and the design and inspection requirements of insurers result in a system that is, at a practical level, "comparable with building control systems elsewhere." Silver (1980, 5).

39 Quebec and Newfoundland have systems of rent regulation that predate those introduced by the other eight provinces.

40 The best empirical evidence for this is that even real pre-tax short-term interest rates were negative throughout the early and mid 1970s.

41 The rise in interest rates was at first an increase in nominal interest rates and only subsequently an increase in real interest rates. However, for institutional reasons (notably the design of conventional mortgage instruments and the application of traditional loan evaluation techniques), both types of increase have a contractionary effect on the private provision of new rental housing units. High nominal interest rates that translate into low after-tax real rates for developers have, because of lending practices, translated into higher lender requirements for initial cash flow incomes from rental projects. Higher real rates have resulted in higher required economic rates of return for the developer.

42 In Ontario the 1979 Act revising rent review explicitly provides that increases in operating costs, amortized capital expenditures, and changes in financing costs may all be used to justify a rental award by the Residential Tenancy Commission.

43 Changes in financing costs arising from a change in ownership have not been allowed as a basis for rent increase awards in Ontario since December 1982; this change resulted from the much publicized sale and resale at much higher prices of 11,000 Toronto rental units owned by a single company in transactions whose sole purpose was to provide the basis for an increase in capital costs and hence financing charges.

44 Changes in Ontario legislation (notably the implementation of a rent register) in 1986 reduced this incentive and made rent regulation tighter.

CHAPTER EIGHT

Building Technology and the Production Process

James McKellar

THE RESIDENTIAL construction industry in Canada has evolved since 1945 with a minimum of capital investment, little standardization, varying skill levels in the labour force, an aversion to technological innovation, and a reliance on a myriad of subcontractors, suppliers, and material producers. It is an industry that has a complex organizational structure, is fragmented, is subject to major cycles in the economy, and is regional in character. It may be argued that these are also its long-term strengths.

Housing stock producers have seldom operated nationally, and the regional and local variations in the way housing is built across the country are more significant than apparent similarities. The local nature of the industry hinders attempts, such as this, to understand the broader picture as it has emerged over time. Many pertinent facts have likely been overlooked or omitted. From owner-builder in Moncton, New Brunswick, to small housing manufacturer outside Montreal, to mobile home manufacturer in Red Deer, Alberta, to merchant builder on the outskirts of Toronto, those who build or manufacture housing across the country may fault the level of generalization that the current data and information can sustain.

The housing stock production process has not significantly changed over the last thirty years, nor will it likely change in the years ahead. Yet, this is a process that is relatively robust and efficient, adaptable and responsive to changing consumer needs and demands over time.

The changes in the residential construction industry through the years have been small and incremental. The industry has not been well studied and much of what is documented is anecdotal. Much residential construction-specific data is lacking. The available literature explains why certain changes might or should take place, but few data exist on what actually happened. Quantitative information is scarce, particularly time series data on the component costs.[1] Investigations of the production process are further hampered by a lack of research and development activity within the industry itself and an almost exclusive reliance on government statistics with which to measure industry performance. Thus, any attempt to study housing production and the technical changes in the industry must be qualitative and judgmental.

Post-War Origins of the Industry

The residential construction industry, as we know it today, arose out of the exigencies of the war, and its birth coincided with the founding of CMHC on 1 January 1946. In the words of CMHC's first president, David Mansur, the "primary duty" of the new corporation would be "finding ways and means for private enterprise to look after needs in the economic (housing) field." He asked that success be measured "by the amount of activity not undertaken" by government agencies "in the public housing field."

CMHC's Integrated Housing Plan initiated the post-war era of single-family detached housing (CMHC 1970, 12). In this scheme, speculative builders undertook to sell houses at a price previously agreed upon, and in turn, CMHC undertook to buy back unsold houses. This program also forged CMHC's close ties with building materials producers and suppliers since one of CMHC's early responsibilities was the issuing of priority certificates for the use of critical building materials such as cement and plumbing fixtures. Some CMHC branch offices even held bulk supplies of nails to be sold directly to priority builders.

This "integrated" plan gave encouragement, security, and confidence to the many small builders who were entering the house-building industry for the first time. Enthusiasm, and a few hand tools, were convenient substitutes for skills and experience. In 1947 and 1948, as many as 491 builders took part in this plan, and each year produced more than 5,000 units or nearly half of all NHA-financed housing in those years. Although annual starts were about 90,000 units, success was short-lived. The effect of the Korean War was reflected in housing starts that plummeted to 68,000 units. The number of NHA builders in Toronto fell from 500 in 1950 to 170 in 1951 (CMHC 1970, 15).

Development of the Industry

Chartered banks were authorized to make NHA loans starting in 1954. This initiative was coupled with NHA amendments that introduced "insured loans" to guarantee against default risk. These actions made insured mortgages more available in many small and remote communities where banks had branches. Such actions reduced the risk of speculative building; borrower affordability and accessibility to the housing market was improved, and financial risk to builders was diminished. The era of the small builder, even in remote locations, was launched. The fortunes of the residential construction industry now rose and fell with interest rates.

Parallel to its initiatives in housing finance, the federal government sought to encourage housing construction more directly through the 1954 NHA amendments. In administering the Act, CMHC promoted national construction standards and sought to raise the quality of construction with its own force of building inspectors. A legacy of these initiatives was improved material performance for a form of residential construction that was to endure right across the land. Today "wood frame platform construction," utilizing the nominal two by four (and now the two by six) is as firmly entrenched as it was three decades ago.

An industry that began with government assistance to small builders through WHL, the Integrated Housing Plan, and the 1954 NHA amendments came to include, in the 1960s and early 1970s, large diversified real estate companies capable of undertaking land development as well as both single detached and multi-family housing. These firms focused on less expensive homes, offering more space at less cost through standardized designs and increased market share. Names such as Bramalea, Markborough, Cadillac-Fairview, Nu-West, Genstar, and Campeau dominated the residential land and construction scene in certain regions to the extent that critics hinted of oligopolies and called for intervention. Diversification for many of these large companies was both geographic and product-oriented. Nationally-recognized names emerged in the home-building industry. These same names began to appear on apartment sites, office developments, regional shopping centres, and industrial buildings in Canada and the United States with increasing frequency throughout the 1970s. [2]

Neither industry nor its forecasters, ensconced in the building boom of the late 1970s, foresaw the immense changes that were about to occur; 1976 was a watershed year; a drop in activity followed and the recession of the early 1980s caused a trauma that was still being felt in the mid 1980s in certain major market segments, particularly in western Canada. Many of the large companies withdrew entirely from residential construction and land development business, leaving the field to myriad small home-builders who had survived on larger custom homes and the home renovation market. Land was left in the hands of the banks or the pension funds that had bank-rolled the developers. In Calgary, there were few holdovers by 1985 from the 1981 list of top ten house builders. The new list had many names that were not even in business in 1981. The industry changed radically right across the country. [3]

Since 1976 there has also been a steady increase in the amount of total residential renovations and a progressive decline in the real value of new residential construction. Total spending by Canadian home owners on repairs and improvements increased 23% over the period 1982-4 to an estimated $9.7 billion. [4] Total dollar spending in seventeen metropolitan areas jumped from $1.7 billion to $4.1 billion in just six years (1978-1984), and 65% of this spending was for improvements. Contractors accounted for 71% of this value. Renovation, which includes both improvements and repairs, has emerged as a big business in Toronto (29% of this 1984 total), Montreal (24%) and Vancouver (15%). However, the majority of renovation spending is still done by home owners beyond these seventeen metropolitan areas (58% of the 1984 total).

It is only a matter of time until the value of renovation construction exceeds the value of new residential construction (CMHC 1985a). Figures already show that renovation accounts for a larger proportion of the total residential construction employment than new work, although new construction still generates greater total employment because of the materials requirement.

In the future, large builders will likely play a reduced role in residential construction and, after winding down their land inventories, will likely concentrate on specific regional markets such as Toronto and Montreal. Even in these markets, profits have not reflected the early 1980s inflation in house prices, and this raises further questions as to the advantages, if any, of size.[5] The residential construction industry is expected to become one made up increasingly of small to medium-sized firms. Innovation will never be a trademark of this group; it will be a reactive industry, treading cautiously on new ground, adopting changes in building practice only when absolutely necessary and pursuing an approach to building that could be termed "business as usual." Its main lines of activity will split into those firms which concentrate on new construction and those which specialize in renovation.

The Industry in 1984

A 1984 survey of the home-building industry found that only 63 firms had annual revenues of $10 million or more; these firms accounted for 25% of total home building industry revenue and averaged thirty-three salaried employees (Clayton Research Associates Limited 1987). Ontario had 62% of these large firms. There were 409 firms with revenues from $2 million to $10 million, and Ontario and Quebec shared 68% of this category. These medium-sized firms averaged only five salaried employees. The 1984 survey also found that average before-tax profit revenue was just 3.6% of the total, down from 8.0-9.5% in the mid 1970s.

Within the industry, residential general building contractors (RGBC) have continued to specialize in single-family detached housing: on average, 71% of output from 1977 to 1982. Since this group is the single largest sub-component of the industry, it gives rise to the claim that the residential construction industry is primarily oriented toward the production of low density, single-family housing by small companies. Other sectors of the industry do not threaten this predominance. For example, the activities of real estate developers declined from 50% of housing starts in 1977 to 24% in 1982 (in step with the decrease in multi-family housing starts). In spite of the prominence of small firms (builders averaged just 9.9 homes in 1985), the 241 medium and large-size builders (4.8% of all builders) together accounted for one-half of all houses built by builders in 1985 (Clayton Research Associates Limited, November 1986).

A significant portion of new houses built every year in certain provinces and regions are owner-built or owner-contracted. In rural areas of the Maritimes, almost all houses are owner-built; in Saskatchewan, approximately 30%; in Sault Ste. Marie, 60%; in Prince George, British Columbia, roughly 80% (CMHC 1985a).

The Housing Production Process

Despite the diversity of firms and their local scale of operation, residential construction is practised according to a few common styles throughout much of Canada. While it is difficult to describe a Canadian residential construction industry, even on a regional basis, it is not difficult to describe the means of housing production and the building technology upon which this production is based. This description will focus on the construction process for new construction and concentrate upon low-rise wood frame construction. That is not to say that high-rise construction and renovation are not important. High-rise construction is not going to command the importance it once had in the 1960s and 1970s, and we are only on the threshold of technological innovations in the renovation field. Renovation will see an increasing volume and an increasing sophistication in firms taking on this work.

The housing production process used for single detached and low-rise, multi-family wood frame housing can best be described as the "assembly line in reverse." As opposed to the traditional assembly line process whereby the product moves past stationary workers, the residential builder has kept the product stationary and scheduled the flow of tradesmen past the house. This process has been ideal for large-scale tract housing that has characterized so much of the industry since World War II. As a process, it stands in marked contrast to the "systems builders" of Europe who rely on factory-built housing predicated on the traditional assembly line process.

The traditional building practices upon which the early merchant builder thrived were refined but not replaced during the period of war-time housing. Shortages of materials and labour, and the need for quick erection time, led to significant improvements over pre-war housing techniques. The house plans and sequential production practices that were introduced by WHL to meet production quotas mark the beginning of the modern residential construction era in Canada.

The quintessential Canadian home is perhaps the "Type C" unit, a storey-and-a-half model used across Canada by WHL in the period 1941-5, in the Veterans' Rental Housing program after the War, and by the early NHA builders. The Veterans' Rental Housing program produced 25,000 of this particular unit in the three years following 1947 (CMHC 1970). The Type C house utilized platform wood-frame construction in an approach that still accommodated a high proportion of site-applied, traditional materials such as lumber, brick, cement, and plaster. The innovation had more to do with arranging and deploying the constituent elements of labour and materials to cope with production schedules. Platform construction was nothing more than the sequencing of the building trades required to complete a house: the innovation being to first complete the sub-floor assembly upon which the walls could be accurately framed horizontally and then lifted into place. There was no major breakthrough in the use of materials, and these first houses had virtually no pre-assembled or factory-built components other than the plumbing fixtures and the convection air furnace.

In the late 1940s and the early 1950s, a good builder used from 1,500 to 1,700 person-hours on-site to build a typical wood-frame, wood-sided bungalow (Scanada Consultants Limited 1970). By 1970 this labour component fell to 920 hours for a comparable house with better landscaping, cabinetry, and finishes. These savings can be attributed primarily to the building materials industry and the introduction of an increasing number of "high factory content" materials and "value added products" such as plywood, gypsum wall board, floor tiles and carpeting, window and door assemblies, two-coat paints, transit-mixed concrete, prefabricated kitchen cabinetry, and light weight structural roof and floor trusses. In addition, power tools, fork lift trucks and truck-mounted hoists, job management and "assembly-line" sequencing of trades have yielded increasing efficiency on the actual job site.

However, these on-site efficiencies are not uniform across the country and reflect the local availability of building products and differences in skill levels and wages. For example, the same bungalow that averaged 920 hours of labour required 1400 hours in the Maritimes; seasonality can further distort this range. Labour practices are difficult to track, but from 1949 to 1969 the average annual productivity change was in the order of a 1.0 to 1.5% increase (Scanada Consultants Limited 1970). Productivity lagged during the 1970s, and labour and materials costs soared.[6] Over the decade 1971-80, the increase in output per person employed (labour productivity) averaged 0.8% per annum in the construction industry overall (Clayton Research Associates Limited 1983). This was similar to productivity gains recorded overall by non-agricultural industry in Canada, but less than productivity gains in manufacturing and goods-producing industries.

In examining the cost breakdown for a new single-family home, a 1982 study confirmed for the US market what also seemed true of the Canadian housing market during the 1970s: between 1970 and 1980 in the US, the cost of a serviced lot increased dramatically, by 248%, to consume 24% of the total cost of the house (Merrill Lynch 1982). Labour and materials for the house itself both fell as a percentage of total cost in the same period, from 19% to 16% for labour, and from 37% to 34% for materials. Financing costs for the builder rose from 7% to 14% and overhead and profit dropped from 18% to 14%.

The picture in the 1980s has been different for both Canada and the US. In Canada, the rate of increase in labour, materials, and land costs all dropped significantly through the early 1980s. In the mid 1980s land was approaching a 10% annual increase; labour held relatively steady at 4%; and material costs had dropped to the same range. In the period 1981 through 1986, according to Statistics Canada (1987), residential construction material prices rose nationally by 29%; union wage rates rose 32%; conventional mortgage lending rates dropped 39%; and land costs increased 8%.

However, parts of the country have experienced dramatic decreases in building costs. Construction costs for a typical bungalow in most centres of Alberta in

1986 were close to those in 1981 (0.4% drop), in nominal dollars (*Alberta House Cost Comparison Study* 1986). In terms of inflation-adjusted 1986 dollars, construction costs in Alberta have fallen dramatically (22%) since 1981.

Innovations in the Process

Construction techniques, equipment and materials have certainly changed since the Type C unit was first introduced, but caution has prevailed. There is a perception in the industry that the firm that entertains minor variations from the norm courts disaster. This shared view is a strong disincentive to the advocates of new ways. Two deviations are worth mentioning – one unsuccessful and the other quietly successful. The first was in response to the growing enthusiasm in North America in the mid 1960s for European "system building" techniques. In the US, this prompted "Operation Breakthrough," a federal program designed to encourage industrialized building of multiple dwellings as a means to solve the affordability problem. In Toronto, from 1968 to 1970 four companies were founded, two using proven European concrete slab systems and two adopting more experimental concrete box module approaches. A fifth company acquired the rights to a Swedish panel system, but later abandoned the venture.

This was a period characterized by a rapid expansion of apartment construction, with starts more than doubling from 1963 to 1968. Profits were high, the stock market decidedly bullish, and the overall economic outlook bright. There was concern that, at least by North American standards, conventional construction techniques had reached the limit of their efficiency. For a number of reasons, some of the largest Toronto developers ventured into large-scale systems building. Market conditions could not have been better.

In summarizing the experience of these companies, it can be succinctly stated that, in the years 1968 to 1974, system building failed in the Toronto market (Barnard 1974). Of the four companies that entered the industry, none demonstrated the capability to compete with the conventional industry. Two reasons can be cited: first, despite generally higher quality and somewhat faster erection times, none of the companies showed the overall cost savings that were originally expected; and second, none of the companies was able to obtain sufficient orders to maintain continuous production, thus losing economies of scale and adding to the cost problems.

The second deviation deals with the little known success of factory-built housing by two companies in Calgary, Alberta that, between them, dominated regional markets for single detached housing during the 1970s. Engineered Homes was founded in 1959 and grew out of Muttart Homes that got its start in 1943 through WHL.[7] Engineered Homes first offered its wood panel, factory-built system in 1960. From 1971 to 1976, the peak years, it shipped products to fifty-two dealers throughout British Columbia, Alberta, Saskatchewan, and Manitoba, and as well as to Wales, France, and Germany. By 1974 it had produced a total of 50,000 houses. The company was purchased by Genstar in 1972, and when Genstar closed the plant in April 1984, it had recorded some 70,000 units over its twenty-four year life span. The factory was virtually a self-contained operation, producing windows and doors, trusses, wall sections, kitchen cabinets, and stairs. The only site work was drywall, roofing, mechanical and electrical systems, plumbing, and finishing. This was a total package that went from factory production to land development and retailing. Its final demise was caused by the drastic downturn in residential construction during the 1980s.

A local competitor was Qualico, a privately-owned company from Winnipeg that entered the prefabrication business in 1959 and operated factories in both Calgary, to serve the Alberta market, and Winnipeg, to serve Saskatchewan and Manitoba.[8] Production was for their own consumption, exclusively, and although the Calgary factory peaked at 1,250 units, including multiples, it normally ranged from 250 to 450 units annually. The production system was characterized by its simplicity, its astute attention to cost and inventory control, and its reliance on long-term employees, some with over twenty years experience in the same plant. The Calgary plant was closed in 1984.

In contrast to the Toronto experience with "systems" building, Engineered Homes and Qualico exemplified a highly sophisticated form of "mixed mode" technology that depended more on the organization and management of the total building process than on factory production *per se.* They combined factory production of value-added components with on-site construction, and they had the capability to vary the mix according to market conditions. The factory work merely duplicated field practice under one roof. Neither Genstar nor Qualico sought technological breakthroughs in their production processes; they simply adapted standard practices to more cost effective production techniques.

From the foundation to the roof, the quality of home-building components and the assembly of these components has improved considerably over the decades. In the early 1950s pre-assembled aluminium sliding windows were introduced and eventually led to the various pre-finished energy-efficient windows in the market today; pre-assembled kitchen cupboards with post-formed plastic laminate counters appeared in the late 1950s, and gypsum wallboard overcame initial trade resistance to gain widespread use by the mid 1950s. Roof trusses also appeared in the 1950s, and floor trusses in the 1960s, although neither substantially penetrated the market until a decade later. Plastic plumbing lines followed the introduction of copper after the war and became the standard in the 1960s for drain, waste, and vent plumbing. Concern with energy efficiency in the 1970s led to insulation improvements including introduction of 2x6 framing to accommodate more insulation and promotion of the R-2000 standard in the 1980s for energy efficient construction.

New construction practices were made possible by such labour saving devices as hand-held power tools, the pneumatic nail gun, and the concrete pump. Concern with the overall performance of the building enclosure led to studies to control condensation in walls and attic spaces, the use of polyethylene vapour barriers, and the airtight drywall approach (*Alberta House Cost Comparison Study* 1985). Some innovations thrived and others failed to realize their full potential. Waferboard was a great success (Salomon Brothers 1986). This was the first alternative structural panel to plywood and was developed and produced in Canada starting in 1966. The Government of Alberta, in looking for ways to capitalize on its abundant, but previously unused, Aspen wood source encouraged the development of waferboard technology. By 1985 over 150 different panels had received performance ratings from the American Plywood Association, and waferboard and oriented strand board (OSB) sales increased more than tenfold from 1980 to 1985.

Permanent wood foundations, on the other hand, have never threatened the viability of masonry or concrete foundation walls. In spite of their economic competitiveness, proven performance in preventing sub-surface leakage, and acceptance of plumbing, wiring, insulation, and drywall for finished basement spaces, they have not gained the confidence of the industry. Thirteen years after their initial introduction, permanent wood foundations are a good example of the resistance of home building trades to materials and techniques that fundamentally depart from current norms (Shaw 1987). For example, it is not only trade resistance that one must contend with. Wood foundations are a good example of the strength of consumer resistance to unfamiliar practices. The substitution of wood for concrete underground is not something the consumer can readily accept.

The industry has a history of moving cautiously toward innovation. Innovation must be compatible with existing building practices, must require relatively little or no capital investment, and must simplify the task at hand (Shaw 1987). For example, pre-hung doors were disseminated into the industry within three years, while roof truss systems, involving greater complexity, took some twenty years to be fully disseminated. David MacFadyen, president of the National Association of Home Builders Research Center in the US, estimates that "for a construction innovation, the mean time to adoption is typically 15 years." The experience in Canada is not likely very different since the countries share similar information sources and networks.

Research and Development in Residential Construction
The highly-fragmented and regional nature of the residential construction industry is not conducive to the type of industry research expected of a consumer products manufacturer. It is highly unlikely that the small builders who dominate the field will even collect market data or systematically gather information on new products, let alone support a research program. Most housing research in Canada, particularly of a technical nature, has been carried out largely by, or with, the financial assistance of the federal and provincial governments. Even the trade associations have turned to governments to fund their research agendas.

The federal role in technical research was prompted by the 1944 NHA, specifically Part V, that came under the jurisdiction of CMHC in 1946. Even today, Part V provides federal funds to conduct research into the social, economic, and technical aspects of housing and related fields, and to undertake the publishing and distribution of the results of this research. Part V refers to research undertaken external to CMHC, and through this external research program, resources have been directed at technical subjects covering almost every aspect of housing performance.

The NRC Division of Building Research established an international reputation for its technical research directed at component performance and did much to improve the understanding of wood frame construction. A combination of the NRC, CMHC, and the Forest Products Laboratory first undertook to develop and promote roof trusses in Canada. The Division of Building Research had the critical task of updating the National Building Code and therefore played a crucial role in promoting change and innovation through its regulatory influence. In its present form, as the Institute of Research in Construction, this organization has redirected its efforts to contract research on behalf of industry and government and has foregone some of its traditional role in code-related research topics.

In the 1980s there has been a marked increase in the activities of various provincial agencies with a direct interest in technical research and development. The Ontario Research Foundation supports building-oriented research; the Ontario Ministry of Housing has promoted research in codes and regulations affecting renovation, assessment of renovation strategies, and forms of direct industry assistance; the Saskatchewan Research Council, a field station of the NRC, is noted for its work in energy efficient housing, notably the R-2000 house; and Alberta's Department of Housing has stressed product development. Fortunately, there is still the impetus for federal-provincial collaboration at various levels. For example, the degradation of concrete garages resulting from salt action is now the focus of a coordinated effort involving provincial governments, CMHC, NRC, the Department of Public Works, CIPREC, and the Concrete Manufacturers Association, among others.

Technical research within the private sector is almost exclusively the purview of the building products industry. A notable exception was the "The Experimental Housing Program" of the National House Builders Association, Technical Research Committee, started in the 1950s.[9] Under this program a series of experimental houses were constructed, with the support and cooperation of CMHC, NRC, Department of Forestry, Canadian Wood Council, and the Plywood Manufacturers of British Columbia The "Mark" series spanned the years 1957 through 1968 and were primarily directed at introducing new materials, new methods of erection, and changes to residential building standards and building regulations.

The Mark I, constructed in Hespler, Ontario in 1957, prompted changes in 2x4 wood framing, reduced the thickness of sub-floor, exterior wall, and roof sheathing, and introduced the heated crawl space in lieu of a basement. Two years later, the Mark II built in Calgary, Alberta, introduced lightweight 2x4 wood roof trusses and 1/2 inch thick drywall throughout the interior. In 1961, the

Mark III, built in Ottawa, utilized preserved wood foundations and plastic supply and waste pipes, and was followed in 1963 by the Mark IV that had the first
all-wood basement to be built in Canada. The Mark V was a departure from its
predecessors in that it sought to reduce costs by using different construction
methods within the realm of existing building regulations. Observations of the
Mark V house showed that materials accounted for 74%, labour 24%, and
equipment rentals 2% of the on-site costs. The labour content of individual
operations varied from 77% for painting to 14% for the installation of electrical,
plumbing, and heating services. As a point of comparison, the 1986 Alberta
House Comparison Study indicates a breakdown of 25% for labour and 75% for
materials for Calgary and Edmonton (*Alberta House Cost Comparison Study*
1986). [10]

The final house in the series, the Mark VI, was built in Kitchener, Ontario in
1968, and as the only two-storey unit, it reflected the concern with increasing
land cost. This experiment demonstrated numerous materials and practices not
yet approved under existing building codes, including precast concrete basement walls and footings, steel floor joists, electric cable heating integrated into a
drywall "sandwich" panel, a prefabricated bathroom unit and plumbing "tree,"
and the use of vinyl siding, soffit, fascia, eavestrough, rain water leaders, and
shutters.

Unfortunately, this type of applied building research, combining both private and public initiatives, did not continue in the same systematic fashion
through the 1970s. The 1970s became the era of the home warranty program and
energy efficiency. Attention turned more and more to issues of land development and subdivision regulation as the price of residential land continued to
escalate. Beyond product research, much of which takes place south of the bor-
der in the laboratories of such companies as Boise Cascade, W.R. Grace, Owens
Corning Fiberglas, or Weyerhaeuser, the thrust of research in the private sector
in Canada is directed at evaluating, establishing, and revising government policies and regulations affecting the industry. An exception is the activities of the
wood industry in Canada, much of which has now been consolidated under
Forintek Canada Corporation. [11]

The Products of the Industry

The storey-and-a-half, Type C, unit of the 1940s provided two bedrooms and a
half-bath upstairs and a living room, bedroom, full bath and eat-in kitchen
downstairs, all in approximately 88 square metres (Doherty 1984). This particular unit became the backbone of immediate post-war housing production and
typified the issues of the day (Galloway 1978). New housing prototypes appeared
by the late 1950s, along with more sophisticated attitudes to community planning and design.

The much-favoured three-bedroom "ranch style" bungalow that appeared in
the late 1950s, closely resembled the two bedroom bungalow, Type B, produced
during the war. By the early 1960s, the "split-level" and "bi-level" were in

wide-spread use and afforded better use of the basement space. In the late 1960s, with the increasing cost of land, two-storey units enjoyed considerable success. For the first 25 years after the war, most single detached housing was a variation on these three generic types, the bungalow, the split-level and the two-storey unit, in two, three, and four bedroom versions.

While average house size increased with the years, neither house type nor size indicate the immense change in lifestyle that accompanied these years. Significant qualitative changes were better reflected in the myriad of housing designs and floor plan layouts that began to appear. Emerging lifestyles required first a single and then a double garage, sodding and fencing, landscaping, washer and dryer, complete interior finishing, carpeting throughout, larger lots, forced air heating, and sometimes air conditioning. Inside the home, two new rooms appeared: a utility room next to the kitchen and opening to a side or backyard, and a family room to house the television set. The kitchen expanded to accommodate an eating area and an extension for children's play. Walls separating the living room, dining area, and kitchen were eliminated to create a more "open" plan and to introduce new activity areas. The picture window and then sliding glass doors gave the interior a brighter appearance, and consumer demands for more amenities led to laundry chutes, linen closets, built-in appliances, and other features that were well beyond necessity and the need for shelter. Many of these features were factory made or assembled and therefore placed few demands for new skills on the on-site labour pool. Pre-assembled kitchen cabinets in various styles introduced a new luxury into the kitchen, and bathroom features followed shortly thereafter.

Changes were not only reflected in the design and production of the house itself. The dramatic rise in the cost of serviced lots was driven, in part, by the consumer's interest in better neighbourhood planning with improved street lighting, curbs and gutters, sidewalks, underground utilities, parks and play-grounds, bicycle and walking trails, and community centres. Local municipalities adopted neighbourhood standards that made many of these "amenities" the norm and provided the engineering standards that would ensure the longevity of this infrastructure. These costs were assigned to the cost of the serviced lot, except in Quebec where they showed up as a local improvement charge on the home owner's annual property tax.

The approval process no longer dealt only with lot size and setbacks but now required provision of school sites, more sophisticated methods to handle storm drainage, hierarchical road networks, public transportation routes, landscaping and irrigation of dedicated public open space, and designation of sites for neigh-bourhood facilities ranging from daycare to convenience retail. More of the real costs of growth, including provision of police and fire stations, new sewer and water trunk lines, upgrading of vehicular routes, regional park systems, and school buildings, also found their way into the cost of the serviced lot through off-site charges assessed on an acre basis against lands being subdivided for new development. The approval process became more complex with citizen participation and environmental review, and the lengthening of this process also carried a cost.

The new communities and planned neighbourhoods of the 1970s and 1980s reflected a different set of consumer expectations from those that shaped the housing tracts of the 1950s. The cost of the serviced lot reflected the charges for meeting a broader set of consumer expectations, as well as the municipality's charges for incorporating the lot into the fabric of the then-existing community. The complexity and cost of land development was now far beyond the capability of most home builders. This gave rise to a segment of the industry, mainly large companies, whose product was not only houses but also the retailing of serviced lots to other builders.

Through the 1970s and the early 1980s, a multiplicity of detached housing types, forms, and styles evolved. Such architectural features as two-storey spaces, loft spaces, and cathedral ceilings were in great demand and made possible by more variety in truss designs such as the "scissor" truss. Bathrooms and kitchens became the focus of attention in number, size, and design, and they were increasingly being customized to suit particular segments of the market. By the late 1970s builders discovered the move-up market, the luxury market, the "empty-nester," and the young professional market, and product differentiation was noticeable. [12] A limited range of stock plans gave way to in-house design departments, and by the late 1970s Canadian builders were making the annual pilgrimage to the convention of the National Association of Home Builders in the US, with a side trip through the California show homes to see what trends were shaping the market place. Show homes, sales pavilions, consumer preference studies, and media campaigns became part of the business. The industry took on new dimensions in marketing and sales.

Changes in single-family detached housing were still modest when compared to the introduction of fundamentally different housing forms to accommodate the growth of multiple-unit rental housing in the 1960s and 1970s. The majority of new apartment units in the Toronto market were built by some half dozen developers, companies which were fully integrated in land, construction, sales, and property management, and often with annual volumes in excess of 1,000 units. The key to their success was the emergence of highly efficient subcontractors (Barnard 1974), particularly in the structural trades of concrete forming, reinforcing bar placement, and concrete casting.

During the 1960s these building trades benefitted from a new sophistication in high-rise building techniques and a large influx of immigrant labour. The advances in high-rise residential construction techniques fall into two categories, standardization and innovation (Barnard 1974). The products of these apartment builders were highly standardized in design, unit layout, structural systems, finishes, and even overall appearance. This in turn allowed components such as partitions, kitchens, and bathrooms to evolve to standard or near standard designs. Prefabrication was made possible by economies of scale, and both on-site advances in standardization and off-site prefabrication were seen as the most effective means to stabilize costs.

Competition among these few large-scale companies promoted capital-intensive new processes in the search for efficiency and, on one of the rare occasions, innovation in construction techniques was aggressively pursued. Two results were the refinement of the "flying form" by Tridel in Toronto and the introduction of the "climbing crane." The flying form technique, involving a pre-assembled, pre-engineered forming system, streamlined the process of on-site concrete pours and achieved construction rates of over one floor per week. The climbing crane, a European feature, increased the versatility of on-site labour crews and allowed greater prefabrication and preassembly of larger, heavier components. Innovations from this period soon became industry standards throughout North America and the "flying form" was to high-rise residential construction what "platform construction" was to single detached housing.

However, high-rise buildings did not account for the majority of multi-family units built across the country. The production of townhouses, duplexes, semi-detached, triplexes, and three-storey walk-up apartments (wood frame, or "combustible" construction) has always exceeded mid to high-rise apartment construction ("non-combustible" construction). Similarly, single-family detached housing starts have traditionally exceeded multi-family starts, although in 1964 for the first time, more apartments (60,435) were built in Canada than detached houses (50,475). Detached starts declined, as a percentage of total starts from 69 in 1973 to 63, in 1982, but this figure has risen through the 1980s and is estimated at 66% for 1987.

This decline in multi-family starts is reflected in the collapse of new rental housing construction in the 1970s. From 1963 to 1970, 85% of all rental starts, which in turn accounted for 47% of all housing starts, were private, non-government-assisted starts. A decade later, by 1980-1, private rental starts had fallen to 10% of total starts. Four structural shifts precipitated this collapse of new rental construction: first, a drastic change in the tax incentives which had encouraged private rental housing; second, a gradual shift in emphasis to government-supported home ownership and social housing; third, a growing view that inflation would persist and even accelerate; and finally, the impending threat of rent controls (Smith 1983). The impacts of this decline were most severe on the large apartment builders who dominated high-rise construction.

One product that has not met with continued success in the Canadian market is the manufactured house, particularly the mobile home. [13] The volume of manufactured housing in Canada dropped from 49,000 units in 1974 to less than 10,000 units in 1984 (CMHC 1985a). Canada's 21 home producers shipped just 3,191 units in 1984, according to Statistics Canada, a 25% decline from 1983. The remnants of an industry which once produced more than 25,000 units annually (1973) are centred in Alberta and British Columbia where nearly half of the country's producers are located (Clayton Research Associates Limited, November 1986). The "prefab" industry that factory-produces sections, complete units, or components for on-site erection consisted of just 81 producers in 1984, down from 87 in 1983 and 97 in 1982 (Clayton Research Associates Limited, January 1987). Total shipment of larger firms amounted to 4,694 units in 1984, down from 4,694 the previous year. Only four firms had more than 100 employees, and these four accounted for 36% of all production. The average firm employed 25 workers and shipped $2.3 million of product.

For an industry that is locally or regionally-based, often fragmented, cautious to introduce new ways of building and new building materials, and usually dependent on many small and medium-sized builders and sub-contractors, it has shown a remarkable ability to adapt to changing consumer demands and preferences over the years. The industry has the capacity to react, but slowly. It still has the capability to produce a wide range of housing options in response to evolving aspirations, changing population mix, and different patterns of consumption among buyers.

Prospects for the Future

The social system of the home building industry is comprised of the material supplier, the home builder who assembles these materials, the home-owner who purchases the value-added product, and government that regulates the process. Technology and the production process must be viewed within the context of this system, and any prospects for the future are likely to be bound by the system as we know it today. Opportunities for technological change and innovation within the residential construction industry are most likely to be found in, and integrated with, current industry practices and approaches.

There will be new incentives on the demand side, particularly the need to find more cost-effective ways to provide affordable housing and to address a growing home renovation market. However, these incentives will apply to improving conventional forms of building, and consistent with historical trends in the industry, changes will be evolutionary, not revolutionary. We are not likely to see major breakthroughs in cost savings as a result of any one particular technology, and as in the past, conventional forms should prove as cost effective as factory-built housing systems in the long run. Most research and development motivated by the desire to reduce costs will be directed to improving these conventional systems with the basic intent of improving reliability, simplifying on-site erection, and reducing the on-site labour component.

The demands of increasingly differentiated markets for greater choice and improved quality will define new product opportunities for factory-produced components and assemblies. The factory environment can encourage systems of quality control, the introduction of more diverse materials and specialized trades, and greater variety of product type. In the past, so many of the value-added components have been wood products such as pre-hung doors, floor and roof trusses, kitchen cabinets, and stair assemblies. These products have capitalized on cheaper labour sources and have been a means to effect labour savings, particularly in periods of labour scarcity. New value-added products may address the need for more customized interior and exterior finishes and trims, reduced maintenance, lower operating costs, increased flexibility, home security, or retrofitting of existing space. These products are likely to be directed at a discretionary market and not necessarily at lowering costs.

Opportunities for research and development of technical innovations can be categorized in five areas. The most obvious is the increased development of value-added prefabricated components and systems, particularly composites. Related to this is the development of engineered structural and non-structural components and systems derived from plastics and reconstructed wood products. Promotion of new materials can be expected from companies not traditionally viewed as construction industry suppliers, particularly in the area of synthetics. There is certain to be increased interest in manufacturing and process technology, including the development of prefabricated and factory mechanization process and equipment using computer-driven controllers. Finally, product manufacturers will continue to pursue the increasing utilization of construction coatings, sealants and adhesive materials.

In the US, the National Association of Home Builders anticipates a gradual shift from traditional craft skills to industrial-type assembly skills as prefabrication of building components intensifies. Prefabrication, including larger basic building components, will expedite the construction process and increase labour productivity among more specialized and skilled construction workers. Several trends are foreseen: new sources of supply will be available in the electronics, plastics, and fabrics industries not previously oriented toward construction; joints and connections will become increasingly easy to install and more maintenance-free; computer-controlled, factory-produced components will compete with conventional assemblies in an increasingly diversified and customized residential market; craft skills will be in highest demand in a limited market for retrofitting, conversion, rehabilitation, and historic preservation; building components will carry a manufacturer's warranty; and plug-in, zip-in, and hook-up connections for telephone equipment, plastic plumbing, heating and electricity will reduce maintenance and servicing needs (NAHB 1985).

The demand is growing for even more energy-efficient building systems and equipment. Energy-efficient products that may find their way into the home include: coatings that can both absorb and reflect heat; window glazings with adjustable thermal and lighting characteristics; devices to reduce indoor air pollution in tightly sealed houses; and various thermal and structural components integrated with microcomputer control systems to provide optimal levels of energy efficiency. Product development may even extend to some of the more common problems of moisture damage in wood frame houses, deteriorating concrete parking structures, and cracked basement walls.

The renovation market could be a fertile area for innovation, and it is the market with perhaps the greatest growth potential (Clayton Research Associates Limited, May 1986). But there are caveats. Barriers to entry into this field are largely non-existent. This has important consequences for the competitive position of firms that try to adopt sophisticated techniques – customers will have considerable choice from among contractors with varying levels of skill. This is also a market comprised of relatively small jobs, scattered throughout existing stock, and often contracted on an informal basis. While structural additions to existing homes are estimated to be an $850 million dollar business today, it will not be an easy market to access with new products and processes (Clayton Research Associates Limited, January 1987). The renovation market is still relatively immature in spite of its apparent size.

Unlike the automobile industry, agriculture, or for that matter, many of the consumer markets, quantitative data on the residential construction industry are scarce. Time series data that could provide a better understanding of how the industry has performed in relation to the business cycle, rising costs, competition, market demands, or regulation and housing policy are difficult to assemble. The lack of housing research by other than government agencies has not placed the industry on a strong footing to meet emerging opportunities. Lack of data on renovation activity is a prime example. In spite of the high volume of expenditure in this segment, there is scant evidence that the industry is cognizant of this opportunity, nor that it has a strong interest in capturing the major share of this construction activity.

Government could play a key role in removing or alleviating impediments to the acceptance of worthwhile innovations. The fragmented nature of the industry, particularly the renovation sector, is a barrier to the effective identification, development, evaluation, and dissemination of innovation. It is a barrier that industry participants on their own, or even through their trade associations, are not likely to overcome. Governments, and particularly the federal government, could provide a national forum to promote innovation and the transfer of technology within the industry. Without this assistance, the industry will be increasingly susceptible to the information sources associated with the larger markets to the south and off-shore producers, as evidenced by Canadian interest in Swedish prefabricated panel production systems and Japanese prefabricated housing.[14]

At present, it is difficult to advance a strong argument in support of some of the foreign prefabricated building systems that are catching the attention of the trade press (McKellar *et al.* 1986). The Japanese prefabricated house, the Swedish wood panel system, or the US mobile home industry have limited applicability in Canadian markets. There are not the profit margins to justify the initial capital outlay for plant and equipment and to support the on-going overhead during downturns in the building cycle (Gietema and Nimick 1987). These foreign competitors, even today, seldom match the level of sophistication that Qualico and Engineered Homes brought to the Canadian market. In fact, Mitsui and

Mitsubishi, two of the foremost names in Japanese industry, have embarked upon house building operations that closely parallel the operations of these two Canadian companies. Both Japanese producers utilize the Canadian approach of wood frame platform construction using on-site erection of "2x4" walls (McKellar 1985).

Future prospects for the Canadian residential construction industry lie neither with foreign competitors nor with neighbours to the south. The post-war residential construction industry in Canada has evolved over the years in response to the demands of this country, and there is every evidence of solid ground upon which to improve the production and delivery of housing to meet future needs. The "assembly line in reverse" has served the residential construction industry well since its introduction in the mid 1940s and, through incremental improvements, it has become in its own right a highly sophisticated building system. No doubt, it will be necessary to embellish or supplement this "assembly line" with advancements in product research and development, factory mechanization and control systems, or new managerial skills.

The residential construction industry has proven its ability to adapt, to shed the lure of foreign approaches to "systems" building and, in fact, has devised a building technology and production process that is highly regarded and now emulated in other countries. If there is one future or one frontier yet to be explored, it is the international market. Perhaps it is now time for the Canadian residential construction industry to meet the challenge of the Japanese or the Swedes in off-shore markets where long-term potential may be the greatest.

Notes

1 A large information base is available, both nationally and regionally, dealing with the output and production processes of the housing industry over the entire post-war period (Statistics Canada, CMHC, and the National Research Council). As is often the case, there is greater information available for the later years. Considerably less, and more fragmentary, information is available in statistical or published form, on the topics of industry structure and cost, land and technological change and transfer in the housing industry.

2 The international trend was perhaps a logical extension of the success of these firms through the late 1960s and early 1970s, and not unrelated to the subsequent decline in Canadian demand and rising opportunities in the US south-west and elsewhere. Also, large-scale commercial real estate and land development began to dwarf the residential activities that had given birth to many of these companies.

3 By mid 1987 the stage had once again been set for a cycle that last occurred a decade ago. The pace of housing production in 1987 surpassed all expectations and the warning signs are out. Ontario and Quebec continue to dominate the industry with 70% of all starts in 1987, and single-family starts are providing the principal momentum for a market estimated at a seasonally adjusted rate of 262,000 units. See Clayton Research Associates Limited (August 1987). The housing industry, in spite of past experiences, is once again

about to ignore the fundamental economic and demographic factors that shape demand. It is not an industry that believes in numbers; it reacts to what it sees, when it sees it, and not before.

4 When allowance is made for renovations to the rental stock, the total amount of renovation spending in 1984 exceeded spending on new residential construction. See Clayton Research Associates Limited (May 1986).

5 Annual reports for two prominent Toronto-based publicly traded home builders show little or no increase on gross profit margins from 1985 to 1986, a period of unprecedented increase in housing starts and "exploding" prices in the Toronto region. See Clayton Research Associates Limited (April 1987). The number of single-detached house builders in Canada's twenty-four CMAs increased in 1986 after falling in 1985, a pattern that reflects the increase in single-detached starts. See Clayton Research Associates Limited (November 1986). There were 4,989 firms in 1985; 80% constructed fewer than ten houses that year. These 3,976 firms accounted for just 21% of all houses built by the industry.

6 The average annual rate of increase in building labour costs in Canada from 1971 to 1977 was near 12%; worse than in the US (6%) and Sweden and Switzerland (10%); comparable to Finland and Denamrk, and better than Portugal (14%), the United Kingdom (15%), and Austria (17%). The average annual rate of increase in building materials costs in Canada from 1971 to 1977 was near 9%; worse that in West Germany (4%) and Austria and Japan (8%); comparable to the US and France, and better than Asustralia (12%), Belgium (16%), and the UK (17%). Data are drawn from the *Annual Bulletin of Housing and Building Statistics for Europe* and the *Monthly Bulletin of Statistics* (various years), both published by the United Nations Economic Commission for Europe.

7 Interview with Gordon L. Magnussen, former president of Engineered Homes. Calgary, Alberta.

8 Interview with Maurice Chornoboy, former Senior Vice-president, Qualico Developments Ltd. Calgary, Alberta.

9 Interview with William M. McCance, former Director Technical Research, National House Builders Association (NHBA), Toronto. The "Mark" series of experimental projects is summarized in a small brochure "The Experimental Housing Program of the National House Builders Association Technical Research Committee," (not dated).

10 The *Alberta House Cost Comparison Study* has been carried out annually since 1979 and serves as an excellent source for consistent construction cost data, over time, for selected urban centres in Alberta.

11 Forintek Canada Corp. was formerly a Crown agency but was subsequently transformed into a for-profit corporation and is now supported by research income.

12 Product differentiation and its impact on design and construction is examined in a research publication *Design Preferences and Trade-Offs for Moderately Priced Housing In Alberta.* Alberta Department of Housing. November 1985.

13 See Bairstow and Associates Consulting Limited (1985) for an examination of the opportunities for manufactured housing in Canada.

14 See McKellar (1985) for an in-depth analysis of the Japanese prefabricated house and its applicability to other markets.

CHAPTER NINE

Net Changes in Canada's Post-war Housing Stock

A. Skaburskis

DESPITE THE data shortcomings, Canada's housing stock and flow statistics have improved over the past four decades. The completions series now better reflect additions to the stock. Demolitions and conversions are now being counted and monitored at local levels and summarized by Statistics Canada. Quality control procedures have become more sophisticated.

This chapter reviews the evolution of census definitions and procedures in enumerating the housing stock. Also considered are the methods used by Statistics Canada to measure the annual flows of units into or out of the housing stock. The chapter then considers determinants of stock changes focusing on demolitions, conversions, and abandonments. Finally, the chapter considers progress since 1945 in the measurement of stocks and flows and discusses future data needs.

Stock Definitions and Statistics

A census takes a snapshot of the stock of housing in Canada at a particular date. In doing so, Canadian censuses divide the stock between private and collective dwellings. Private dwellings are "structurally separate sets of habitable rooms with a private entrance ... giving access ... without having to pass through another dwelling." The difficulties in applying the definition yield errors that are small in comparison to the numbers depicting the total size of the stock. The errors, however, are large in comparison to the number of units added each year through completions or conversions. They are large in comparison to demolitions and, when combined with other errors, they overwhelm attempts to estimate stock losses by census and completions reconciliation methods. A 2% error in classifying building type will swamp attempts to measure events affecting 0.2% of the stock.

The 1986 Census identifies 9,515,930 private and 19,800 collective dwellings. The former includes 469,000 unoccupied private dwellings (but does not include unoccupied seasonal or marginal dwellings) and 55,265 dwellings occupied only by temporary or foreign residents. This chapter is concerned exclusively with the stock of private dwellings. The Census distinguishes unoc-cupied private units that are available for rent or sale from those that are not available.

Canadian censuses also categorize dwellings by type of structure: for example, single detached dwellings, double houses, row houses, other attached houses, duplexes, apartments, and movable homes. Unfortunately errors in past censuses and changes in the magnitude of these errors mar the usefulness of available building type statistics. In 1951,

> ... there had been some misunderstanding on the part of the enumerators as to the application of the definition for single-attached dwellings. This misunderstand-ing appeared to be such that in 1941 too large a number of single-attached dwell-ings were reported. In 1951, the reverse appeared to be true, with enumerators tending to report too many apartments and flats (*Census of Canada 1951*, 10: 362).

The switch to self-enumeration in 1971 introduced new procedures and new sources of error as indicated by the 1976 caution:

> ... owing to a significant response bias in the 1971 figures for structural type of dwelling, particularly in the larger urban centres of Quebec, the comparability of 1976 and 1971 data should be viewed with caution. Different studies were under-taken to evaluate the extent of the error. It was found that in the identified problem areas in the Montreal core, the 1971 figures for "single attached" were overstated at the expense of the "apartment" category, which was underestimated by 36% (*Census of Canada 1976*, 3.1: 31).

In 1976 census, enumerators identified the building type on the outside of indi-vidual census forms. While the procedure appears to have given reliable mea-sures, it was not replicated in the 1981 census. A return to 1971 procedures led to response bias and the new warning. [1]

> From the structural perspective the counts for apartments in buildings with five or more storeys are believed to be relatively accurate. Counts for other types of dwell-ings in multiple unit structures (for example, apartments in buildings of less than five storeys and row houses), on the other hand, may contain varying degrees of error. For these dwellings there have been two types of misclassification. First, there are misclassifications among various types of the multiple unit structures. For example, apartments in buildings of less than five storeys have frequently been classified as row houses and semi-detached. Second, there are some misclassifica-tions between multiple and single structures. For example, a duplex may have been misclassified as a single detached.

Both the magnitude of the errors and the variation in their size are important. Their large size precludes accurate estimation of the changes occurring as a result of conversions that add units within the existing stock. Thus, we are least able to measure changes in the stock most often occupied by the housing-disad-vantaged.

Table 9.1
Conventional housing stock by province†: Canada, 1941-1986
('000s of conventional dwellings)

	Canada	Atlantic provinces‡	Quebec	Ontario	Prairie provinces	British Columbia
All dwelling types						
1941	2,638	243	659	932	578	226
1951	3,512	376	882	1,204	703	348
1961	4,725	437	1,243	1,695	866	475
1971	6,247	515	1,684	2,299	1,060	676
1981	8,416	666	2,243	3,053	1,466	969
1986	9,516	773	2,516	3,357	1,184	1,160
Single-family dwellings						
1941	1,904	194	311	681	523	196
1951	2,362	296	354	845	581	287
1961	3,086	340	486	1,171	704	379
1971	3,704	381	662	1,395	790	467
1981	4,883	501	978	1,728	1,026	636
Other dwellings						
1941	734	50	348	252	54	30
1951	1,150	81	528	359	120	61
1961	1,638	97	757	524	163	95
1971	2,543	135	1,022	904	270	209
1981	3,533	165	1,265	1,324	439	332

SOURCE Statistics Canada: *Current Investment Indicators*, Science Technology and Capital Stocks Division.
† Data are given for census years. 1986 counts by dwelling type are not available.
‡ Data unavailable for Newfoundland prior to 1949.

Table 9.1 presents the best estimates of the conventional housing stock by region available at this time. [2] Conventional housing consists of dwellings whose foundations form an integral part of their structure. These differ slightly from census counts because they exclude mobile homes.

Flow Definitions and Statistics

In addition to census stock counts, Statistics Canada uses other survey data to estimate the stock flows that change the size and character of the housing stock. Four types of flows add units to the stock: completions, conversions to residential use, recovery (from abandonment, temporary loss, or temporary conversion), and reclassification (including temporary to permanent dwellings, and definitional changes). Other flows result in losses (for instance, temporary mergers, conversion to non-residential use, damage that makes a dwelling uninhabitable, and demolition). Net change in the stock can, in principle, be estimated by adding the flows that increase its size and subtracting losses. In practice, net change is also measured by comparing consecutive census counts. If data on all flows were available, then both should yield the same numbers. That they do not points to errors in the counts and differences in coverage or definition. These discrepancies are easily observed when we attempt to reconcile stock and flow data.

Statistics on completions are developed monthly by CMHC. Their *Starts and Completions Survey* (SCS) collects building permit summary statistics monthly from all urban areas with more than 10,000 people. The rest of the country is sampled quarterly to complete the data set. Excluded from the SCS definition of additions are mobile homes, tents, trailers, and the other miscellaneous building types where counted by the census as the occupants' usual place of residence. The SCS does not cover conversions or additions to existing buildings. It excludes temporary dwellings and seasonal homes, some of which may become usual places of residence and hence be included in census stock counts. The CMHC survey yields good counts of permit-obtaining completions by building type, as presented in Figure 9.1, but excludes dwellings built without permits.

New conversions are a separate category of additions. These are units built intentionally for immediate conversion. For example, the "Vancouver Special" is a large single-family dwelling (SFD) built in a neighbourhood zoned for single-family houses which may be converted immediately after the last building inspection takes place. With minimal effort, the house is turned into a duplex. In Calgary, new permit-obtaining duplexes have been turned into quadruplexes. In Toronto, one-bedroom apartments became a larger number of "bachelorette" units. New conversions are often completions done without permit and should be treated as unreported additions. Using only permit-obtaining completions statistics understates the number of new units added to the stock.[3]

The housing stock can also be augmented by physical conversion within the existing stock or by conversions from the non-residential stock.[4] For example, single-family houses are converted into duplexes and triplexes. In so doing, the habitable space of such houses is typically increased – for example, by finishing off basement or attic areas. Such activity expands the physical stock and increase its value, but often is not recognized in published flow series. Many large old single-family houses in the inner city were converted to rooming houses during and after World War II and later reconverted to small apartments as the housing preferences and needs of small households changed. The extent of the physical conversion that has taken place in the inner areas of large cities, and its relative importance in providing low-income housing, has led housing analysts to recognize the multiple converted dwelling (MCD) as a distinct building type. The flexibility offered by MCDs is an important means by which the supply of

FIGURE 9.1 Annual dwelling completions by type: Canada, 1951-1986.

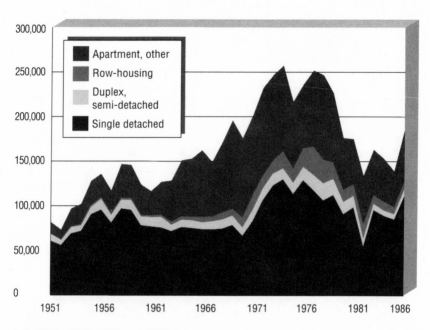

SOURCE: *CHS*, 1976 and 1981.

low-income housing can be adjusted and expanded. Loss of these units, through demolition or deconversion to single-family houses, reduces the housing stock available to low-income people. That their loss can fuel future housing problems of this kind makes it important to monitor this component.

The conversion of seasonal dwellings to year-round use, the sale of a second home, and the transfer of group housing or foreign-owned dwellings to a private household can show up in the census as an addition to the stock. Previously-vacant "temporary dwellings" are added to the census count when they become someone's usual place of residence. If not taken into account in stock and flow reconciliation, they introduce errors and show up as large residuals. The residuals also contain the errors mentioned earlier and cannot therefore shed light on the extent of the net changes from any one particular kind of flow.

One can also distinguish between permanent and temporary losses. Permanent losses are caused by the demolition and abandonment of buildings because of fires and other hazards and disasters. Retrievable losses may result from a temporary merger of units or from the conversion of dwellings to non-residential use. Reclassification of a building's use can also appear as a component of stock change. The conversion of a single-family house into a group home will,

Table 9.2
Completions, conversions and demolitions of conventional dwellings:
Canada, 1946-1985

| | Net stock | | Completions | | Conversions | Demolitions | |
	Total	% Single	Total	% Single	Total	Total	% Single
1946	2,998,018	68	60,500	76	6,868	8,424	71
1951	3,563,665	67	81,059	74	2,868	10,112	68
1956	4,159,168	66	135,700	70	5,242	12,362	66
1961	4,794,776	65	115,608	66	2,706	11,157	57
1966	5,469,090	62	162,192	46	2,655	16,064	52
1971	6,375,126	59	201,232	41	2,751	14,299	54
1976	7,500,633	57	236,249	54	1,931	11,888	69
1981	8,521,485	58	174,996	56	2,112	14,252	64
1985	9,090,936	58	139,106	61	6,307	9,680	71

SOURCE Statistics Canada: *Current Investment Indicators*, Science Technology and
Capital Stocks Division.

for example, show up as a loss in the census count of private housing units (and
an increase in the count of collective dwellings). The loss may be temporary as
the boarding house may eventually be changed into a MCD or converted back to
its single-family use.

Table 9.2 presents the Statistics Canada estimates of stock size for each year
since 1946. It also lists statistics on permit-obtaining completions, demolitions,
and conversions. The annual stock estimates were developed with the help of
population statistics and estimates of average household size. Table 9.2 shows
single dwellings falling as a proportion of the total stock (and as a proportion of
completions) during the 1950s and 1960s. The estimates of recorded demolitions
show no systematic change in type of building lost since 1946. The proportion of
demolished single dwellings within the series, however, is suspiciously high. In
case studies carried out in Vancouver and Saint John, most demolitions
recorded as single dwellings actually contained more than one dwelling and
should properly have been registered as MCD losses.[5] That Statistics Canada
demolitions series are usually underestimates is partly a result of units not being
properly recorded on demolition permits, and partly a result of the pervasive
confusion of MCDs and single dwellings.

Table 9.3 summarizes the flow data and reconciles them with census counts.
The table shows the "residual," that is, the mismatch between the two sets of
data. The residual is a net difference; it understates the magnitude of the errors
in each of the stock flow series that are caused by unrecorded gross demolitions
cancelling out unrecorded gross conversions. The magnitudes of these errors

Table 9.3
Reconciliation of conventional stock and flow statistics:
Canada, 1941-81
('ooos of dwellings)

	Canada	Atlantic provinces†	Quebec	Ontario	Prairie provinces	British Columbia
1941 Census stock	2,638	243	659	932	578	226
1941-51 Flows						
Completions	604	39	169	216	109	70
Demolitions	85	8	22	30	19	8
Conversions	49	3	14	18	9	6
Residual	233	26	61	68	25	53
1951 Census stock	3,512	376	882	1,204	703	348
1951-61 Flows						
Completions	1,118	62	324	4,146	200	123
Demolitions	117	13	31	43	23	12
Conversions	32	2	10	11	4	6
Residual	133	10	57	77	9	10
1961 Census stock	4,725	437	1,243	1,695	866	475
1961-71 Flows						
Completions	1,558	93	400	603	260	203
Demolitions	160	18	54	45	24	19
Conversions	27	3	11	7	2	4
Residual	94	0	84	41	12	14
1971 Census stock	6,247	326	1,684	2,299	1060	676
1971-81 Flows						
Completions	2,285	166	508	818	467	325
Demolitions	136	11	39	37	26	23
Conversions	24	3	11	5	1	3
Residual	-12	-8	78	-33	-37	-12
1981 Census stock	8,416	666	2,243	3,053	1466	969

SOURCE Statistics Canada: *Current Investment Indicators*, Science Technology and Capital Stocks Division. Residual was computed by taking the difference between the 1951 stock and the (1941 stock plus completions-demolitions plus conversions for 1941-51).
† Data unavailable for Newfoundland prior to 1949 unification with Canada.

can be revealed only by special surveys using trained interviewers. The large errors obscure those changes in the existing stock that affect the least well-housed Canadians. Knowledge of the size of this stock component can be improved through monitoring procedures such as those carried out in the US Annual Housing Survey.

Table 9.4 presents the best estimates of net stock losses available at this time. Presented in Table 9.4 are estimates of "net losses" – the losses within the existing stock, net of the additions within this stock, that are the result of physical conversion. These loss statistics were developed by comparing the relevant period-of-construction (POC) data from the census with the total number of units counted in an earlier census. These estimates seem low; note for example the especially small losses estimated for Quebec and Ontario.

Net stock loss rates corresponding to the dwelling counts presented in Table 9.4 are presented in Table 9.5. Stock losses in the 1970s were generally higher than in previous decades. An average of 47,000 housing units were lost each year across Canada, representing an annually-compounded stock loss rate of 0.8%. Losses and loss rates increase as the housing stock ages and as cities grow larger and are less able to accommodate further growth without demolishing older buildings. The future rate of stock loss is expected to increase as a progressively larger proportion of the stock enters its last phase of useful life. Table 9.5 shows the highest loss rates to be in the prairie provinces. This illustrates the importance of rural losses and helps to explain part of the difference between the net loss statistics presented here and the demolition series listed in earlier tables. Stock losses in rural areas are rarely recorded on demolition permits. Rural losses often occur through the abandonment of buildings that are no longer needed.

The Determinants of Stock Change

Case studies for Vancouver, Calgary, Toronto, Montreal, and Saint John help to explain the direct causes of demolitions, conversions, and abandonments (Vischer Skaburskis, Planners 1979a).

DEMOLITIONS

Current demolition and conversion rates in Canadian housing are affected by the factors that influence the effective demand for new housing.[6] The determinants of demolition rates are the forces that increase the value of inner city land. The factors affecting the demolition of particular buildings are those that give rise to their owners' expectations of more profitable land uses. At more general levels, the rate of demolition is affected by city growth rates, city size, age of stock, and local development policies. It is affected by government housing policies that stimulate new construction. Case studies in Vancouver have shown one unit is lost for every four new units built in the inner city. Demolitions also result from fires and other hazards.

Public works have brought down many old buildings since 1945. Approxi-

Table 9.4
Permanent stock losses: Canada, 1946-1981
('ooos of dwellings)

	Canada	Atlantic provinces	Quebec	Ontario	Prairie provinces	British Columbia
1945 stock						
In 1961 Census	2,633	285	687	964	466	230
Actual	2,824	314	696	978	588	244
Net loss	191	29	10	13	122	14
1960 stock						
In 1971 Census	4,437	402	1,191	1,641	754	444
Actual	4,692	434	1,231	1,681	864	473
Net loss	255	33	40	40	110	29
1970 stock						
In 1981 Census	5,779	471	1,575	2,178	931	615
Actual	6,245	519	1,671	2,285	1071	684
Net loss	466	50	95	107	142	70

SOURCE Vischer Skaburskis, Planners (1979a, Table 2).

Table 9.5
Annually-compounded net stock loss rates:
Canada, 1945-1981 (%)

	Canada	Atlantic provinces	Quebec	Ontario	Prairie provinces	British Columbia
1945-1961	0.5	0.6	0.1	0.1	1.6	0.4
1961-1971	0.6	0.8	0.3	0.3	1.4	0.7
1971-1981	0.8	1.0	0.6	0.5	1.4	1.1

SOURCE Calculated from data in Table 9.4.

mately 40% of all demolitions in Montreal during the 1960s and 1970s were the result of public works projects. Transportation projects caused much of the loss – 6,822 units made way for improved road systems. Urban renewal in Saint John removed some 300 units a year throughout the 1960s; some of these dwellings had earth floors (Vischer Skaburskis, Planners 1979a). Calgary cleared much of its downtown low-income stock to make way for public buildings, parks, and future development sites; no count of losses was taken. Vancouver inner-city neighbourhoods had large tracts of land cleared to make way for public projects, and again did not count lost units. Urban renewal as a factor affecting stock loss rates has decreased in importance since the early 1970s.

City policies also affect stock loss rates. A drive to enforce by-laws can lead to the removal of dilapidated buildings and thereby help preserve the neighbour-hoods and keep other buildings from following suit. Fire and health by-laws, property standards by-laws, minimum maintenance ordinances, and building codes have been enforced to keep up standards. Fire and health by-laws appear to have been effective in dealing with dangerous and unsanitary dwellings, but their enforcement tends to lead to building closure and eventual demolition. Case studies have shown that property standards and minimum maintenance by-laws are not effective tools for stock maintenance (Vischer Skaburskis, Planners 1979a). City officials have been cautious in trying to enforce such habitability standards, being uncertain of success in legal challenges. Vancouver planners have concluded that the enforcement of property standard, health, and fire by-laws do not, by themselves, do much to help conserve the existing stock.[7] In Montreal, subsidies were used to restore and upgrade some 7,000 housing units during the 1970s.[8] Attempts to reduce loss rates by instituting demolition control by-laws have usually led to the conclusion that legislating the preservation of economically obsolete buildings is futile.[9]

Municipal governments have tried to reduce the incentives for demolishing low-income stock by means of zoning legislation. Smith and Tomlinson describe Toronto's experiment:

> In an attempt to prevent conversion by demolition and rebuilding; the City of Toronto responded by imposing land-use restrictions on land that previously contained rental apartment units. In October 1980 the City passed a by-law limiting every site in the City occupied by an apartment building 20 years of age or older to a maximum building density of one times lot area and a maximum building height of 11 metres (37 feet). For most sites, this meant replacement buildings were substantially down-zoned, which reduced land values for alternative uses and reduced the likelihood of demolition applications ... (1981, 110).

They conclude that the long-term maintenance of this policy would have had a significant adverse effect on land-use efficiency. Silzer updates this policy initiative:

The City's efforts to save rental apartments from the wrecker's ball began in 1980, when council passed the first of several restrictive by-laws. Council's intent was to discourage demolitions by limiting the size of building that could be built on sites occupied by rental apartments. The courts quashed this by-law on the grounds that the Provincial Planning Act does not give local councils the right to use their zoning authority to preserve existing buildings. In effect, councils are only allowed to delay demolitions until building permits have been issued. They cannot withhold permits if plans conform with zoning regulations (1985, 8).

CONVERSIONS

Tight housing markets make the conversion of basements and attics into "in-law" suites more attractive to home owners. Housing shortages create the demand for converted units. This demand, in turn, creates the revenue potential that allows some home buyers to add suites to help them pay for the high-priced housing. These additions help low-income people by expanding their stock of housing. While economic factors explain the motives for physical conversions, a number of other considerations affect the rate at which conversions occur. Restrictions imposed by city hall, neighbours' attitudes, and financial constraints can restrain conversion activity. Organizational factors, building characteristics, and stock conditions also affect conversion rates. [10]

While conversions add low-priced housing units, they may also add low-quality units that raise health and safety concerns. Neighbours often dislike the congestion created by the increase in local traffic and parking problems. Cities have tried, often unsuccessfully, to reduce the extent of conversion activity. Building codes and zoning ordinances appear to be relatively ineffective means for reducing conversion activity. City councils have been reluctant to ban low-priced housing units during periods of housing shortage. While the City of Vancouver Planning Department tried to monitor some 6,000 illegal units during the 1970s, they could do little to change the situation. Neighbours prefer the illegal status since they have recourse to threats of legal action if the tenants become a problem. [11]

The deconversion phenomenon associated with the gentrification of inner-city neighbourhoods is explained by the economic and demographic factors that bring wealthier households back to the inner city. [12] Gentrification is a land-use change that keeps the exterior building envelopes intact. It changes the housing services offered by existing stock and reduces the need for replacement construction in inner city neighbourhoods. Gentrification, of course, reduces the low-income housing stock.

Attempts to preserve low-income stock have also failed to achieve their ends by encouraging deconversion activity. In the mid 1980s, staff in the Toronto City Planning Department observed that rent controls increased the rate at which this city's inner-city housing was being changed back to single-family use. Silzer (1985) saw owners trying to evade rent controls by converting their units to hotel accommodation, and notes: "Conversion to tourist accommodation accounts for almost another third of the rental apartment losses."

The issues raised by conversion and deconversion activities are serious and multi-faceted. They pose questions as to the efficiency of land use, the consequences of constrained housing markets, and the equity consequences of functioning markets. They touch local political nerves. They can affect the growth and form of the city. The issues surrounding conversion activity continue to be a topic on the political agenda in Canada's larger cities. The ability of a city to respond to these issues might be improved by developing statistics that show and monitor the number of its converted dwellings and their units' and occupants' characteristics.

ABANDONMENTS

Urban abandonments are usually the result of a reduction in housing demand (for example, the migration of middle-income families to the suburbs) aggravated by the aging and deterioration of the stock. Under-maintenance brings down their physical quality and the trend affects the neighbourhood and discourages investors. The lack of demand for housing keeps rents and resale opportunities low. Eventually faults develop in the buildings, and these bring building inspectors: the by-law citation is often a key step in the abandonment process. Owners are likely to try a quick sale unless they have already moved to another city. When the sale does not materialize, and if the building inspectors are persistent, then the owner may give the building to the city or simply let it go.

Factors affecting the rate of abandonment cause a reduction in housing demand. Other factors include such considerations as the owner's knowledge of rehabilitation potential and ability to engage contractors. The Saint John study identified abandonments that were the result of elderly owners being surprised by a code violation, unable to look after or pay for repairs, and unable to sell their buildings after a half-hearted attempt.[13] RRAP was the saviour of many houses in the inner-city of Saint John: "60% of the stock would have been lost had it not been for RRAP" was the view of one building inspector. RRAP also helped weed out the worst buildings and was reported to have brought about a few abandonments. The better structures could be rehabilitated with RRAP assistance, and their owners would then offer the improved housing at rents that would attract tenants from buildings that were in too poor a condition to be rehabilitated. The reduced revenues from the increased vacancies in those buildings discouraged the owners into abandonment when they found that their buildings could not be sold and that they could not maintain a positive cash flow.[14]

Post-War Progress in Stock and Flow Statistics

Stock and flow statistics, when linked with data on housing needs and demands, play a role in motivating, directing, analyzing, and evaluating the policies that favour housing progress. Progress in the sense of a regular pattern of change is

Table 9.6
Net additions and completions
to the conventional housing stock:
Canada, 1881-1981
('ooos of dwellings)

	Conventional net stock	Completions between censuses	1945 stock plus completions	Surviving 1945 stocks	Residual
1881	797				
1891	950				
1901	1,080				
1911	1,478				
1921	2,137				
1931	2,373				
1941	2,677				
1945	2,804	2,804	0		
1951	3,519	418	3,222	2,633	407
1961	4,724	1,266	4,488	2,632	408
1971	6,247	1,614	6,102	2,366	504
1981	8,416	2,285	8,387	2,016	817

SOURCE Vergès-Escuin, 1985, Table xii. CMHC *CHS* 1945-81.
Residual = Net Stock - Surviving 1945 Stock - Completions.

clearly illustrated in the Table 9.6 stock statistics. The first column of numbers shows the housing stock to have increased since 1881 at a growing rate. The reconciliation of past stock and flow statistics is an indicator of progress in the accuracy of past data-gathering efforts. The completions statistics for Canada are given in the second column of Table 9.6. The third column adds these to the 1945 stock and yields a shortfall: the 418,000 recorded additions would yield a 1951 stock of 3.221 million. The 1951 census enumerators, however, found 297,000 units that had not been recorded in the additions series or had been missed in the earlier census.

The fourth column of Table 9.6 presents the counts of surviving 1945 stock identified in each successive census. The period of construction statistics developed in the 1961 and 1971 censuses are accurate because the enumerators were trained to identify the age of buildings in their assigned neighbourhoods. While the switch to self-enumeration in 1981 is expected to have increased the number of errors, they are still not expected to be serious: not many buildings were constructed during the 1940s, and the war created a memorable event which helps people identify the pre-1946 dwellings. The fourth column of Table 9.6 shows that the pre-1946 stock has shrunk 600,000 units since the war. This number understates the total loss from demolition, abandonment and conversion by the number of units that were added within the older existing stock through physical conversions. Nevertheless, the net loss estimates yield useful information on the extent to which past completions were under-counted.

The last column of Table 9.6 presents the difference between the census stock counts and the surviving 1945 stock, plus the recorded permit-obtaining completions. By 1981 the census stock counts exceeded the 1945 surviving stock plus completions by 816,652 units. This discrepancy is not explained by improvements in census coverage because improvements would also increase the count of the surviving 1945 stock. One explanation for the large discrepancy in the stock and flow statistics suggests the under-counting of completions during the early post-war years. A part of the residual can be explained by the rehabilitation and conversion of older stock to the extent that its residents report the building as being new.

The reduction in size of the residuals over several decades is due to improvements in data quality, reflecting progress by showing how both our ability to measure the size of the stock and our care in monitoring the stock have increased. The large size of past errors points to areas for further historical research. Were they the result of sampling error? Or do they represent a healthy informal economy producing non-permit-obtaining completions during the early post-war period?

Progress has also been achieved in the definition and monitoring of stock changes. Since 1945 a more detailed classification of building types and occupancy patterns has been introduced. Vacant units are explicitly considered and classified according to the reasons for their vacancy. The inclusion of seasonal and temporary houses, the concern with mobile homes, and distinguishing residential units that are attached to commercial structures all demonstrate a growing awareness of the diversity of the stock and, indirectly, may signal progress in our concern for stock-related issues. Statistics Canada's decision to gather and monitor demolition and conversion statistics, as well as its concern with ongoing stock estimation processes, confirm the view that progress is being made in government's ability to deal with stock-related issues; the agency's newly-created Technology and Capital Stocks Division is to monitor housing stock changes, illustrating the importance that is now being attached.

Future Data Needs

The data needs of future policy makers will be varied. Information needs change as new problems are identified, as policy options develop, and as programs are designed and evaluated.

During the initial phases of problem identification, an important purpose of information is to motivate people to form a constituency that will seek the political commitment to deal with the problem. Aggregate and summary statistics generally do not do this unless they are accompanied by good narrative. Instead, studies showing pictures and telling stories may be more important in the problem identification stage. Improvements in national summary statistics will not reveal the drama behind the apartment owner's sledge-hammering walls as he tries to reduce unit counts on a demolition permit application. Improvements in central data-compiling procedures, even by accident, will not explain why a city, apparently intent on stock preservation, has had to order the demolition of boarded-up houses. At the same time, stories alone are not enough; anecdotes can be shrugged off.

Local studies of stock conditions and rates of change are more suitable as we move to develop the context for the policy problem. Studies of stock conditions can demonstrate the extent of the problem, showing that it is broad enough to merit policy, rather than ad hoc intervention. While studies can illustrate the magnitude of the problem, more broadly-based information is needed to determine whether the problem is sufficiently manageable to be recognized on the policy agenda. A data-generating process is even more valuable when it helps to identify policy alternatives. Local counts of demolitions can easily generate information, for example, on the determinants of local stock losses. The direct determinants of stock losses discussed earlier in this chapter, for example, were developed by means of local studies. [15]

More comprehensive views of the factors affecting stock changes need data that are more broadly based. Cross-sectional analysis of data from all major Canadian cities can best reveal and measure the effects of indirect factors. Estimates of net stock losses attributable to shifts in housing demand, for example, require the examination of patterns of change within a broad range of cities experiencing different types of pressures. An expanded and routinely administered Survey of Housing Units can be of value in developing the data needed to explain the determinants of stock change and the knowledge of likely impacts of alternative housing policies. Repetition of the survey, however, is essential to develop data that show how the stock changes over time.

The federal government's increasing focus on collection of accurate local statistics will help policy development at all levels of government. [16] Improved accuracy helps the government anticipate, for example, the problems brought about by rising stock loss rates. The increased federal demand for accurate local statistics can, indirectly, help local officials develop their own awareness of issues and opportunities for their own policy interventions. The demand for better statistics changes the nature of their work and shifts prestige and power. Federal use of summary statistics shows local government officials who develop the data that their work is important and that its quality matters.

Accurate statistics describing the nation's housing stock and flows are essential to the design of efficient and fair programs. Broadly-based statistics developed by the federal government are important in determining the spatial targeting of programs and in allocating government funds. The drawback of routinely gathered data is the impossibility of tailoring the definitions used to specific policy problems. [17] This drawback is compensated for by gains from the data's uniform quality. Problems of systematic errors can be overcome by using the stock and flow statistics as indices showing the relative size of the stock or level of flows in different parts of the country. If national or provincial programs are to address concerns related to stock losses, the relevance of census and CHS data helps ensure the equitable distribution of program benefits.

Post-war policies gave rise to programs that would expand the housing stock available to low and middle-income Canadians. Correspondingly, the focus in data gathering was on description of the stock and of the process that added new units. The next generation of data improvement will recognize the increasing importance of redistribution consequences. The ability to monitor these can be increased by spending more on developing accurate counts of dwellings by building type and by POC. Improvements can also be made by identifying MCDs as a distinct building type and by ensuring that the census statistics reflect the definition. This will probably require a redrafting of current data collection procedures.

As the complexity of housing markets becomes more broadly recognized, the importance attached to reliable stock and flow statistics will increase. Statistics will be needed to monitor changes occurring at local levels and trace the effects of policy and of market processes. The ability to compare the local area housing statistics developed during consecutive censuses can greatly advance our analytical capabilities. Improvements in the statistics used to identify and explain the redistributions and transitions that are taking place within the stock will yield increasingly greater payoffs. The housing industry's increasing reliance on rehabilitation and renovation work will increase its demand for stock statistics. Policy makers will want to know of possible housing adequacy problems and may want the capability to recognize emerging problems. Future analysts will want to monitor housing progress, and will need the data to explain the changes that have taken place.

Notes

1 "Cautionary Note on Data Quality – Structural Type" attached to *Census of Canada 1981. Occupied Private Dwellings: Type and Tenure.* Catalogue #92-903.

2 The numbers were made available in 1985 by Robert Couillard, Chief of the "Current Investment Indicators Section" of Statistics Canada.

3 Their prevalence, however, causes the same kind of neighbourhood problems that are brought about by physical conversion of the existing stock, with their effect of increasing the demand for parking currently being the most significant.

4 A church in Montreal was recently changed into a condominium; schools have become elderly housing projects; warehouses have been turned into luxury suites.

5 Case studies comparing estimates of losses to losses recorded on demolition permits show the summary series on demolitions to understate true losses by 0.96 in Montreal, 1.5 in Toronto, and 2.7 in Calgary. See Vischer Skaburskis, Planners (1979a, Volume 2). The unusually low Montreal ratio may be explained by the fact that a large proportion

of recorded losses was incurred through fires and other events not requiring demolition permits.

6 Architectural history shows that most major monuments that have been demolished were removed during the periods of a nation's greatest self-confidence when it was certain of its historic role in developing a new world. Old Rome was demolished, stone by stone, to provide the material for Renaissance and Baroque accomplishments. The forces generating demolitions are closely related to the events that mark progress.

7 Dr. Ann McAfee, senior housing planner for the City of Vancouver, has often advocated the use of a "stick and carrot" approach that links code enforcement drives with rehabilitation subsidies.

8 The Saint John building inspectors interviewed in 1976 believed that 60% of their inner city housing would have disappeared had it not been for the Rental Rehabilitation Assistance Program. See Vischer Skaburskis, Planners (1979a, Volume 2).

9 A broken water pipe can bring down a vacant apartment building in one cold winter day; the broken pipes have made questionable structures in Saint John not worth repairing. Fires in Montreal are the legendary means of avoiding difficulties with demolition permit applications. A Victoria landlord was known by that city's building inspector to have rented his house to students who agreed to "feel free" on the premises and soon had the neighbours petitioning the Municipal Council for the demolition permit that had previously been refused.

10 The growth of rehabilitation/conversion sectors within the building industry was included by Damas and Smith Limited (1980) as another factor affecting conversion activity.

11 Dr. Ann McAfee has described the incentives that help maintain the illegal stock in the underground economy. Their illegal status keeps the tenants from demanding too many improvements lest the unit be closed.

12 Stock characteristics also affect conversion rates. Jim Anderson, Director of the Calgary Housing Department, observed that Calgary conversions used to take place primarily in inner-city neighbourhoods that had large houses. Much of the city's older stock was seen as not being convertible because of the units' small size and "unromantic" appearance. The planners expressed concern over the long run effect the demolition of large inner-city houses would have on the flexibility of future stock. If most of the old houses were to be replaced by more efficient new buildings, the stock would be less able to respond to sudden changes in housing demand.

13 Abandoned buildings were seen by Mr. Sid Lodhi, Chief Building Inspector in the City of Saint John, as a determinant of abandonment rates. Abandoned buildings were quickly destroyed by vandalism and the cold winter weather. They destroy the immediate environment, and unless they are taken down by a city-sponsored demolition, they infect their neighbours and cause a rash of local abandonments.

14 In the Saint John study, other causes of abandonment were found to include: an increasing prevalence of lease-held land; conversion to rooming houses during the 1975-6 overbuilding; a property tax policy that penalized owners of rental accommodation; acceptance of reduced housing quality by owners and tenants, leading to more deterioration; reluctance of financial institutions to mortgage inner city property; increasing age and

death of owners; absentee ownership of inner city buildings; increasing cost of heating; increasing cost of demolition; RRAP-induced speculation in deteriorating buildings; ARP-inspired starts elsewhere and the flooding of the rental market; turnover in tenement ownership and changes in the owner's expectations and plans for their properties; lack of rehabilitation incentives prior to RRAP; the domino effect of abandoned, deteriorated buildings; and code enforcement drive triggering the finding that abandonments of buildings by their owners are increasing.

15 However, the studies were funded by CMHC as part of a national study of demolitions, conversions, and abandonments.

16 The demand for local information has also been sparked by Statistics Canada's starting to compile demolition and conversion series.

17 Also, federal program administration and evaluation requires that statistics of equal quality be developed for all parts of the country.

CHAPTER TEN

Measuring Transitions in the Housing Stock

E.G. Moore and A. Skaburskis

NEW HOUSING was just 50% of all housing investment in Canada in 1986, down from 69% in 1951; the percentage attributable to major improvements was 33% in 1986, up from 24% in 1951 (Table 5.1). As modifications to the existing stock of housing – whether conversions or deconversions, depreciation or renewal, or additions or improvements – become ever more important in changing the stock, so too does public debate regarding its management. Conversions and deconversions alter the size and number of dwellings in the housing stock. Depreciation lowers the quality of dwellings, while normal market transactions and renewal often shift patterns of ownership. Investments in additions and improvements slow down or reverse the deterioration of the stock.

This chapter discusses "transitions" in the stock. Transitions refer to events which change the status of a particular physical entity. To narrow the focus further, we consider only those transitions which occur to structures and dwellings in the existing stock. For a stock distributed over categories defined in terms of tenure, quality, or value, the flows into and out of each category in a given period of time produce the observed change in overall distribution.

The importance of changes within the existing stock increases as population growth subsides, and the need for net additional built space declines. Rising incomes generate demand for different housing attributes, the growth in two wage-earner households increases the desire to live in the inner city close to work, and the postponement of childbearing increases the significance of access to downtown amenities. The pressure to modify the existing stock increases and the number of transitions will grow through such actions as the purchase of older houses, deconversion and addition. Transitions are also sensitive to broad shifts in the patterns of demand arising from demographic change; for example, the growing number of elderly households (who typically cannot afford to spend much on property maintenance) increases the demand for government support for maintenance of the quality of the housing stock (in 1981 the average age of urban recipients of RRAP loans was over 58 years).

Transitions define how the stock changes over time. Individual flows identify the particular types of transaction (such as own-rent conversion in row houses)

Table 10.1
Changes in tenure of individual properties:
City of Toronto, 1976-1985

| | | Tenure of unit in 1985 | | | |
Tenure of unit in 1976	Owner	Owner-tenant	Tenant	Vacant	Total
Owner	57,193	2,704	4,635	1,229	65,761
Owner-tenant	10,132	11,069	3,104	328	24,633
Tenant	6,192	2,388	14,431	1,044	24,055
Vacant	1,746	295	1,335	401	3,777
Total	75,263	16,456	23,505	3,002	118,226

SOURCE City of Toronto Planning Department special tabulations, 1986.

that contribute to change, and thereby inform public debate. For example, Table 10.1 describes the changes of status of individual properties in the City of Toronto for 1976 and 1985. Of the 65,761 properties which were owner-occupied in 1976, 57,193 were owner-occupied in 1985, while 2,704 had changed to mixed owner-tenant occupancy. Of the 118,226 properties which remained in the stock over this decade, 30% were in a different tenure in 1985 from 1976. The net effect was a significant increase in owner-occupied units and a decrease in owner-tenant units.

Table 10.1 shows that, between 1976 and 1985, 41% of owner-tenant properties had become solely owner-occupied and 13% had become solely tenant-occupied. Such transition rates can be manipulated by public intervention. For example, the imposition of rent controls may increase the transition rate from rented to owned properties and reduce the rate of transition from owned to rented. However, the net effect of these changes will depend on the number of dwellings which start out in each of the tenure categories as well as on the rates.

The role played by these rates is also important to analysis of future distributions of attributes of the housing stock. If we assume that the rates remain constant over successive time periods, we can project future distributions of housing attributes (ownership, quality, size, and composition) using standard markovian methods (Emmi 1984). [1] If, however, we can identify the ways in which interventions such as the imposition of rent controls affect transition rates over time, we can create future scenarios characterizing housing attributes using simulation models incorporating plausible shifts in transition rates.

Post-War Transitions
In examining the trends in post-war transitions, each of the three different types of status change are considered in turn. Numerous problems of data adequacy

FIGURE 10.1   New units from conversions as a percentage of new construction: Canada and selected cities, 1964-1982.

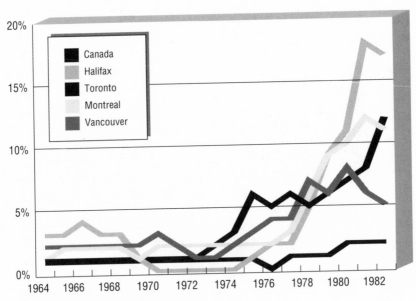

SOURCE: CMHC special tabulations.

discussed in the next section, and often the available data only permit statements to be made about the aggregate number of transitions rather than their detailed structure.

PHYSICAL CONVERSIONS

Migration during World War II brought many single people from central Canada to the coastal cities. Their housing needs were met by the conversion to rooming houses of the larger houses in the deteriorating inner neighbourhoods of these cities. Studies of such conversions show many thousands of units being added to the "collective" housing stock of cities such as Vancouver and Halifax. This stock remained through the 1950s and became the subject of code enforcement drives in the 1960s. Much of it was removed through urban renewal projects, public works, and the enforcement of fire and safety by-laws.

Change in income level and lifestyle made the old rooming houses functionally obsolete even though more adults were living outside family units. More young people came together so they could afford to rent more-traditional housing. Unrelated people formed households and generated demand for the smaller units found in central cities. The demand for small, less expensive dwellings in Vancouver increased tremendously during the 1960s, while the demand for

FIGURE 10.2 Dynamics in the occupied housing stock: City of Toronto, 1976-1984.

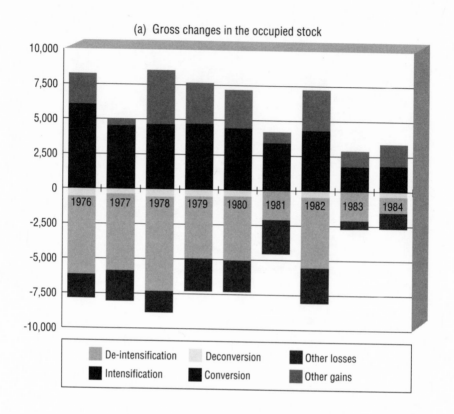

(a) Gross changes in the occupied stock

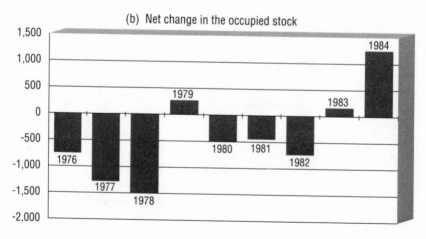

(b) Net change in the occupied stock

SOURCE: City of Toronto Planning Department special tabulations.

rooms virtually disappeared. The inner city stock underwent a major transition during this period in response to changing norms and increasing affluence.

Nationally, how important have conversions been in the overall stock? One measure is provided by the national building permit series (Statistics Canada, Catalogue 64-001, 1963-83) which identifies the relation between conversions in the existing stock and new construction. Figure 10.1 indicates this relation for Canada as a whole and for the metropolitan areas of Halifax, Montreal, Toronto, and Vancouver for the twenty-year period. While only providing a general guideline, it does indicate the dramatic increase in importance of conversions during the late 1970s, especially in the older metropolitan areas where housing of pre-war vintage is concentrated. It was truly an urban phenomenon, with the rates for Canada as a whole increasing only modestly during this period.

An indication of the degree to which conversions impinge on the stock is provided in Figure 10.2 which is based on data from the City of Toronto for the period 1976-81. While there has been a consistent net loss of dwellings during this period, the total number of transitions has been substantial in each year. Only small proportionate changes in specific transitions would be sufficient to produce large shifts in the net change components.

The most detailed study of conversions comprised five case studies in Calgary, Montreal, Saint John, Toronto and Vancouver (Vischer Skaburskis, Planners 1979a). That study demonstrated that conversion activity was strongly dependent on the character of the local housing stock. In Calgary, for example, the importance of conversions from the existing stock was limited by the fact that most of the older properties were small and unsuited to subdivision, in sharp contrast to the situation in parts of the City of Toronto. The implication is that the initial impetus for analysis needs to come at the scale of the individual housing market; aggregation across markets, at least in the first instance, may obscure rather than clarify impacts of specific public interventions.

At the substantive level, after much conversion activity immediately after World War II involving the transition of older large homes to lodging houses (particularly in Vancouver), the frequency of conversion fell dramatically in the face of the massive efforts to increase new construction. It was not until the 1970s that conversion activity increased again; upward pressures on housing prices created both a need for some home owners to pass on the costs of ownership to a tenant and a situation in which inner-city redevelopment became cost competitive with new construction.

A major problem in Vancouver and Calgary in particular is the occurrence of "new conversions" in which newly-constructed buildings are converted to multi-unit structures prior to occupancy; this phenomenon reflects an incompatibility between current zoning and owner/developer desires for more intense land use and, hence, higher rates of return on investment. Not only are inspection and enforcement problems raised, but it takes time for corrections to be made to the property inventory in the city.

With respect to the existing stock, the problem again is primarily one of the incompatibility between local zoning or other by-laws and current market demands which leads to illegal conversions and poor information regarding the adaptability of the stock. In Saint John, the fact that rented units are taxed at a higher rate than single-family owner-occupied units is a disincentive to report conversions. In Toronto, the growth of the illegal "bachelorette" reflects both a reaction to stringent rooming-house controls in the face of strong demand for singles accommodation and a desire to avoid the parking requirements associated with many inner city neighbourhoods. At the same time, as many units are added to the stock, however, these units also become targeted to a specific socio-demographic group, and the larger units they replace become unavailable for other low-income groups such as childless couples and small families. This discussion suggests, therefore, that some measure of dwelling size might be particularly important in assessing the effects of conversion activity on the availability of housing for specific groups within the city.

TENURE TRANSFERS

Many conversions or other modifications or improvements are associated with shifts in tenure. Much of the early post-war conversion activity involved transitions from previously owned single-family structures to multiple-unit rented apartments and, in some cases (especially in Vancouver) to lodging houses. A subsequent upsurge in conversion activity, however, has seen an increase in transitions in the other direction, from rental property back to owner-occupancy. Between 1971 and 1981 there was a small, but widespread, decline in the proportion of rental units, particularly among single family dwellings.

The small net shifts in rental proportions, however, are the outcome of a more dynamic situation as indicated above for Toronto in Table 10.1. In that sample, 30% of the units changed tenure status, and significant shifts were experienced from owner-tenant status to solely owner-occupied. As Steele and Miron (1984) have argued, the growth in condominium conversions and other transfers from rental to owner-occupancy reflect the increase in incomes, particularly among urban two-earner families, and the associated desire for ownership, tax advantages from home ownership that increase with inflation, the imposition of rent controls, and general market pressures for redevelopment. To pursue the analysis more rigorously, we would have to show how specific transition rates changed as a result of any one factor. Without a systematic accounting procedure, it is difficult to assess the direct and indirect effects of public interventions. While it is possible to show that direct imposition of condominium conversion controls, for example, constrains some kinds of transitions, it is difficult to identify how the stock might adjust in other areas or sectors to consequent market pressures.

CHANGES IN QUALITY

Changes in housing quality arise from four factors: the addition of new units of higher quality, the demolition of units of lower quality, the progressive deterioration of the stock with age, and investment in improvements. Unfortunately, there is little empirical evidence to measure such transitions. For example, it is often claimed that urban renewal failed to remove the poorest housing, but no data exists to assess the rates at which different quality units were removed from the stock as a whole in any city. The existing data does focus on two issues: the net effects of all four factors and the amount of investment in the existing stock, whose effects are assumed to be positive.

With respect to the first issue, the evidence on overall improvement in quality is clear. Although there was debate over measures of quality in the Census returns of 1951 and 1961 (see Dennis and Fish 1972, 42), the 1971 review of housing by CMHC showed that the number of units in need of major repair had declined from 545,000 in 1945 to 118,000 in 1970. Similar improvement was reported with respect to the percentage of units with basic amenities such as piped water, flush toilets, and separate bathrooms and kitchen. Numerically, the major impact on aggregate quality stemmed from additions and removals rather than investment in upgrading. However, as the major deficiencies were reduced, the more subtle battle against progressive deterioration became joined.

Since 1944 the National Housing Act has been "an Act to Promote the Construction of New Houses, the Repair and Modernization of Existing Houses, and the Improvement of Housing and Living Conditions." However, actual spending under the act was dominated by new construction programs for a quarter of a century and even urban renewal provisions were primarily interpreted as simply meaning slum clearance (Dennis and Fish 1972). The main federal rehabilitation initiative was in providing guarantees for home improvement loans from approved lenders. Among provincial governments, only Quebec showed interest in developing its own rehabilitation initiatives.

Direct entry into the rehabilitation grant and loan arena did not occur until the implementation of RRAP in 1973. RRAP expenditures grew rapidly through the 1970s, peaking in 1983 and then declining (Figure 10.3). Unlike conversions, publicly-subsidized rehabilitation became disproportionately represented in small towns and rural areas, particularly in Quebec. In 1975, the first full year of the program, 91% of both owner-occupant and rental loans were in urban areas; by 1981, 67% of owner-occupant and 22% of rental loans comprising 58% of the total loan value were to rural areas. In the dominant owner-occupant sector, 44% of loans were made in Quebec. Even though the RRAP program emphasizes the rural component, the amount of renovation activity is likely to be underestimated to a greater degree in rural than urban areas because of the greater incidence of sweat equity in rural compared to urban areas.

The degree to which transitions to better quality have been achieved by these programs is open to some debate. While, in general, there is no question that the money has been allocated to the appropriate target groups and spent on work of acceptable quality, half of the owner-occupant RRAP units and more than 40% of the rental RRAP units still possessed substandard elements after work was completed (CMHC 1986c). Since the primary recipient group is the older home

FIGURE 10.3 RRAP units by urban/rural areas and tenure: Canada, 1974-1985.

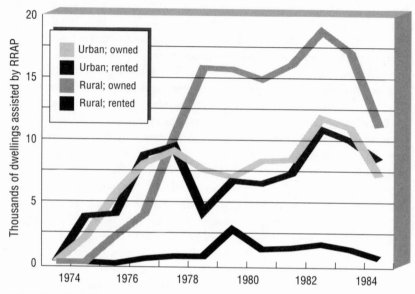

SOURCE: CMHC special tabulations, 1986.

owner, the question also arises as to whether the loan expenditures constitute real improvements in the expected life of the unit or merely concentrate future maintenance expenditures at a particular point in time. Evaluation of these longer-run consequences depends on building appropriate data series with a longitudinal component.

In contrast to the transition to higher quality is the progressive deterioration of the stock. Many earlier studies of changes in quality were concerned with the concept of filtering which comes from the "observable process in which a dwelling occupied by a higher income household depreciates and is passed down to a household with a lower income" (Davies 1978). The concept thus embraces both change in condition and change in occupancy and is perceived to have positive redistribution benefits (Sharpe 1978). However, few have examined the changes that actually occur in the buildings, namely the deterioration, obsolescence, and transitions in value that accrues to particular buildings in particular locations. Often the idea of filtering has been accepted uncritically and analysis pursued with simulation models based on inadequate data. During the late 1960s studies of filtering were popular, as analysts attempted to depict and explain the process by which the stock deteriorated with associated changes in occupancy from higher to lower income groups. The studies tended to be marred by the lack of empirical evidence and often resorted to an aggregate, ecological approach using census data on housing value (Maher 1974). This lack was in part due to the cost of establishing a proper accounting system, in part to the time needed to develop the data series as well as the commitment of the analysts and their client policy makers to market processes and allocation systems. Filtering had to work. It was the only way the market would supply housing to most people, and if it did not work, the virtue of housing markets would be suspect. An ideological commitment to the concept of filtering encouraged the theoretical study of its efficiency.

Now there is strong evidence that even the dominant downward direction of the transition process embedded in filtering theory lacks empirical support. Rising incomes, declining household sizes, and changing locational preferences have led to the deconversion of multiple converted dwellings and the regeneration of old neighbourhoods. No longer is filtering producing an "excessive supply" of low-priced housing. The reversal of the process in many neighbourhoods (Laska and Spain 1980) illustrates the durability of the stock and its adaptability; at the same time, it reduces the housing supply for lower income households as both rich and poor engage in a predetermined battle over whose dollars will move the market the most. The ideologically-inspired view of filtering, showing how the poor get better housing through the operation of the market, is no longer tenable and forces us to re-examine past assumptions.

The Data Issues

Discussions of how to measure "progress" in the housing field or how to evaluate the effects of public interventions are replete with comments regarding the inadequacy of available data. A similar case can be made, only more strongly, in regard to transitions. As has been argued elsewhere (Moore and Clatworthy 1978; Moore 1980), there is a fundamental relation between the types and quality of data collected and the substantive questions which can be addressed in a planning or policy context.

Prior to the development and use of property-based information systems, data for studies of housing dynamics were obtained from three main sources: (1) the decennial and mid-decade censuses which could be used to provide estimates of conversions using "residual" methods (Skaburskis 1979); (2) event data – primarily building permits, rehabilitation loans records, and sale-resale data – which were used to identify change data directly (for example, Statistics Canada Catalogue 64-001 Series; the CMHC Annual Housing Statistics Series; Morrison 1978); and (3) assessment files (McCann 1972) and city directories (Peddie 1978). Each source has limitations in terms of reliability, bias, and coverage (Vischer Skaburskis, Planners 1979a), although some sense of dynamics can be provided from a judicious use of these sources.

A good example of data concerns is provided by the Vischer Skaburskis, Planners (1979a) study of conversions in five major cities. The authors encountered serious problems in trying to measure conversions in each city, although they tend to be compounded in higher density, inner city environments.

Table 10.2
Conversions from single to multiple units and residual error
in federal housing statistics

| | | Residual error in housing counts | | |
	Conversions	Single units	Multiple units	Total units
	(a) By province, 1980			
Newfoundland	44	1,386	-713	673
Prince Edward Island	0	3	-142	-139
Nova Scotia	209	906	-1,088	-182
New Brunswick	141	635	304	939
Quebec	1,072	9,871	-3,222	6,649
Ontario	576	9,740	-12,730	-2,990
Manitoba	5	695	-1,219	-524
Saskatchewan	11	1,585	-2,461	-876
Alberta	94	5,892	-4,444	1,448
British Columbia	376	2,955	-3,955	-1,000
	(b) For Canada, 1971-1980			
1971	1,605	-6,598	8,447	1,849
1972	2,321	-13,222	12,134	-1,088
1973	2,441	-15,759	13,202	-2,557
1974	3,417	-17,678	12,168	-5,510
1975	2,073	-15,934	11,577	-4,357
1976	1,931	28,843	-25,684	-3,159
1977	2,242	49,061	-50,471	-1,410
1978	2,137	42,062	-52,368	-10,306
1979	2,442	44,122	-42,245	1,877
1980	2,528	33,668	-29,670	3,998

SOURCE CMHC special tabulations 1971-1981.

Building permits constitute the prime source of data; in many cases, the work to be done is not set out on the permit, making it difficult to tell if new units are being created. Inconsistencies exist between dwelling-oriented and property-oriented records in different cities. Finally, the fact that many conversions, particularly the addition of basement apartments, are illegal erodes the confidence in many published counts of conversion. For example, in 1978 Vancouver had a list of over 6,000 properties which were identified as having illegal conversions; Calgary had many illegal duplexes and quadruplexes which were being created immediately after completion of new units built for lower densities. This recording problem will continue to be of concern and its impact needs to be assessed in the context of any data development scheme, whether automated or not.

At the aggregate level, published statistics provide little useful insight into the dynamics of change. Table 10.2 provides examples from CMHC statistics on conversions by province. A simple equation should link the stock in successive years (S_t, S_{t+1}) to the number of completions (C), demolitions (D) and conversions (V); in other words, $S_{t+1} = S_t + C - D + V$. Yet, when this is done, there is a residual error (R) which indicates the degree to which the equation is not reconciled in each year. Table 10.2 shows not only that this residual error is large relative to the recorded number of conversions but that the relative error increases when the stock is separated into single and multiple unit structures. Furthermore, even the direction of the error has changed during the 1970s, with the number of single units overestimated prior to 1976 and underestimated afterwards as the momentum of deconversions increased.

The issue here is that without a greater commitment to measuring the transitions in the stock, as opposed to marginal distributions, how are we to identify the mechanisms which contribute to change? The further implication is that we cannot assess the longer-run consequences of current dynamics or the impacts of interventions whose primary function (as opposed to intent) is to change transition rates rather the resultant distributions.

The Development of an Accounting System

In pursuing the goal of improving the measurement and monitoring of transitions, two related challenges have to be met: (1) to establish appropriate procedures for measuring transitions such that we can document change at different spatial scales and for different segments of the housing market within a time frame appropriate to the analysis of public policy impacts; (2) to build a theoretical structure for integrating processes of physical conversion, tenure change, and investments in residential improvement within a more general theory of housing supply-demand relationships.

These two challenges are interdependent for, on the one hand, one cannot design appropriate measurement and data collection procedures without some idea of the theoretically relevant categories for which data are needed while, on the other hand, theoretical development is dependent on the availability of data on the nature and magnitude of transitions in the housing stock.

An important step at the scale of the local jurisdiction is to establish a set of housing accounts which identify the stocks and flows within the housing market (Byler and Gschwind 1980). The motivation for the establishment of accounts at this scale is that mismatches between supply, which is largely immobile in the short-run, and demand occur at the local level and "making progress" is essentially defined by the ability to promote more effective utilization of the stock (Merrett and Smith 1986).

Many of the effective instruments for generating change in the local housing stock, such as zoning by-laws, building permits, and inspections, are exercised locally and with some specificity with regard to the neighbourhood, structure type, and tenure. Assessing the effectiveness of specific local interventions requires that changes in the stock be monitored both before and after intervention; in particular, an evaluation is often needed of the types of transition being generated by new sets of controls, regulations, or incentives. The ability to undertake such evaluations is dependent on building appropriate data bases capable of documenting the relevant transactions.

An important corollary of this perspective on local intervention is that housing accounts need to be built from the "bottom up"; the foundation of the accounting system must comprise effective local accounts which contain the capability of monitoring local change. The ability to produce aggregate accounts for larger regional entities depends on the consistency of accounts across lower-level jurisdictions. The role of central agencies, whether federal or provincial, is vital since consistent aggregation depends on rigid definitional standards being established for a wide range of measures (Byler and Gschwind 1980). The central agencies must take an active role in setting these standards.

The ability to build a coherent set of accounts is a task which can only be contemplated with modern information system technology. The number of annual transactions even in a modest-sized housing market is beyond the scope of most local authorities to handle without recourse to computing facilities. To combine accounts across jurisdictions requires not only access to further computing resources but also demands that a wide range of definitional issues are reconciled. This is unlikely to occur without significant leadership from a central agency such as CMHC.

The basis of any accounting framework is the identification of the units of observation and the categories into which the units can be placed at any given point in time (Byler and Gale 1978). In the case of housing, there are a number of possibilities which reflect different questions of concern. For example, analyses of quality change are most readily undertaken using the dwelling as the basic observation, while physical conversion is defined most appropriately at the structure or property level. Two factors favour using the property as the basis for accounting. First, most investment decisions are made with respect to the property rather than the dwelling, and therefore, this approach has greater theoretical appeal. Second, the building block of local information systems is the assessment or property file.

The basis for future accounts is beginning to emerge. A number of cities and regions have developed computerized property information systems (Mclaughlin et al. 1985; McMaster 1985; Nuttall and Korzenstein 1985). To date, however, the use of these systems has been directed primarily at administrative as opposed to planning functions as these are seen to generate the most immediate returns. The design of each system, while taking advantage of technical innovations of a more general nature, is tailored for local needs. As yet, there are few widely accepted definitions to secure the comparability of data across jurisdictions and form the basis of sensible aggregation. More importantly, however, the requirements of administrative systems do not necessarily guarantee the quality of data for planning purposes. Only through the demonstration of the utility of data obtained from various sources as inputs to decisions can reliability be improved. The development of effective accounts is not just a question of commitment to technology but also to the infrastructure needed to provide reliable inputs to that technology.

Why are we concerned with more detailed accounts at all? Surely, the mountain of statistics currently produced by federal, provincial and municipal agencies is sufficient for decision-making purposes. As long as the emphasis is on increasing the supply of housing to meet growing demand under expansionary economic and demographic conditions, existing statistics are probably sufficient. This growth scenario dominated most of the 1950s and 1960s; however, in the 1970s, and particularly towards the end of the decade, changes in the demographic character of the nation, coupled with a slow-down in the economy, produced conditions in which adjustments in the existing stock became more important. This was especially evident in metropolitan areas and was important not only in absolute terms but also, and perhaps more significantly, in the public policy domain.

As population growth has slowed, household growth has been sustained by increased propensities for both young and old to live alone, coupled with increasing rates of separation and divorce (Miron 1988). At the same time, as demand for smaller dwellings has grown, the rise in two-income families has produced an increased spread in the nation's income distribution; average income has risen, but so has the proportion of the population in financial difficulty. We see, therefore, both more demand for specific forms of ownership such as condominiums and greater pressures to provide sufficient housing for low-income households.

The variation in housing conditions from one locale to another means that aggregate statistics for the region or nation often provide a poor guide to what is happening in specific communities; they need their own data to inform local decision-making. Not only have the demographic and labour force pressures been felt predominantly in metropolitan areas, but rates of growth, increases in house prices, and the adaptive potential of the housing stock varies appreciably between cities. For example, newer urban areas such as Calgary have far fewer old large properties capable of being converted to multi-unit structures (Vischer Skaburskis, Planners 1979a). Meanwhile, as is illustrated in Figure 10.3, the rural stock is also undergoing change although its character is somewhat different from the city with a greater input of sweat equity.

The essence of measuring progress is more subtle that in earlier growth periods. National progress is essentially a collage of local successes and failures, for an aggregate gain assumes that a loss or deficiency in one locale can be offset by a gain elsewhere. At least in part, progress must be measured in terms of the development of policy instruments which facilitate the better adjustment of detailed housing supply and demand in specific locales, as market conditions change rapidly and mobility cannot resolve states of disequilibrium.

Progress in the sense defined above depends critically on developing a better understanding of the relation between stock flexibility or adaptability and supply-demand relations in individual housing markets. At present, the ability to assess whether specific intervention strategies, such as CHIP or CHRP or provincial rental conversion programs are being effective is largely informal; if planners are to determine whether particular by-laws, zoning ordinances, or other interventions are having desired (or undesired) effects, there must be a commitment to more critical levels of measurement over longer periods of time.

Public Policy and Housing Accounts

Local governments must cope with changing housing needs and mismatches between local supply and demand. On many policy issues, net measures are simply misleading. Local governments need a sound base for monitoring transitions in the stock and for identifying the contributory courses of change. Furthermore, modifications to the existing stock are becoming progressively more important, although to different degrees in different jurisdictions. We need to understand better the processes of physical conversions and investments in improvements, and the types of interventions that will promote effective stock use. Even large-scale surveys, such as that undertaken by the Community Development Strategies Evaluation (1982) of the US community block grant program, have been less than conclusive, for the truth is that surveys just cannot identify the richness of local environments which interact with the wide range of possible local interventions.

One path to the future is to make a stronger commitment to the integration of monitoring, evaluation of program impacts, and detailed analysis of housing dynamics in individual markets. This also implies sufficient standardization to permit extraction and aggregation of selected accounts. A number of jurisdictions have already made a start at this. What is needed is a central set of guidelines with respect to issues, such as common definitions of housing categories, methods of linking dwelling and property data, and procedures for improving the quality of building permit data. Only when comparability is established can the relations between interventions and account parameters be analyzed across jurisdictions and over time.

To be better able to cope with localized mismatches between supply and demand, local governments must be able to identify the conditions under which specific controls (such as zoning changes or rigid code enforcement) or incentives (such as rehabilitation subsidies) are effective. This is beyond the ability of existing theory or of survey methods to achieve in a cost-effective and practical manner. Improved measurement has had an impact on theory and practice in most areas of scientific activity. Economic arguments about household behaviour, for example, were revolutionized by large-scale micro-level household surveys of the 1960s that led to advances in the modelling of disequilibrium (Hanushek and Quigley 1978), while the development of the Survey of Income and Program Participation (SIPP) in the US had impacts on theories of family and household behaviour. To improve theories of housing dynamics and the impacts of stock intervention, analysts need data on housing accounts: the same data needed by local governments for day-to-day stock management. For provincial and federal governments, the primary concern should be to improve the allocation of public funds and to ensure that policies and interventions are appropriate to the stock management problem.

Central agencies can play an important role in the development of housing accounts. They can provide support to municipalities in system development, help integrate the adoption of common definitions and standards, and favour submissions for program funding by local governments that make use of a systematic accounting of local housing conditions and change. It is to be hoped that such initiatives will be taken by federal and provincial agencies in Canada.

Notes

1 In the Toronto case, if the structure of transitions were to remain constant for the indefinite future, eventually 71% of properties would be owner occupied, 18% tenant-occupied, 9% owner-tenant and 2% vacant.

CHAPTER ELEVEN

Housing Form and Use of Domestic Space

Deryck W. Holdsworth and Joan Simon[1]

POST-WAR CHANGES in the housing of Canadians are manifested principally in three respects: the changing size of the suburban single-family house and its varying layout in subdivisions; the increasing proportion of residents living in high-density, and increasingly high-rise units; and changing programs of social housing that have similarly varied in scale, massing, and density. Each of these has facade variations across the country, but generally innovations and standards developed in the Toronto region have provided the model for national plans (with the exception of a "west coast" variant that percolated from Vancouver onto the Prairies, and the persistence of a Montreal style of duplex housing).

These changes in housing form can be seen in four broad time periods: a period of housing crisis extending from 1945 to the early 1950s; a phase from the early 1950s through until 1961 in which the modern residential construction industry became established; a period through the 1960s in which the industry was confidently meeting goals; and a phase of retrenchment beginning in the early 1970s. In outlining the distinctive and salient changes to form and appearance among three types of housing in four periods, it is intriguing to see the extent to which (1) British, American and Canadian models provided inspiration over time; (2) the institutional/bureaucratic framework of CMHC has enhanced, prescribed, or prohibited changes in form; and (3) demographic and lifestyle changes are reflected in or influenced by changes in housing form.

Salient Aspects of the Pre-1945 Stock

The heritage of housing stock and attitudes related to housing earlier in the twentieth century is an important prelude to consideration of post-war progress. Additions and changes do not occur in a vacuum, either locally or in the broader international context. Dominating the stock of urban housing in pre-Depression Canada is a series of increasingly suburban environments for family life in detached houses (Doucet and Weaver 1985; Holdsworth 1977, 1986; Marson 1981; Spelt 1973). These could be small cottages and cabins or elaborate four-bedroom, two-storey houses; they could be on small lots of 7.6 metres

(25 feet) frontage or verdant 18.3 metres (60 feet) lawns; they could be close to workplace or separated by streetcar or automobile journey. Notwithstanding the tenements, bunkhouses, and rooming houses of central city areas, and a stock of Victorian housing that was already becoming inadequate through overcrowding or conversion, most Canadians had an image of the ideal house as a cosy cottage set within its own garden surrounds. The home thus had a physical and a psychological moat. Cheap land on the fringes of towns and cities that was accessible through an evolving road network made attainment of such an ideal feasible for most – even if, for some, the housing was minimal and completion required an investment of sweat equity (Harris 1987). Only a few Canadian urbanites embraced the notion of high-rise apartment living as an acceptable alternative. Most occupants of the apartment blocks of the 1910s and 1920s were single, and likely clerical workers working downtown; the apartment zones of Vancouver, Toronto, and Montreal were encroachments on former Victorian elite districts that have been abandoned as newer suburbs grew up (McAfee 1972).

In rural areas, generations of folk and vernacular housing had accumulated into an architecturally rich and varied stock prior to 1945. The Prairie region had the most marginal housing; many homes there remained little more than the pioneer house awaiting more prosperous times that never came. Company housing was the norm in resource extraction areas and – given the ephemeral nature of many of these settlements – was rarely innovative in style or comfort.

Throughout the twentieth century, Canadian housing design was driven by the residential construction industry's focus on the job of subdividing raw land into lots and building housing. An important but minor theme has been the attempt by reformers to get Canadians to build better communities for all and better dwellings for the poor. Theories about housing design and site layout were tied to health concerns that had both physical and social dimensions. Before 1960 the physical health hazards of slum housing dominated arguments for housing reform. As the worst housing was torn down, water and sewage servicing became almost universal, and medical advances wiped out tuberculosis and other contagious diseases, social health questions moved to centre stage.

Housing reformers attempted to establish minimum acceptable standards for Canadian workers and their families in the early twentieth century. In Toronto, the campaign to attack the environmental evils brought by industrialization and urbanization was initiated by Dr. Charles Hastings, the city's Medical Officer of Health. G. Frank Beer, a prominent Canadian clothing manufacturer, sold the idea of housing reform to his fellow businessmen on the basis that bad housing affected the physical health of workers, reducing their output; hence good housing was good for business.[2] Under Beer's leadership, the Toronto Housing Company was created, and the limited dividend concept was introduced to Canada (Spragge 1979). Their first project, Spruce Court, was built in 1914.

In the same year, Prime Minister Borden enticed Thomas Adams, the leading town planner, to move from England to Canada to advise the Commission of Conservation. Adams was a housing reformer and Garden City advocate who linked housing design and site layout. This design approach maximized the use of the available space and it is one of the visual features which initially distinguished social housing from market housing.[3] This Garden City influence can be seen in Toronto's Spruce Court, Ottawa's Linden Lea, Halifax's Hydrostone development, and in some company towns during the 1920s (Saarinen 1979; Delaney 1991).

The Toronto Housing Company's Spruce Court units are the earliest definition of minimum acceptable standards for workers' housing. Given the average monthly wage of a skilled worker, not everyone could benefit from such a scheme; nor given the cost of land in the inner city, could developers deliver such housing on a broad scale. From this earliest attempt until today, housing reformers have been plagued by the dilemma of creating adequate decent new housing at a price affordable to the low-income earner for whom the housing is intended.

Adams was instrumental in establishing criteria for the design of housing for veterans under the first national housing program – the $25 million Federal-Provincial Housing Loan scheme in 1919. These were intended to be houses which workers could afford to buy, but the spirit of patriotism that fuelled the desire to provide adequate "Homes for Heroes" was dampened by economic realities. Rural soldier settlement schemes were more numerous than urban projects and were typically built on poor land that farmers had ignored to that point. The image of a house on land was paramount, rather than higher-density urban schemes.

During the Depression of the 1930s, the question of housing adequacy and affordability came to the fore. High unemployment and poverty focused attention on the slum areas of major cities. Many observers were reluctant, however, to come to terms with the broader implications of socially-assisted housing, recognizing on the one hand housing as but one element of a broader package that included medicare and minimum incomes, and on the other hand, that the Canadian social model did not necessarily have to embrace the idea of state-provided housing. The emphasis was more on assistance to the private sector, assisting private mortgage companies and owners with joint loans (as for example, in the 1935 DHA provisions). The block unit, in the fashion of British, German, and other experiments of the 1920s, clashed with the North American ideal of single-family housing. Ironically, New Deal block housing schemes in Pittsburgh and Cleveland were to provide important models for Canadian post-war architects and planners.[4] The Garden City principles were re-introduced after being transformed by the American greenbelt experiments of the 1930s.

Wartime Housing Limited (WHL) introduced a new standard in Canadian housing design. The crown corporation drew the country's leading architectural, engineering, and construction industry talents together to design and build houses for the average working man. They applied scientific principles to the planning and manufacture of temporary housing for war workers that could be removed at war's end when the war factories were no longer needed. At the same time, they had to alleviate the municipalities' fears that the housing would deteriorate into slums. The simplified "1930s Cape Cod" designs were considered by many to be progressive, experimental, and distinctive, but by others to be "packing cases" and eye sores. These standard two or four-room, one-and-a-half storey houses created a distinctive "Canadian House" which can still be found from coast to coast (Wade 1986).

The drab vistas of rows of "little boxes"[5] were dictated by the existing servicing to the rented lots, by provincial laws requiring that houses must face onto a 20.1 metre wide street, and by setbacks that were usually fixed by local zoning. Lot widths were usually a uniform 12.2 metres to provide a 4.9 metre fire separation between houses. WHL prohibited the erecting of fences or garages, causing consternation among the tenants and contributing to the bleakness of the site.

The Post-War Housing Crisis

Veterans' housing became an urgent concern in the immediate post-World War II period, but this was only a part of the million houses thought to be needed in the decade after the war. A large residential construction program was also seen as an important source of employment by a government fearful that war's end could lead to higher unemployment and a return to the Depression. WHL was wound down and CMHC, created in 1946, began to take over responsibility for housing; a number of the staff moved to the new crown corporation. The building program quickly shifted from rental houses to home ownership. The wartime designs continued to be built, but now on (unfinished) basements because these were permanent dwellings. The "Cape Cod" strawberry boxes were similar to the architect-designed "dream homes" for affluent clients and could be related to either a French or British heritage (Page and Steele 1945). CMHC contributed to a longstanding tradition of influencing mass design by pattern-books when they developed and promoted a series of plans for bungalows and ranch-style houses in the late 1940s.

The popular "house and garden" magazines and the technical architectural press envisioned modern houses built with new materials and new industrialized technologies – scientifically lit, heated, and air and sound conditioned – but the houses actually built were traditional bungalow designs constructed in traditional ways. The demand for houses was strong, and subtlety of design was not a dominant concern among buyers. Although the houses were small, the families were soon large by today's standards. The baby boom had begun and suburbia, away from urban overcrowding with ample light and air and at least a promise of trees in the future, was seen as the ideal environment for child-rearing.

The Wildwood development in Winnipeg was an exception to this cautious approach. Wildwood was a model community that used design principles advocated by Stein and Wright and demonstrated in their Radburn, New Jersey, development (CMHC 1986d). Pedestrians and vehicles were separated so children could go to school and play without parents worrying about traffic; local services, grocery shopping, and recreational facilities were within walking distance of everyone (CMHC (1986d, 36). Veterans were surveyed to determine their preferences for size (58% wanted 3 bedrooms), style (42% preferred one and a half storeys), and heating (73% requested forced warm air) (CMHC 1986d, 42).

As the housing shortage eased and families became more prosperous, units became larger, and there was more scope for innovation and novelty. By 1951 house builders such as the Shipps in the Kingsway area of Toronto were becoming land sub-dividers, and before the end of the decade, they would be part of the new development industry. They purchased a 101.2 hectare orchard and started to build Applewood Acres just west of Metropolitan Toronto. Within four years, 800 families were living in homes built from one of the eight architect-designed plans available on MacIntosh Crescent, Russet Road, or Greening Drive. Their typical buyer was a 38-year-old salesman with an annual income of $6,600, a wife, and two children, aged 10 and 6 years. This was the second house they had owned; they had made a downpayment of $55 and were carrying a $11,000 mortgage. They owned one car (although 20% of their neighbours owned two cars), and a television set. The house was designed to appeal to the wife. White, enamelled-steel kitchen cupboards with a maple chopping block incorporated into the countertop, a stainless steel sink with a pull-out attachment for rinsing dishes, vinyl tile floors, and the hood for the kitchen stove made these "state of the art" kitchen designs. The four-bedroom brick house had a full basement and an attached garage. The 18.3 metre wide lot had been sodded by the builder and came complete with an apple tree. The new owners were likely to put in a patio and back yard barbecue. The back yard had become an outdoor summer living room. The suburbs' two schools were jammed to bursting, and the one church had to conduct Sunday services on shifts. The suburban lifestyle was in place (Fillmore 1955).

In less well-planned subdivisions, rows of bungalows continued to march across the agricultural landscape on the outskirts of most Canadian cities. But as the housing shortage eased, home buyers began to demand neighbourhoods rather than stark tracts. CMHC used its leverage as mortgage approvers to get smaller builders to improve site design. Layouts were recommended to improve the visual quality of the housing areas and also traffic safety, especially for children playing (Kosta 1957). This was an era when municipal planning was still in its infancy. CMHC sought long-term solutions to the quality of planning by encouraging the establishment of departments of urban and regional planning in universities. To fill the gap, British planners were encouraged to immigrate and their ideas and preferences would invariably influence subsequent design work.

There was little government concern for those who could not afford to become home owners. A few small apartment buildings were built by non-profit sponsors. In 1948 the citizens of Toronto voted to allow the City of Toronto to finance the construction of Regent Park North in an area which had been identified as a slum in the Bruce report of 1934 (Rose 1958). When the Section 35 provision for rent subsidies was added to NHA in 1949, the stage was prepared for an ongoing program of assisted housing. In 1951 Newfoundland became the first province to complete a project. Both Regent Park North and the pioneering scheme in St. John's, Newfoundland used a nineteenth-century design solution which had virtually disappeared in Canada after the First World War: row housing. In Toronto, the row houses were introduced into a walk-up apartment scheme because the residents of the area had complained that they were being moved out of dwellings which had front and back yards into apartment buildings. Also many of the families to be rehoused were large: five to ten children were not uncommon. Two-storey houses were the most compact design solution, and row houses were more economical than detached dwellings.

Regent Park North also introduced the notion of the "superblock" into Canadian planning. The existing development pattern was considered to be overcrowded, and overcrowding was seen as the root cause of slum housing. The clearance of large areas had been recommended in the City of Toronto Plan of 1944. Redevelopment, it was argued, had to be organized so that large areas of "blight" would be removed. The Toronto planners, supported by federal financing under the new NHA urban renewal program, sought to create a "modern type of residential development" which would have permanent amenities (Rose 1957). The requirement to retain the arterial roads which form the site boundaries was recognized because people needed to move easily across the city and to and from the site. Existing small streets were eliminated to remove the hazards of through traffic from an area designated for families; closing streets would add to the open space for play. Trees and grass were virtues needing no explanation. When it opened, Regent Park was the symbol of reform success: a new and green landscape which would help people build new and successful lives for themselves and particularly for their children, despite the wise voices who recognized that housing was only one of the social problems facing the tenants. The new housing did improve the physical health of the inhabitants, but it was also touted by some as a quick fix for social problems. The social experiment was well studied, the lessons documented, but few listened. Within a decade this redevelopment would become the symbol of failure.

Housing Mix in Community Development

During the mid 1950s the residential construction industry got going, and CMHC turned its attention to questions of quality and design. CMHC stimulated public and professional awareness of planning as a social concern by underwriting the creation of the Community Planning Association of Canada. They sought to improve the architectural design of houses by encouraging architects to get involved in house designs, and by creating the Canadian Housing Design Council to give awards for good design, they tried to raise public awareness of design. The Building Research Division of the National Research

Council was charged with improving the quality of the material and techniques used in house construction (Bates 1955).

In the private sector, Toronto's Don Mills set the standard for suburban development across the country. A neighbourhood concept was introduced with higher density dwellings near the core. Initially, these were three-storey, walk-up apartments, but as development proceeded the scale of the apartments grew. The combination of apartments, townhouses, and detached housing in Don Mills was instrumental in fabricating a distinctive Canadian suburban landscape. Unlike in the US where the work was done by engineers, subdivision layouts became a standard part of professional planning practice.[6]

Don Mills Developments also acted as the matchmaker between architects and builders of tract housing. It was a traumatic encounter for both groups. Builders had to learn the virtues of design, and architects the reality of the marketplace. The marriage, when successful, resulted in a standard of environmental quality in mass housing that was unusual in North America.

The single family housing component of Don Mills was laid out on lots that were wider and shallower than current practice, with the result that there is a sense of spaciousness, enhanced by retained vegetation and enriched by extensive planting. With increasing affluence, the standard bungalow grew larger and split sidewards. The garage was moved out of the back yard to free the space for outdoor living. Wide lots allowed room for the ranch house plan with all the rooms on a single floor as well as space for an attached car-port (or later an attached garage) beside the house. Inside, the space opened up: dining rooms became dining areas and were used for multi-purpose activities. Kitchens were "strategically" located to oversee children's play, and they increased in size to make room for the new appliances which were touted as making housework easier. Utility rooms were the latest feature with space for the new electric washers and dryers. Because oil had replaced the coal furnace, the basement space was free to become the "rec-room" for the growing children. Gradually, the rec-room crept upstairs and was renamed the family room. This new feature aided parents in their search to maintain the living room as an adult-only space which was needed as suburbanites struggled to attain a "gracious living" lifestyle.

An equally influential source of design ideas in this period came from CMHC itself, whose deliberate intentions of promoting high standards of site design to establish role models led to some intriguing mixed-density projects. In Toronto's Regent Park South, the federal extension to the city-initiated slum clearance program and in Warden Woods and Thistletown on the fringe of the metropolitan area, the British row-house model was combined with high-rise apartments. The planning and design was in the mainstream of the International Tradition of modern architecture which drew upon Le Corbusier's ideas of how to rebuild Paris for the promised new world after World War I. The public housing projects had a British flavour; indeed, many were designed by British-trained architect-planners imported by CMHC.[7] Westwood Park in Halifax and Jeanne Mance in Montreal are similar examples, in materials, scale and proportion, and appearance. On the west coast, Vancouver's Little Mountain project also used the row house, but the 224 units were stucco-clad in contrast to the brick used in the east.

In the private sector, the work of Murray, Fliess, Grossman, and later Klein and Sears fabricated a Canadian vernacular for higher-density units that was distinct from the British model. The Canadian version was designed for the private rental market. As a consequence the space norms were higher, the groupings smaller and more diversified in design, and the amount of car parking provision much greater. As early as 1955, Rogers Enterprises had identified a market composed of several groups: those who wanted to live in a house but were not prepared to buy, older people with grown children who had left home, and families with three children who could not squeeze into the standard two-bedroom apartment. Rogers had visited Chatham Village in Pittsburgh and came away convinced that this beautifully landscaped 1932 development planned by Wright and Stein would be attractive to Canadian tenants. Accordingly, Murray and Fliess designed Southhill Village in Don Mills in 1955. The split-level row house plan was a North American first. All units had their own private gardens, three bedrooms and many had a bath and a half (Bowser 1957).

A Change of Scale and Pace

By the early 1960s the established residential construction industry was poised to open a new era of construction and development across Canada, extending its expertise in the suburban market and also diversifying to take advantage of the new limited dividend (rental) schemes brought in after 1962 by CMHC. The pace of change was dramatic.

In the suburbs, big houses on big lots were the norm, although through the 1960s and into the early 1970s the constraints of land prices and servicing costs began to squeeze lot sizes. As lot sizes decreased, the garage (now frequently for two cars) began to edge forward. Semi-detached houses allowed builders to construct two houses on one lot, which was usually 15.2 metres wide. Because house buyers preferred detached dwellings, many semi-detached were split above grade, but continued to be joined at the foundation level. Escalating house prices actually contributed to the growth of the dwelling itself: people wanted to feel they were getting their money's worth. The early row-house schemes provided parking in corral areas. When the provincial strata title and condominium acts made it possible to own a townhouse, buyers demanded that the car had to be next to the house. Townhouses began to sit on top of garages. McLaughlin's schemes in Mississauga, Ontario are typical of suburbia of the time, while Bramalea, Ontario was a harbinger of the newer high-density row housing. Both were essentially offsprings of the Don Mills model, where later quadrants of the development were also adjusted to the new cost constraints and houses were located on smaller lots. In Vancouver, the growth of the house took on a slightly different appearance: sidestepping zoning by-laws, contractor/developers erected barn-like duplexes, the lower floor being justified as a unit for the mother-in-law; in reality, the house was two units, and the upper unit typically had an apartment-style balcony across the street exposure. These "Vancouver specials" filled much of the lot, and since they were usually infill housing on a vacant lot or a replacement house, they rarely harmonized with the smaller bungalows and cottages around them.

Under limited dividend schemes, developers were also locating blocks of six-storey rental apartments in areas beyond the inner city. Initially, these were rather stark and with minimal decoration on the exterior, but as CMHC gave "bonuses" for amenity items, balconies began to develop as the dominant feature of these structures. Adjustments by CMHC to the condonable size of a balcony that would qualify as an amenity make it possible to date many of these rental apartment blocks. At first, the balcony was barely wide enough for a potted plant; then, after 1977 balconies had to be at least two metres wide and partly recessed in order to qualify.

Perhaps the most dramatic aspect of this new development phase, and one that would eventually raise alarm bells in community and political arenas, was the clustering of these apartment blocks. No longer six-storey clumps but twenty-or-more storeys high (facilitated by the development of the hammer-head climbing-crane), the resultant densities soared to 300 persons per acre in Vancouver's West End and Toronto's St. James Town. These new buildings were aimed at a new market, the "swinging singles": young adults in this era wanted to get out of the suburban family home and live on their own, preferably near urban attractions. Many newly-weds were attracted too by the amenities and location before starting a family. Over the years the formula was refined, and the apartments were marketed to a more mature and affluent population. As the apartments soared and densities increased, community open space, shopping, and transportation facilities became over-loaded. Many municipalities began to question the economic cost to the community of this form of development. Multiple-unit starts surpassed family-housing starts, and many citizens began to question this form of living, with concern being particularly strong about the effect of high-rise living on children. Internationally, it became the agreed norm that young children should not live higher from the ground than they could comfortably climb: four storeys. As a result, the public sector moved to curtail the construction of high-rise family apartment buildings, but the private sector continued to build them.

Apartment layouts were dictated by car widths. The structural grid was determined by the most economical dimensions for the parking garage, and these were then projected upwards: apartments were slotted into the projections of the parking spaces. Living and dining-room "L's" were wrapped around galley kitchens; the only variation seemed to be whether the two bedrooms and bath were to the left or right of the apartment door.

Throughout the 1960s apartment buildings continued to be sited on large lots and well back from the street. In lower-cost projects, the uncovered land was often a parking lot, but in more luxurious developments this space was used for increasingly elaborate recreational facilities and extensive landscaping. Frequently, the only features which distinguished the buildings from one another were the details on the balconies and the design of the fountain by the front entrance. The physical design on buildings changed little over the next decade, but the economics changed radically. It became increasingly difficult for private developers to build rental accommodation to serve the low end of the market. Some developers turned high-rise buildings into condominium projects and tried to attract a luxury market. For example, the Shipps, still building in the same area of Toronto, built a 442-unit high-rise tower called Applewood Place. They were still selling a lifestyle concept, but now to singles and childless couples. The closed-circuit television and sophisticated electronic security system was a selling feature. Residents had access to hobby rooms, party rooms, separate card rooms for adults and young adults, two health clubs, a roof-top swimming pool, whirlpool, sauna, gym, and sundeck. Outdoor recreation facilities included three tennis, two badminton, and two shuffleboard courts.

In the public housing arena at the same time, CMHC pushed for a high standard of design. Good design was seen as a way of overcoming public opposition to this form of development. Some outstanding projects, such as Malcolm Park in Vancouver or Alexandra Park in Toronto were created. However, there was a basic ambivalence. Many, including those working in public housing authorities, believed that poor people should live in poor housing. Public housing could not look good – row housing and point blocks rather than single dwellings were inevitably the norm – and cost constraints limited attempts to provide attractive settings for such shelter. The scale of redevelopment projects triggered community debates about the appropriate density of new housing as well as the preservation of existing neighbourhoods. In Toronto, Trefann Court, the only remnant remaining uncleared from the two massive Regent Park projects, came to symbolize the public's reaction against the bulldozer approach and a transformation of emphasis from renewal to rehabilitation (Fraser 1972). On the fringe of the city, large-scale, high-density point blocks at Jane and Finch brought what social planners had so far labelled as "inner-city" social problems to peripheral locales (Social Planning Council of Metro Toronto 1979).

The Search for Vernacular Conformity
From this decade of rampant growth and some excesses of scale and juxtaposition, the period of the 1970s began a phase of retrenchment. Inflation began to nibble away at the suburban dream. Between 1971 and 1976, the price of the average house doubled in a number of Canadian cities. The land component of the house price was identified as the primary cause of the price escalation. Suburban sprawl was attacked as wasteful, and smaller lots were advocated to produce more efficient subdivision layouts.

To keep alive the middle-class expectation of home-ownership, both federal and provincial governments experimented with assisted home ownership schemes. Individual families responded to the higher prices by housewives returning to the paid labour force. The new family lifestyle spurred the return to urban living and encouraged the creation of infill housing schemes.

Socio-political as well as economic-political pressures brought forth a set of smaller-scale, neighbourhood-sensitive schemes, almost the only things tolerated in the face of earlier excesses. New federal programs created cooperative and non-profit housing programs. Heritage and tenant groups in the Milton Parc area of Montreal combined forces to stop the march of Concordia Estates' high-rise complex, La Cité. Using mortgage funds provided by CMHC, 700 dwellings were bought from developers who had planned to demolish the buildings. Resident-owned and controlled cooperatives rehabilitated the tenements and built new small-scale, in-fill projects with the goal of retaining the existing population in the revitalized inner-city neighbourhood (Helman 1981).

In Vancouver, the architectural firm Downs/Archambault attempted to adapt the west-coast shed style that had become popular for expensive North Shore houses into clusters of social housing (for example, Champlain Heights). On the South Shore of False Creek, the city used CMHC land-banking funds to transform a derelict railroad and industrial area into one of the city's most desirable neighbourhoods. The pods of stacked townhousing and medium-rise apartments, interspersed with parkland, are arranged along a sea-wall promenade. Cooperative housing for low and moderate-income families is juxtaposed with luxury condominiums (Hulchanski 1984).

The St. Lawrence neighbourhood, Toronto's post-industrial redevelopment scheme, also used federal land-banking, as well as the cooperative and non-profit housing programs, to revitalize a decayed industrial area adjacent to the city's core. Mews townhouses and mid-rise apartments, built at twice the density of False Creek, line streets laid out to echo the traditional grid street pattern. The new neighbourhood has attracted young working professionals as well as the low and moderate-income families targeted by the government housing programs.

Under the non-profit program, community groups across the country have been challenged to meet the needs of special groups. During the 1970s a new type of accommodation appeared, such as Transition or Interval houses that provided refuge for women and their children from violent family situations while they re-establish their lives in a new permanent home. Finding a decent place to live for women, especially those with children, has become problematic as vacancy rates in many Canadian cities have approached zero. Cooperative housing has been particularly attractive to single parents to cushion the downward mobility which is frequently associated with divorce (Simon 1986). Housing specially designed to meet the needs of the disabled have also been created using the federal housing program and the needs of those requiring the more supportive living environment of group homes have been addressed.

High-density low-rise became the fashion, but rising urban land values imposed increasing densities. The miniaturized versions of traditional housing which resulted (smaller units, smaller gardens, reduced neighbourhood amenities) continued the trend to internalize and privatize space which traditionally had been open to the whole neighbourhood (Simon and Wekerle 1985). As of the mid 1980s some schemes went back to the high-rise form but without the 1960s landscaping that went out of fashion in the architectural profession. One of the most visible facade changes was a transition from the "bonus balcony" to an enclosed "solarium" (named a "Florida" or "Hawaii" room depending on the local sun-destination), perhaps a sensible acknowledgment of the reality of the Canadian climate. The extended use of glass was accompanied by other glittering details inside and outside the unit. Private developers also rediscovered the city centre was a place in which couples wanted to live. Infill townhouse projects as well as condominium construction housed the more affluent. Gentrifiers were upgraded to a new rehabilitation and restoration industry as individual sweat equity gave way to more widespread and well-financed transformation of urban neighbourhoods. The Anglo-Irish slum district of Cabbagetown was the first Toronto neighbourhood to be gentrified. Typically, interior walls were removed and spaces opened up, then refitted with European-style kitchens and spacious bathrooms as stage sets for conspicuous consumption. As the well-heeled repossessed the central locations, the poor were pushed out into less-central locations. The process has been repeated across Canada in pockets of desirable and historic (at least in local terms) neighbourhoods such as the South End in Halifax or Kitsilano in Vancouver (Ley 1986).

Gentrifiers were not alone in renovating inner-city Victorian housing. Postwar immigrant groups, such as the Italians and Portuguese in Toronto and Montreal and the Greeks in Vancouver, contributed to the retention and upgrading of earlier housing stock. They too altered the fabric, both inside and out, in making a more appropriate setting for home and neighbourhood life. Thus, the sweat equity approach to house building and house renovation, so important to rural Canada, is now also an important segment of the urban mix. Indeed, it helps to define the vernacular quality of this phase.

As of the mid 1980s there was the beginning of a *nouveau* suburban revitalization, with big two or three storey houses filling small lots in inner suburbia (Dunbar Heights in Vancouver and East York in Toronto, for example). Again, the process involves a mix of individuals (often contractors exploiting informal sector labour) and more organized property capital. Like the earlier Vancouver specials, these new houses added a jarring feature to a relatively homogeneous low-scale suburban streetscape.

Suburbia did not decline. Erin Mills, the offspring of Don Mills, together with its clone, Meadowvale, continued to set the standard for well-planned suburbia. The lots got smaller and the garages doubled. Now the streetscape is a 6.1 metre expanse of paving in front of a two car garage, with a token strip of planting to remind one of the traditional front yard; the front door is almost in an alleyway. Inside, the number and size of bath rooms have proliferated. When houses were small, careful planning for the efficient use of space was an important criterion. Today, dream fulfilment seems the design determinant; curving staircases link expansive front hallway to the private domain of upstairs bedrooms, and route visitors to elaborate kitchen/dining rooms, rec rooms, and living rooms. New housing in suburban tracts of Richmond, near Vancouver or O'Brien's Hill on the edge of St. John's, Newfoundland and any city in between exhibits the same garage-dominated mass; the same revivalist tudor/Spanish/colonial details are thinly added to the remaining front facade.

Rehabilitation and renovation of older housing within the city have returned older neighbourhoods to families. The appeal of urban attractions combined with the schedules of working wives made city-living popular. Condominium construction in the inner city has assumed a strong neo-vernacular flavour as part of the post-modern fad invoking various streamlined and glittering moderne/art deco signifiers. This is noticeable especially on the Fairview slopes above Vancouver's False Creek and in Toronto's Harbourfront development.

The Future [8]

The single-family suburban house is still undeniably a goal for many Canadians. That new households have to work their way up the housing ladder from starter-homes (that are now likely townhouses rather than the strawberry box semi-finished house of the early 1950s) and that they may trip through mortgage overload or divorce (and stay forever in a townhouse or apartment complex) does not deny the fact that the industry and consumer alike still view the home as castle. Agricultural land is still being transformed, suburban "communities" are manufactured, and streets of new housing slowly develop the subtle signatures of occupancy rather than mere purchase. Insofar as several decades of mixed-density and mixed-tenancy models have evolved for Canadian suburbia, in the wake of Don Mills, among others, and that the current restructuring of metropolitan workplaces has developed nodes of office and manufacturing employment near the urban fringe, then the housing future may well be evolutionary rather than revolutionary. More of the same is in fact more of a mix, and the intersect of private industry production and local planning facilitating can mean that many Canadians will be able to think they have successfully attained home-ownership – or at least sovereignty over some semi-private space. The embryonic resurgence of the bungalow and other single-storey house forms in some suburban tracts might be a hint that builders will again "feed" the starter home buyers in more conventional means, rather than just build link-detached townhouses. As such, the limits of miniaturization and internalization may well have been reached.

Even so, the other aspect of the Canadian housing mix, social housing, is likely to get short shrift. After a decade or more in which assisted housing in the form of cooperative schemes of various kinds has been perceived in some quarters as being subsidized housing for trendies in inner-city locations, the broad political capital for low-income housing has been largely exhausted. What assisted housing there will be in the near future is likely to be cheaply built and almost deliberately designed to look cheap. Only when a generation of nascent slum areas develops, and with it a consciousness of indignation or fear, will there likely be another round of reformist design principles that could energize new housing programs. As such, the redefinition of basic adequacy has to be reconstituted anew. That extreme statement needs to be modified regionally, of course, and as the Canadian space-economy adjusts differentially to booms and busts in resource extraction and resource processing, what will define the social debate in municipal and provincial arenas will vary considerably. Space and massing will likely revert to earlier solutions. The hardening of attitudes towards social and tenancy mix is further reflected in less liberal postures towards children in adults-only buildings, or children near seniors' complexes.

Finally, it is likely that the renovation and rehabilitation industry will continue to grow and that existing streets and neighbourhoods will see dramatic piecemeal renewal. The densification of suburbia – replacing strawberry box bungalows with large four-bedroom-and-winding-staircase houses – will accelerate, and probably create as much tension as that which accompanied the invasion by townhouses and apartments a decade or so ago. Those apartments and townhouses will likely need a significant input of rehabilitation money both for the public sector and the private sector stock. Rent control has likely created the need for massive repair programs, since incremental repair work funds dried up or were deliberately withdrawn. The longer those repair funds are withheld, the more likely it is that a new generation of slum housing will develop, necessitating a crisis response from various levels of government. And existing single-family housing will attract continued attention from renovators. One challenge will be to create a set of regional restoration vernaculars within the technical information industry, so that owners and developers can modernize while maintaining the historical and regional nuances of older housing stock.

Notes

1 This essay was drafted in 1986, prior to Joan Simon's untimely death. Through several months of discussion, and especially through our joint interviewing of some leading contemporary practitioners in the Canadian housing design field, we evolved the perspective that is presented here. Our evidence is different than that preferred by some data-driven social scientists, but our intent from the outset was to look at housing form and housing design in a more humanistic manner. Joan had only just begun to insert her own considerable practical perspective into this essay, and I have been loath to tamper with it in ways that would detract from what we believed an essay in such a volume should look like.

2 A half century later, when non-profit and cooperative housing programs were introduced, the responsibility for social housing would again devolve to the private sector.

3 Until recently, more expensive housing continued to use space on the lot to compensate for design deficiencies; rising land costs changed the picture.

4 Interview with Henry Fliess, architect and planner, Toronto, April 1986.

5 This and other aesthetic opinions expressed throughout this chapter draw on the

opinions of architects and planners as recorded by the authors during interviews in the winter and spring of 1986.

6 Interview with John Bousfield, architect and planner, Toronto, March 1986.

7 Interview with Wazir Dayal, architect, Toronto, January 1986.

8 Joan Simon, always pragmatic and rooted in the problems of the present, balked at predicting the future. Her comment was:

> Cheap and nasty for the poor, the only people to have targeted housing, tax-incentive schemes for the middle income and the on-going fascination with revivalist castles for the rich will explore the wonders of art deco, tudor, georgian colonial. Canadians will use the new technology to remain in the hot tub 24 hours a day with sushi delivered.

===

CHAPTER TWELVE

Substandard Housing

Lynn Hannley

HOUSING ONCE considered standard may now be considered substandard. To identify what is substandard requires some widely-accepted standards against which a dwelling can be compared. A historical approach contributes to an understanding of how and why standards have changed, and hence the perception of the substandard housing problem.

Early settlement patterns were primarily based on private initiative and not constrained by publicly-defined standards. As a result, housing was sometimes developed with little regard for public health, safety, or comfort. Shanty towns developed around most major urban areas, some by owners on an incremental basis depending upon their resources. Surface sanitary facilities were often in close proximity to the outdoor wells that provided drinking water. This soon created problems that affected the general public. In Winnipeg, for example, unsanitary conditions and outdoor privies were partly responsible for a typhoid outbreak in 1904-5.

An urban reform movement matured in the period from 1905 to 1920, with a focus towards housing health and safety standards. Municipal expenditures on infrastructure and services rose quickly with the rapid urbanization of Canada's population in the first two decades of the twentieth century. Urban centres in western Canada developed from nothing, while those in Ontario and Quebec expanded on existing infrastructure.

The extent of municipal expenditure on services in western cities is illustrated by Artibise:

In 1906, Alberta's cities spent only $2 million; in 1912 they spent $16.6 million. The total surged to a high of $36.5 million in 1913 (1982, 136).

Much of this expenditure was anticipatory, funding the servicing of subdivisions that were yet to be built. However, in part expenditures were also made to improve conditions in existing built-up areas; in Winnipeg, for example, the 6,500 outdoor privies blamed as a cause of the typhoid epidemic in 1904-5 had been reduced to 666 by 1914. By the end of World War I, sanitary conditions and services were improved within the urban core of most municipalities.

During the 1930s determination of housing standards shifted from the private to the public domain. With the passage of the DHA in 1935, the federal government became actively involved in housing issues and after the passage of the NHA in 1938 began to work on a national building code. Indicators of substandard housing focused on the physical aspects of housing; vermin, the lack of appropriate sanitary facilities, water within the unit, appropriate ventilation, and electrical power. In addition, a household's unit was considered substandard if they were living under crowded conditions. The Bruce Commission in 1934 developed two minimum sets of standards; one to determine minimum health and safety standards and the other to establish minimum internal/external amenities.

The 1941 census gathered data on the need for external repairs.[1] Data were also collected on inside running water, bath and privy, electrical power, and the type of heating and cooking fuel. In addition, households with more than one room per person were considered to be crowded. The 1941 census data, as well as the work done by local municipalities in assessing their housing stock, were used in the Curtis Report of 1944 to develop several indicators of substandard housing. One indicator was the need of a unit for exterior repairs (as defined by the 1941 census), a flush toilet, or a bathroom.[2] The Report also recognized that, while a unit's condition may not be substandard, the neighbourhood may be.[3]

As the physical standards of housing increased across the country, expectations about "acceptable" housing changed. In the 1951 census, the definition of "in need of major repair" was modified to include an interior badly in need of repair, that is, large chunks of plaster missing from walls or ceilings. The 1971 Federal Provincial Task Force on a Developmental Approach to Public Assistance proposed indicators of physical adequacy and occupancy standards that were considered necessary for safety, health, social, and personal well-being. Rather than just using the number of persons per room to determine these, the Task Force took into account household type, age of occupants, per capita floor area, minimum floor areas for specific interior spaces, and access to play space by children.

Substandard Housing Prior to 1945

The Curtis Report found that much of the Canadian housing stock in communities of over 30,000 in 1941 was substandard. Using three indicators (needs external repairs, lacks exclusive use of indoor flush toilet, and lacks exclusive use of installed bath), 31% of the total stock were classified as substandard. The stock in centres of under 30,000 was also less than adequate; 25% (of 626,466 dwellings) needed major repair; 31% lacked a flush toilet; and 44% lacked an installed bath. Of 469,247 rural non-farm dwellings, 40% had no electricity; 69% lacked a bath; 67% lacked a flush toilet; and 30% were in need of major repairs. In addition, 39% of 729,744 farm dwellings were in need of major repairs, and only 20% had electricity. While some housing can be brought up to standard through simple renovation or the installation of a bath or toilet, in all 298,000 substandard dwellings required immediate replacement according to the Curtis Report.[4]

Regionally, the incidence of substandard housing was higher in the Prairie and Atlantic regions, parts of northern Ontario, and selected communities in Quebec. In part, this difference was due to a lack of municipal services (that is, piped running water) necessary for flush toilets and other bathroom facilities. In other cases, dwellings were built initially with an indoor flush toilet, but added installed bath facilities only later as the household could afford it.[5] As well, the metropolitan stock was in greater need of replacement.[6] The metropolitan stock was older, originally of poor quality, and some of it had deteriorated to the point of being considered slum housing, while the stock in the smaller communities, much of which had been built by home owners, was newer and of better quality.

The substandard housing problem prior to 1945 was not just one of poor quality units; it was also one of poor quality neighbourhoods, crowding, and unsanitary conditions. Despite progress made prior to 1945 in municipal servicing and the articulation of physical standards for housing, limited residential construction during the 1930s and World War II, as well as the growing demand for housing,[7] left many households living in substandard conditions.

Many Canadian households with low incomes were living in poor quality housing under crowded conditions. The Curtis Report estimated that approximately 50,000 or 28% of low-income renter households were crowded (based upon an occupancy standard of 1 person per room including the kitchen). Some 40% of households with annual incomes of under $499 were crowded, compared to just 12% of households with incomes over $2,000. The Report estimated that 150,000 units were needed to house crowded families (110,000 in metropolitan areas and 40,000 in the smaller communities.) As well, 44,000 units were required to house crowded non-family groups (32,000 in metropolitan areas and 12,000 in smaller communities.

Post-War Housing Policies and their Effectiveness

The Curtis Report proposed the development and rehabilitation of housing to service the needs of all Canadians regardless of income (a plan that included low-rent projects, cooperative housing, and private ownership), the institution of comprehensive town planning, and the introduction of a residential section of NRC's Building Code. It suggested that the provision of municipal services, the setting of building codes and standards, and housing production were the best ways to attack the substandard housing problem.

Although the preamble to the National Housing Act of 1945 indicated that it was "An Act to Promote the Construction of New Houses, the Repair and Modernization of Existing Houses, the Improvement of Living Conditions, and the Expansion of Employment in the Post-war Period," initial post-war programs focused only on production of new housing units. During the early post-war years, the physical quality of housing was affected by four policy directions that were being pursued by CMHC. These included the development and management of public housing, urban renewal/redevelopment, support for privately owned and operated limited dividend corporations, and support for home ownership.

PUBLIC HOUSING

The first housing units developed and managed by CMHC were for veterans and their families. Of relatively good quality for their time, they were seen to set a standard of appropriate housing and neighbourhood development. Although some thought that these publicly owned and operated dwellings should be converted to low-income units once they were no longer needed for veterans and their families, they were eventually sold off. This program was terminated in 1949 in favour of a low-rent housing program shared 75% to 25% (capital costs and operating losses) between the federal government and a province or municipality. The first project under this program was completed in 1951, and only 11,624 units were built altogether from 1951 to 1963. After the program was amended in 1964 to provide for a loan of 90% of the capital costs of the project and 50% of the operating losses paid for by CMHC, production surged. From 1964 to 1978, 145,183 dwellings were built under the 90% federal loan program.

Although public housing was likely more adequate (physically) than the existing stock it replaced, it was not without problems. Public concern was raised about building size and quality, as well as social segregation. Even existing tenants were not satisfied with the product. In a user study of public housing, Martin Goldfarb Consultants identified a number of concerns. As illustrated by some of the points raised by the tenants, a physically adequate dwelling unit still did not adequately meet their needs.

> Alone, shelter itself is not a solution to the housing problem; tenants are looking for a "home"....
>
> People accept public housing as a last resort and expect to stay for only short periods of time. Prolonged duration of residence increases frustrations regarding the lack of a "home." (1968, 37)

Often, from the tenant's perspective, the problem with public housing is that it fostered social dependency, rather than encouraging self-reliance and creativity. This situation is succinctly relayed in the following statement from a study conducted of Vancouver public housing tenants:

> Tenants of public housing, unlike tenants of privately owned houses, are for the most part forbidden to erect fences or ... any other protective device affecting the appearance of their project. [The] individual tenant ... cannot set up store-bought play equipment for the use of his own family. If he did, (a) it would be destroyed by children too large or too undisciplined to be using it; or (b) it would figure in an accident to someone else's child, for which the tenant would be held responsible. If one keeps this in mind, one will avoid resorting to the popular but

unrealistic stand that people are being shiftless and grabby when they ask for essential facilities which they are not free to provide themselves (Adams 1968, 31).

Public housing was meant to be temporary housing; to be occupied by households only until they were able to afford home ownership. In addition, there was a concern that public housing not be built to a standard that would compete with the private sector. This perception of public housing is reflected by the following statement by a senior government official and member of the Board of Directors of CMHC, in response to a proposed public housing policy statement.

... public housing projects should also be at a minimum standard as far as accommodation is concerned, but not as far as external design, siting, etc. are concerned. In other words they should improve the community but only provide a bare minimum of housing for the occupants.... This should be deliberately used not only to achieve economy, but to make clear that we are not competing with private enterprise who we assume will be building a more attractive product intended for those who can afford it (Dennis and Fish 1972, 174).

Surrounding communities reacted negatively to early public housing projects specifically because the projects were designed to provide temporary housing and not to compete with the private sector. As a result, the Hellyer Task Force recommended:

The Federal Government initiate a thorough research program into the economic, social and psychological issues of public housing. Until such a study is completed and assessed, no new large projects should be undertaken. (Canada 1969, 55)

By the mid 1970s a number of provinces, including Ontario and British Columbia, had discontinued production of family public housing.[8] Public housing certainly can be criticized because it addressed only one facet of housing standards (that is, the provision of basic accommodation built to minimum standards) and did not address other goals (that is, social aspects and consumer choice and control); however, it did and continues to provide affordable housing for some of the lowest income Canadians (Canadian Council on Social Development 1977).

The commitment to provide adequate housing for low-income households did not vanish with the end of new public housing schemes; rather it shifted to the development of non-profit, mixed-income communities, developed, and managed through municipal and private non-profit corporations. These projects provided housing to low-income and moderate-income households. Unlike many previous public housing projects, the new non-profits were often family-oriented (much of the later public housing constructed was for seniors) and built to be marketed to households who would pay market rents, as well as to those who would pay on a rent-to-income basis. From 1978 to 1985, 85,041 non-profit housing units were developed (CMHC 1985b). Non-profit housing was built in every region of the country. These projects, smaller in scale and built to fit the existing neighbourhood, did take into account some of the social aspects of housing.[9] In December 1985 the Minister in charge of CMHC indicated that all federal social housing expenditures would be directed to those in need. This policy direction could effectively curtail income mixing within non-profit projects, as the intention of the policy is to house only those below certain income levels.

URBAN RENEWAL/REDEVELOPMENT/IMPROVEMENT

Slum clearance was not a new issue in 1945. The condition of the urban poor had been a concern of the urban reform movement in the 1920s. Much of the activity directed towards slum clearance before and after 1945 resulted in the razing of the existing neighbourhoods and the development of large publicly-owned housing projects. Prior to 1956 federal enabling programs required that land acquired in a clearance area be replaced with low-income housing projects. The inappropriateness of the bulldozer approach to urban renewal was eventually recognized; and in 1956 the National Housing Act was amended to provide federal support for locally originated urban renewal studies and to remove the restrictions on the use of the land acquired in a renewal area. Most municipalities initiated such studies.

Certainly, the housing stock produced under these relocation schemes was better in quality than the original stock, but questions were raised about the other aspects of the community to which redevelopment and relocation were not very sensitive; and by 1968 the federal government imposed a moratorium on all new urban renewal approvals.

In 1973 the concept of neighbourhood improvement was introduced through a cost-shared program between the federal government and the provincial and municipal governments. This approach included the rehabilitation of housing stock within an existing community, through RRAP and the complementary upgrading of local services, facilities, and infrastructure through the Neighbourhood Improvement Program (NIP). In addition, provision was made for the acquisition of land for new social housing within the community. Neighbourhood improvement was a program with great potential that was widely expected to address the failings of urban renewal.

NIP had a short time horizon; the program itself lasted for five years, with a three-year time frame for plan development and implementation.[10] As such, it could not be expected to result in many comprehensive plans for neighbourhood improvement and revitalization. Nonetheless, NIP proved to be a limited success. The improvement approach was considered by most to be more progressive than urban renewal; however, the expectations that many had for NIP were not realized. Critics argue that too much emphasis was placed on the improvement of municipal infrastructure and not enough on housing renovation and redevelopment.[11] This focus of expenditure on infrastructure was in part one of the problems of the old urban renewal program.

There were other problems with NIP. Not all older communities were eligible. For example, the Edmonton inner city area did not qualify because of unstable land uses and some zoning that did not conform to CMHC criteria. That area, which had some of the poorest quality older housing stock, did not receive any federal/provincial or municipal assistance; and in 1986 the area still suffers from a deteriorating housing stock. Another failing of urban renewal that neighbourhood improvement was to address was the involvement of the user in planning and implementation. Here, the degree of success depended on the nature of the indigenous organizations within the community. Effective community involvement requires an organized community group with the skills and resources to make decisions and direct the planning process. Mechanisms are needed to enable community groups to deal with competing objectives and power elites. NIP was a program that required a degree of community consensus, energy, and political sophistication; yet community development was not a program component.

RRAP, NIP's companion program, however, did improve the standard of existing housing stock, albeit in a limited way. RRAP provided assistance to improve the quality of the housing stock in designated NIP areas. During 1974-8, 25,464 owner-occupied units as well as 26,446 landlord-owned units received RRAP loans; an additional 257,773 units (excluding hostel beds/non-profit units) received RRAP assistance from 1979 to 1985 (CMHC 1985b). The policy focus of RRAP was modified in 1986 from one of stock improvement to a social housing program targeted to those in need. While both owner-occupants and landlords are eligible, only owner-occupants in core need are eligible, and to receive the full assistance, landlords must agree to post-rehabilitation rents that are 50% of average market rents. Whether this new policy thrust will serve to retard the rehabilitation of existing stock has yet to be determined.

LIMITED DIVIDEND

Under the Limited Dividend Program, CMHC provided direct loans to private corporations to provide low-rent housing. Much of the initial housing produced under this program was not adequate (Dennis and Fish 1972). A review of the design and layout of a number of projects by CMHC in 1960 indicated that many projects lacked landscaping and play areas. In addition, many projects were large and contained a high proportion of one-bedroom units. Although CMHC attempted to regulate the quality, size (maximum 100 units), and unit distribution (averaging 2.5 bedrooms), the overall quality of these developments did not improve; developers continued to build projects at minimum standards, and some were poorly located. Other problems with the Limited Dividend Program related to the management of the projects, the failure of the project management to prevent under-utilization, and the income verification requirement.[12] In the mid 1970s there was a shift away from entrepreneurial housing for low-income households to provision through non-profit corporations, a trend which had started in the mid 1960s.

HOME OWNERSHIP ASSISTANCE

Post-war housing policy was heavily directed towards private home ownership. The introduction of mortgage insurance and high-ratio mortgages enabled households with little wealth to purchase a home and changed perceptions about home ownership. Historically, many households had built their own homes on an incremental basis as resources allowed. That pattern changed; households now sought to purchase a home that was already fully developed, and municipalities no longer tolerated partially-built units.

The emphasis of early post-war policy was on the production of new units for the more affluent; existing housing would then filter down to households who could not afford to purchase new. Ownership has been favoured by Canadians in part because it provides the consumer with control. In the 1970s new policy thrusts provided effective and previously-unavailable choices to modest-income consumers both with regard to type of housing and tenure. Governments at all levels introduced programs to assist modest income households to afford new housing. These programs included grants for first time home owners, subsidized mortgages for low cost housing for low-income purchasers, reduced lot prices in government land developments, programs to assist owners build their own housing, direct NHA financing when private funds were not available, rural and Native ownership programs and programs for cooperative housing.

While by the 1980s, the majority of Canadians enjoyed physically adequate housing and had a choice both with regard to housing form and tenure, there still remained pockets of households occupying physically inadequate housing and lacking consumer choice; many of them in communities without effective housing markets, for example in rural and remote areas and on reserves. As illustrated in Figure 1.2, in 1981, 23% of Native households on reserves lived in units that needed major repair as compared to 7% of non-Native households; 51% had no central heating as compared to 9% elsewhere. The situation was similar for off-reserve status Indians, non-status Métis, and Inuit. In addition to the lack of basic facilities, such as plumbing and central heating, crowding is also an issue. Based upon one person per room, one in six Native households were living in crowded conditions, as compared to 1 in 43 for the rest of the population. Inuit households faced the worst crowding situation, as 40% of the households had more than one person per room, and approximately 8% had more than two people per room.

In some instances, government intervention to improve the physical adequacy of the housing effectively reduced consumer choice, as George Barnaby of Fort Good Hope indicated in his presentation to the Northwest Territories Special Committee:

> ... there has always been a problem with housing, especially since the territorial government got involved in it. [Before] the government moved North, I mean everybody owned and built their own houses and had responsibility for every-

thing they decided. They did it for themselves. About 1968 or 1969 there was a big push by the government to change everything around.... There was lots of time and money spent introducing a new rental program of housing. At that time people were promised that they would pay a couple of bucks a month and they would have a lower rental unit, that is what they were called. So that was a pretty good deal, you get all your electricity and fuel oil plus the house for two dollars a month.

... So a lot of houses were destroyed, some of them were pushed over with cats, some of these people still do not have houses. Their houses were never replaced. And they would have no choice but a rental house, that means their houses were taken away from them and then they would have to rent from the people who took them away (Northwest Territories 1984, 36).

For these households, living in substandard, crowded conditions, there has been little progress in the post-war period. Housing policies that are predicated on filtering require a market and a sufficient stock of housing; both are non-existent in these communities. Housing programs, targeted to provide ownership opportunities to moderate income households, require households to have an income and to carry at least a portion of the debt service. Many residents in these communities have incomes that would not even cover the cost of operating the housing units. In many communities in the Northwest Territories, for example, the majority of census families in 1981 had incomes under $15,000 per year (*Census of Canada 1981*). Progress has also been inhibited by the fact that these households have had limited political power; they are few in number compared to the total population and not visible for the most part to those who, like the early members of the urban reform movement, would have championed their cause.

The Municipal Role

One indicator used in the Curtis Report to classify a unit as substandard was the lack of indoor flush toilet and bath facilities. By this measure, the policy thrust to install and improve municipal piped services helped ameliorate the substandard housing problem. In addition, rural electrification and the development of individual mechanical systems (for example, septic tanks and cisterns) had a positive impact on the substandard problem in rural areas. There has been a significant overall improvement in the housing stock from the perspective of plumbing facilities; nationally, only 86,000 households lacked flush toilets by 1983. [13]

The development and introduction of building codes and other standards also had a positive impact on improving both the housing stock and the urban environment. The National Research Council of Canada developed a uniform code for building standards in the 1930s, but not until three decades later did such standards became widely implemented by the provinces.

Prior to their use at the local level, uniform building standards were implemented primarily through the efforts of CMHC. In 1947 CMHC produced

Building Standards and Apartment Standards which, while modelled on the National Building Code, included additional requirements thought to be important for good quality residential development. In 1957 the Building and Apartment Standards were transferred to NRC's jurisdiction, which published them in 1958 with the Building Standards renamed as Housing Standards. Until 1965 housing projects that received CMHC mortgage funding or insurance had to meet the requirements outlined in these documents. In 1965 the Residential Standards were prepared by the NRC; these were a combination of the current Housing Standards, Apartment Standards, National Building Code, and some additional requirements considered important by CMHC. However, it did not include site planning requirements. Therefore, CMHC developed their own site planning criteria. The site planning standards, set out in the various CMHC publications from 1966 to 1980, set the standards of residential development for all of the projects CMHC funded and insured. In 1980 these standards became advisory rather than mandatory for all the Corporation's market housing projects. Critics considered this move regressive because it reduced the Corporation's ability to control the quality of residential development. However, by 1980 most urban municipalities had developed their own site planning criteria, some more restrictive than CMHC's; in any case, the void in local controls that existed when CMHC developed its criteria no longer existed.

Mandatory standards in housing and residential development helped ensure that newly-built housing was not substandard. However, many might not agree with the extensive nature of the standards and requirements. It could be argued that increased standards have increased the cost of housing, thereby contributing to the problems of affordability. Furthermore, some owners wanting to convert existing housing into smaller suites or a rooming house, for example, will do the work without obtaining a permit, and as a result the housing may not meet basic fire and safety standards. In addition, in some situations there is a reluctance to have existing health and safety regulations enforced for fear of possible loss of stock.

The Curtis Report (Canada 1944, 16) recommended the initiation of an extensive process of town planning. The committee also recommended the establishment of a Dominion Town Planning Agency "that would be equipped with all the necessary facilities for the promotion and co-ordination of town and community planning throughout the country"; however, since planning, like codes and standards, was under provincial jurisdiction, the realization of such a body was problematic.

The initial post-war challenge was the development of enabling legislation to ensure appropriate community planning as well as the development of a pool of planning professionals. During this period existing planning legislation was revised and the necessary structures were developed to enable town planning; planning schools were established; and associations such as the CPAC and CHDC were organized.

In the early post-war period, CMHC played a role similar to that envisioned in the Curtis Report, albeit limited. CMHC encouraged municipalities to undertake studies and plans of potential urban redevelopment areas and provided grants to determine slum areas in need of urban renewal, required community plans, provided assistance to municipalities through the review of proposed housing developments, and provided information resources to builders through the *Builder's Bulletin* which addressed, among other subjects, ways to achieve variety in new subdivisions at no extra cost. Community planning was not coordinated by a national body; rather planning and plan implementation became a function of local and regional government. The implementation of provincial planning legislation was facilitated by changes in the nature and organization of local government as well the development of regional structures which enabled the development, control, and financing of communities on a planned basis. By the 1960s most local jurisdictions had planning departments in place.

The focus of community planning has evolved since 1945. The initial focus was to ensure orderly and controlled growth. The development of zoning and land-use controls and the provision of municipal services (sewers, roads, and sidewalks, for instance) were the main focus of community planning during the first two post-war decades.

However, in the late 1960s community planning also had to address concerns over quality of life. Continued growth was no longer a common social goal, and citizen opposition to major public and private development was common across the country in the late 1960s and in the 1970s. For example, citizens opposed the Spadina Expressway in Toronto, the Third Crossing in Vancouver, and high-density redevelopment of older neighbourhoods in Edmonton. Municipalities also used their power to expropriate to realize public projects. Toronto, for example, attempted to expropriate five home owners living in an urban renewal area in order to implement a plan whose purpose was to improve areas residents' housing conditions, and Winnipeg expropriated a resident to make way for a sewage treatment plant (Lorimer 1972). This period can be characterized by citizen distrust of planners, developers, and the planning process; publications with titles such as *Forever Deceiving You, Up Against City Hall, Fighting Back,* and *The Revolution Game* were published. By the late 1970s planners began to recognize that planning had to take place in the public arena and that citizen participation was an essential component of the planning process.

Community planning in the 1980s is much more complicated than it was in the early post-war years since it must address a variety of competing goals and objectives from different interest groups.

While some might argue that the planning process is too bureaucratic and the standards have become too high, it would be difficult to argue that community planning has not resulted in overall housing progress. Amongst other factors, the provision of green space within planned urban environments, the transportation networks, and play spaces for children certainly have resulted in an environment that is an improvement over the slums of the past which, as described by the Bruce Commission, lacked basic amenities:

> ... the extensive use of trucks which clutter up the streets and make it unsafe for children. Dwelling units are edged in between junk yards, sheds and commercial buildings.... The sordid appearance of the district is largely unrelieved by trees and grass (Ontario 1934, 29).

Future Indicators of Substandard Housing

Future indicators of substandard housing will be effective only insofar as they reflect society's fundamental goals. Social goals that are important in the development of future measures of housing quality include the following: enhancing equity of opportunity, preserving the dignity and privacy of the individual and the family, promoting diversity, freedom of choice, enhancing health, safety, and quality of life, fostering a sense of community, and preserving the natural environment. Future measures of housing quality should not only take into account indicators of physical adequacy but also social adequacy and consumer choice and control.

INDICATORS OF PHYSICAL ADEQUACY

Macro indicators that have been used to identify substandard housing include need for repair, lack of basic facilities, the physical quality of the residential environment, and crowding. In the 1980s, however, the development of a macro definition applicable across the country is a more challenging exercise. There are three general concerns to take into account in developing such a definition.

One is that a single "macro" standard for the nation may be less appropriate than "micro" standards that vary among regions or by categories of households across Canada. Housing needs and standards vary with climate. They also vary with the social and demographic characteristics, housing needs, and aspirations of the occupying household.

A second concern is the weighting of the indicators to distinguish between units that are physically inadequate and those that are substandard. A comprehensive approach, taking into account both exterior and interior conditions and facilities as well as their level of operation, would be useful for housing assessment in the future. The US Department of Housing and Urban Development has developed criteria for distinguishing between physically inadequate and seriously inadequate. This definition includes both absolute and relative measures and takes into account the number of times the toilet, heating, and electrical systems broke down over a period of time. In addition, cost of operations may be an indicator of physical adequacy that may be significant in the future. For example, a unit that is extremely expensive to heat may be considered substandard.

A third concern is the reliability of any data base. Much of the basic housing data has been gathered through the Census, which currently is self-administered. While hard data on physical facilities and number of rooms are likely reliable, data on the quality of the unit and its overall state of repair may be less so. [14] In addition, it is important to determine the basis on which units will be counted, as the definition of dwelling has an impact on the magnitude of the substandard housing problem. A count based upon the Census definition of dwelling unit as "a structurally separate set of living quarters" will yield fewer substandard units than one that allows living quarters with shared facilities. For example, the Survey of Housing Units indicated substantially more dwelling units with shared toilet facilities than did the 1971 Census. This discrepancy may be a result of the SHU staff being more likely to classify units such as bachelorettes and basement suites as separate units than were Census staff. There are many cases where the definition of dwelling is not clear, including basement/attic suites occupied by lodgers and rooms in a rooming house or a bachelorette. The definitional problem needs to be addressed to ensure data base consistency.

INDICATORS OF ENVIRONMENTAL INADEQUACY

Factors that may be significant in the future are (1) the immediate environment created by a specific housing development, (2) the physical form of housing, its location, and its relationship to other residential and non-residential development, and (3) the overall quality of the municipal infrastructure.

A number of high-density, high-rise developments were built in the 1970s, some of which were publicly-owned and occupied by families with children. Play is considered by many educators/psychologists as a primary medium for development during early childhood. For these children, whose play activities may be constrained by lack of available play space, such housing may be environmentally inadequate.

The overall design of housing environments has an impact on people's lifestyle and behaviour. The idea of planning for defensible space was a response to high crime rates and vandalism in high-density projects. High crime levels as well as a fear by residents of walking their neighbourhood streets are indicators of an environmental inadequacy. While the first indicator can easily be quantified, the second, because of its subjective nature, is more difficult to measure. Subjective measures will be important in the future, and there is a need to develop more techniques that will facilitate them.

Concern over air quality, heating costs, and the potential dangers of nuclear power plants could make solar power more attractive. The attendant need for improved access to sunlight has design implications both for housing and urban form. For example, in many Canadian cities, it would be difficult for the residents to convert to solar energy simply because subdivision layouts make roof orientations inappropriate for conversion. An inability to convert to solar power may well be a future indicator of substandard housing. Lack of access to direct sunlight for a predetermined number of hours a day (a requirement which

currently exists in Sweden) could also be a future indicator of substandard housing. Much of the high-density housing developed in the late 1960s and 1970s did
not consider shadowing of adjacent properties to be significant, with the result
that a number of buildings have very limited hours of sunlight.

The location of housing next to major roadways or industrial developments
has come to be viewed as inappropriate and to be avoided where possible. However, in addition, there are also factors that affect environmental quality with less
regard for location, for example, polluted air. The lack of clean air may well be
considered an indicator of environmental inadequacy in the future, as may
proximity to toxic waste disposal sites.

SUITABILITY

Crowding has been the most common indicator used to determine the suitability of housing. However, future indicators must take into account changing
lifestyles and expectations as well as specific needs of particular groups/individ-
uals. A more precise measure of crowding should take into account household
configuration, including the age and sex of the occupants. For example, a two-
person household consisting of husband and wife could be considered suitably
housed in a one-bedroom, two-room unit, while one consisting of a single parent mother and teenage son would not. The National Occupancy Standards –
used in the allocation of federal social housing funds – recognize in a limited
way the needs of different types of households. For some households, housing
suitability is not just dependent upon built form but also upon the availability of
support services. Many ex-psychiatric patients, for example, will not be suitably
housed unless they have access to such services. Changing needs, expectations,
demographic trends, and cultural patterns will all impact on future indicators of
housing suitability.

CONSUMER CHOICE AND CONTROL

From the perspective of the consumer, lack of effective choice is an indicator of
housing inadequacy. Effective choice does not exist if a household has no real
decision-making power over form of tenure, form of housing, location of housing, and the adaptation of interior and exterior living space to meet individual
needs. As previously indicated, many tenants living in public housing consider it
housing of the last resort – a unit that provides a minimum level of physical
comfort at an affordable price but not a place to call home. Historically, effective
choice was a function of a household's income: more income, more choice. In
the future, however, effective choice may not be so dependent upon a household's income, as the new and emerging forms of owning and renting become
more widely available.

Lack of control and security are also indicators of inadequate housing from
the perspective of the consumer. This concept includes control over a household's expenditure on housing and an ability to have a say in the management of
the home and its immediate environment. For example, although limited by

ability-to-pay, owner-occupants have control over the choice to repair and maintain their unit, a tenant does not. If a unit is in poor shape and the owner will not repair it, a tenant can either choose to live under those conditions or find alternate accommodation. Security of the home includes protection from external forces that would result in the loss of the home by the resident; it could include factors such as an income loss that affected the household's ability to maintain its housing payments, eviction from the home because of a lack of security of tenure, or the development of noxious industry adjacent to the home. While many Canadian households have security of home, some have none; for example, in most jurisdictions, roomers and boarders are not covered by landlord and tenant legislation, and those occupying the temporary shelters have no permanent home.

Future Approaches to the Substandard Housing Problem

Of the housing stock in Canada, 46% is more than twenty-five years old; of this stock, approximately 22% was built prior to 1941 (Statistics Canada 1983). CMHC's own projections, published in 1985, indicate that 75% of the housing stock that will be available in the year 2001 has already been built. An article in the *Toronto Star* in 1986, which addressed current housing problems, indicated that "tens of thousands of people are estimated to be poorly housed – in substandard apartments, public housing projects, rooming houses, hostels, and drop-in centres" (Harvey 1986). While journalistic style has a tendency towards exaggeration, there is a growing stock of units that can be classified as substandard. In part, this physical deterioration of some stock is related to a general lack of maintenance and ongoing replacement of short life items.

If the trend towards polarization identified in Chapter 4 continues, a large proportion of the rental stock will be occupied by the low-income groups. Will the market incentive be there for owners to maintain or improve rental housing without government assistance? Initially, inadequate construction is also another reason for physical deterioration. Some of the early limited dividend projects, as well as some of the earlier public housing projects, will require major repair and maintenance in the near future. In addition, a number of older mobile homes and parks may require extensive renovation or replacement.

In order to address these future problems, a comprehensive rather than a pragmatic approach is necessary. A comprehensive approach is best undertaken at the local municipal level. Within each metropolitan area a process should be put in place to get existing interest groups to work together with the municipal government, first to determine the scope of the problems within the community, and then to establish an action plan that would ensure the equitable distribution of resources. At the same time, the financial support of senior levels of government is needed; municipal governments lack the capital necessary to implement such plans.

While some existing approaches will be used in the future, the scope of policies and programs will probably be broader. Rehabilitation programs, for example, should not only address the structural and mechanical components of a dwelling unit but also its overall design and layout to ensure defensible space. Projects that cannot be renovated to provide defensible space may have to be replaced. It is suggested that provincial and municipal planning legislation and regulations probably will be modified to take into account access to sunlight to permit the use of solar energy. In developing new stock, especially in rural areas and Aboriginal communities, it is essential that environmental and cultural considerations be taken into account and that the end user direct the policy and program approaches. Housing policy should take into account the special needs of households. In addition, our approach to the measurement of physical adequacy should expand in the future from one of lack of basic facilities to one that takes into account performance and cost of operations.

Notes

1 A dwelling that exhibited one or more of the following characteristics was considered to need external repairs: sagging or rotting foundation causing the walls to crack or lean, shingled roofs with warped or missing shingles, chimneys cracked or with missing bricks, and unsafe outside steps or stairways.

2 In developing this indicator, the Subcommittee wanted to determine the proportion of the housing stock needing replacement, rather than rehabilitation.

3 One method considered was developed by the Committee on Hygiene of Housing of the American Public Health Association. That Committee used a detailed rating system (from 1 to 30 points) of a variety of deficiencies, including the need for major repairs (interior and exterior), lack of basic facilities (electric light, flush toilet, and bathing facilities), defective facilities (worn out pipes), totally unsatisfactory building design or construction (completely inadequate fire escapes), infection with vermin, lack of the minimum interior space requirements or outside layout needed for healthy housing, and location within a slum area.

4 This included 125,000 units in metropolitan areas, 50,000 in smaller communities, and 123,000 rural units (100,000 farm units and 23,000 non-farm units).

5 For example, the City of Edmonton, which showed 46% substandard units if all three indicators are considered, showed just 24% substandard units based upon the need for external repairs. The same pattern holds true for Halifax, Saint John, Quebec, Trois Rivières, and communities in northern Ontario.

6 The Curtis Report found that approximately 15% of the substandard stock in major cities needed replacement, compared to 8% in the smaller communities.

7 Although housing construction peaked in the late 1920s, by 1932 the market had collapsed. In Montreal, for example, fewer units were built from 1932 to 1939 than in 1928 (Archambault 1947); while in Calgary, residential permits from 1930 to 1939 were less than in 1929 (Safarian 1959). This situation was common across the country. During the 1930s the birth rate and net family formation declined, and immigration was at an all-time low. There is debate about the number of units built during this decade, and the comparison of some estimates of dwelling starts and net family formation would indicate an over-

supply of units as compared to net family formation during the 1930s. However, supply was inadequate to meet the potential demand for housing; doubling up was common; and many households were living in crowded conditions. In Montreal, for example, fewer dwellings were built from 1932 to 1939, than there were marriages in 1938 (Archambault 1947). The shift in population into war-time centres such as Halifax, Ottawa, and Montreal heightened demand for housing in these communities.

8 Although others, including Saskatchewan, Alberta, Newfoundland, Nova Scotia, Prince Edward Island, and the Northwest Territories, continued to develop family projects.

9 While still tenants, many occupants considered living in a non-profit housing project more acceptable than traditional public housing. Perhaps because this new form of housing received greater public acceptance, the massive studies relating to neighbourhood acceptance and user satisfaction that were conducted on past projects have not been repeated.

10 The NIP Program expired on 31 March 1978, although the approved NIP boundaries continued to define the only areas eligible for RRAP funding.

11 Much of the funding in a NIP Project in Winnipeg went toward sewers, water mains, street reconstruction, sidewalks, and street lighting.

12 The intent of the program to provide housing at low rent to households with limited incomes. While income verification was required upon initial occupancy, it was often not subsequently obtained, and in some instances, households of higher income remained resident in LD projects, although theoretically they were no longer eligible.

13 Such improvements have not been evenly distributed, in that a greater proportion of the overall housing in the Atlantic region lack a flush toilet. This region has approximately 8% of the total housing stock, but its housing stock accounts for 24% of all units in Canada without flush toilets. Although not all housing lacking basic plumbing facilities in Atlantic Canada was rural, one of the factors for their disproportionate number is the rural nature of this region.

14 The Atlantic Housing Survey, which tested alternate means of assessing dwelling conditions through respondent assessment, interviewer appraisals and CMHC inspections, found that the assessments varied considerably between the approaches. As Streich (1985, 17; TEEGA 1983, 63) indicates,

> ... occupant assessments matched inspectors' assessments 76% of the time on structural deficiencies, 45% of the time on general exterior condition. Interviewers' assessments were more consistent with inspectors' appraisals with reference to structural condition than with regard to overall condition.

CHAPTER THIRTEEN

Housing as a Human Service: Accommodating Special Needs

Janet McClain[1]

THE NEED for shelter is often measured in terms of average amounts (such as rooms and floor area) required by age group or size of household. However, such measures do not reflect particular requirements of the "special needs" consumer. In housing, these needs may include ground-level construction with off-street entrances and access ramps instead of stairs, wider doorways, open-space planning in the kitchen and bathroom, and up-graded electrical circuits for health and mechanical devices. Given the limited mobility and perception of some special needs housing consumers, in-house laundry and recreational space are also important. Convenient access to grocery stores, pharmacies, banking, and child care is also important, as well as proximity to neighbours and friends with compatible lifestyles. Access to on-site building maintenance, personal security, and attendant care or assistance is also important. Ideally, housing should be located within an envelope of community support and social services that provides both occasional assistance as well as ongoing service and treatment programs as needed.

At times, these distinctive housing qualities reflect the needs of almost everyone. At one time or another, upon being confined by illness or injury or limited by immobility or income, people join the ranks of special needs consumers, with some staying much longer than others. If the condition is temporary, the lack of choice and the inability to search for alternative accommodation may not be such a hindrance; in that case, the daily frustration of living in housing with more "handicaps" than its residents may be less appreciated. Many special needs consumers are not fully mobile. Their needs – changing as they do with age, household and family composition, economic resources, and health condition – often cannot be accommodated because their present housing is not directly linked to support services. A trade-off may have to be made: new accommodation that is better suited to their physical needs at the expense of proximity to one's support community.

Since 1945 Canada has experienced a growing special needs population. These needs are a result of improvements in health care and longevity, changing social policies, decreasing long-term care in institutions, the desire for more independent lifestyles, and more openness in society about physical and mental disabilities and social problems. As well, reliance on an informal network of family and friends in one's home community, though still the primary base of support, may be less of an option for persons with special needs than it once was.

Outside of an institutional setting, dwellings with a component of care or special services were scarce in Canada prior to the 1960s. Up to 1966 publicly-funded social services were solely the responsibility of the provinces and municipalities (except for vocational rehabilitation programs for disabled persons). The 1966 Canada Assistance Plan (CAP) introduced federal cost-sharing of social services, intended to expand the scope of these services. There was a shift toward prevention and more community and personal development-based services away from protective and institutional care. As a result of CAP, the provincial and municipal role in social services was expanded to include information, counselling, referral, crisis intervention, and family planning. Developmental services were made available such as group homes for children and young adults, rehabilitation and accessible transportation for disabled persons, homemaker's services and meals-on-wheels, halfway houses and Native Friendship Centres, and daycare for young people and adults (Guest 1980, 195). Information, referral and crisis intervention services are made available without charge on request; other services such as rehabilitation or homemaker's services have charges unless recommended as part of an aftercare treatment program. Other services such as daycare and group homes have user charges which are means-tested (Guest 1980, 195)

The level and quality of social services under CAP were improved by the introduction in 1966 of the Medical Care Act and the Hospital Insurance and Diagnostic Services Act. These Acts provided comprehensive coverage of health care needs. CAP paid for health care services in homes for the aged and nursing homes which were facilities not covered by the new medical care and hospital insurance legislation (Hum 1983, 69-71; Hepworth 1975 4, 6). Provincial leadership was also important in the adoption of innovations in preventive health care (Alberta), child welfare (Ontario), and homemaker's services and seniors' welfare programs (Quebec) (Chappell *et al.* 1986, 91-92; Hepworth 1975, 4, 6).

The Hellyer Task Force in 1969 gave public recognition to the critical housing needs of low-income singles and families and, for the first time, the housing needs of disabled persons were acknowledged. Many provincial governments established special commissions and task forces in the mid 1970s to address the concerns of disabled persons and their advocacy groups as well. Special offices such as the L'Office des personnes handicapées in Quebec were established and laws were instituted or amended to assure basic rights of access and self-determination (Québec 1988, 156-8).

Many of the changes in the 1960s assumed that the existing housing stock was both readily available to persons using social services and appropriate for community-based service delivery. However, only when RRAP and the Nonprofit Housing Programs were revised in the mid 1970s was there a firm commitment to the rehabilitation and production of accessible units. In addition, cooperative, municipal and community-sponsored non-profit housing organizations used the new federal government financing under NHA Section 56.1 to expand the stock of special needs housing for senior citizens, disabled persons, and single mothers.

The Need for Supportive Housing

The development of housing for the elderly in Canada eclipsed the provision of all other forms of special needs housing both in terms of unit production and as the focus for much of the housing research and social policy literature. The objectives in building and developing self-contained and hostel housing for the elderly were to lower shelter costs, to provide accommodation more designed for the changes and infirmities of aging, and to promote access, security of tenure, and independence.

Most of the housing literature focused on the needs of an aging population; little had been written on disabled persons, women, and the homeless prior to the mid 1970s. The United Nations International Year of Women (1975) and the Year of Disabled Persons (1981) prompted the funding of research and motivated governments to make public statements about the needs of these groups. During the Year of Disabled Persons, for example, the federal government incorporated into NHA programs provisions to make social programs, facilities, and services more accessible to the disabled community.

SENIOR CITIZENS

From 1946 to 1981 approximately 146,000 senior citizen dwellings were financed under NHA programs. Of this total figure, all but 3% of the units were new construction. Some 47,000 hostel bed places were also constructed primarily by non-profit corporations, municipal governments, and provincial housing corporations. This production makes up the largest proportion of publicly-financed housing stock provided to any client group in Canada. Prior to 1970 a fairly even mix of family and senior citizens housing stock was built by provincial housing corporations. During the early 1970s the production of seniors' housing soared; many of the larger-scale apartment projects and complexes were constructed at that time.

Who needed this housing and why did the federal and provincial governments show preference for the construction of seniors housing? By the early 1970s the need had been well-documented. Several studies had highlighted the surge in Canada's elderly population (Bairstow 1973; Yeates 1978; Stone and Fletcher 1980; Marshall 1980). Similarly, the low-income status, pension coverage and housing affordability problems of elderly widows, singles, and couples had been noted by many researchers (Bryden 1974; Gutman 1975-6; Bairstow 1976; Huttman 1977). Major studies called attention to the housing needs of the elderly poor and emphasized the need for housing rehabilitation and maintenance programs for elderly renters and home owners. The reports also pointed to inadequate OAS/GIS and CPP benefits and the need to improve provincial income supplement and housing assistance programs (National Council on Welfare 1984; Social Planning Council of Winnipeg 1979; and Social Planning Council of Metro Toronto 1980a). Furthermore, various studies had begun to identify a growing population of older women who were living alone without adequate financial resources and few housing options. The needs of this population were identified in the 1970s but were not articulated until the 1980s when national statistical information on housing tenure, cost, and living arrangements was analyzed by gender and household composition (see Stone and Fletcher 1982).[2]

The housing problems of the elderly shaped housing programs and policies throughout the 1970s. Over the last decade this population has aged along with their self-contained public and private dwellings. While a small proportion of the residents in subsidized housing, for example, needed on-site care and social services in the 1970s, now a greater proportion of this population has infirmities and limitations in mobility which increase their requirements for supportive services. This "aging in place" phenomenon (Gutman and Blackie 1986) has generated debate among housing and social service providers.

Institutionalization and Homes for Special Care

Canada has a fairly high rate of institutionalization of the elderly – close to 9% receive some form of institutional care – compared to the United States, England, and Wales.[3] The reasons for the predominance of institutional care are many. Homes for special care were a significant part of the housing stock produced for elderly persons across Canada (Table 13.1). Up to the mid 1970s, the number of rated beds in homes for special care was over 40 per 1,000 population (age 65 years and over) in the provinces of Prince Edward Island, Quebec, Ontario, Saskatchewan, Alberta, and British Columbia. Then, in 1975 the numbers increased significantly to as much as one bed for every ten older residents in Prince Edward Island and Alberta.

In many small towns and rural areas, there are few alternatives to special care homes. Hostel residences with specific levels of care were often constructed by provincial and private sponsors because to offer a particular level of care to a potentially small population base was less expensive than to offer individualized primary support care covering a wide range of needs. Often individuals were placed in special care homes who could have been sustained independently with home help services or through congregate housing that offers a minimum level of on-site services. Older women in rural areas were sometimes institutionalized prematurely because no one was around to look after them (Baum 1974; Cape 1985). Many provinces have now questioned the effectiveness of their evaluation of seniors housing needs in non-metropolitan areas.

Early research found that almost half of the older persons in expensive institutional care in Montreal could lead independent lives. One of the main reasons for institutionalization was the lack of readily available alternatives (Zay

Table 13.1
Number of rated beds in homes for special care:
Canada, 1966-1985
(per thousand population)

	1966		1975		1980		1985	
	20-64	65+	20-64	65+	20-64	65+	20-64	65+
Canada	6.3	47.0	11.4	85.7	11.8	82.9	10.9	74.2
Newfoundland	3.4	28.0	8.3	70.3	8.0	63.4	7.4	54.9
Prince Edward Island	8.9	45.7	18.6	102.4	17.9	99.0	17.1	92.4
Nova Scotia	4.7	29.7	11.5	75.1	13.5	83.7	13.1	79.1
New Brunswick	5.6	36.6	10.4	70.4	15.1	99.9	12.7	79.6
Quebec	5.7	52.7	7.8	65.9	7.7	60.5	7.4	54.8
Ontario	5.7	41.7	12.9	96.2	13.1	90.5	11.8	78.9
Manitoba	7.9	50.8	13.1	82.1	13.8	79.8	14.1	79.1
Saskatchewan	7.0	43.2	14.4	81.4	14.3	79.8	13.6	74.2
Alberta	9.1	71.7	14.3	115.7	13.2	112.2	11.3	97.9
British Columbia	8.0	51.4	12.6	87.4	13.5	87.1	13.4	81.2
Yukon	–	–	22.3	662.5	19.0	316.2	14.2	267.5
N.W.T.	–	–	4.2	84.4	3.0	50.1	3.5	77.7

SOURCE Health and Welfare Canada (various years); Hepworth (1985, 152-63); and special tabulations (1987). "Rated Beds" are rated for cost-sharing under CAP.

1966b, 17). It is paradoxical, at a time when there is general support for community care and helping older people remain in their homes, that the share of government resources going to hospitals and residential care institutions has increased (Townsend 1981, 21-2).

Assessing Levels of Need for Supportive Services
There are many difficulties in measuring functioning and in estimating levels of services and the need for supportive housing (Heumann and Boldy 1982, 28-30). The methodology and practice of assessing physical and mental functioning, mobility, and self care have changed over the last twenty years. There is still debate in the gerontology literature about assessing the ability of older persons to maintain independent households or to remain in their own homes (Gutman 1975-6; Lawton 1976; Wigdor 1981; Connidis 1983). While assessment of the activities of daily living is relatively straightforward when done by experienced persons, the decisions about whether a senior can continue these activities is more complicated. Although a person may be frail, their ability to function and maintain daily activities is contingent on availability of family members and community support services.

In a 1982 study of environmental competence among independent elderly,

72% of the survey's respondents felt that maintaining independent living was important (Morgenstern 1982). Almost half the respondents recognized the possibility of having to move because of need for care as health deteriorated. Despite financial difficulties or loss of a spouse, these respondents were committed to maintaining their own households. Thus, the ability to care for themselves was "crucial to their independent lifestyle" (Morgenstern 1982, 1-2).

A 1982 study of life satisfaction and the living arrangements of older households in Quebec City found that "la vie active" is fostered by contentment with the conditions and quality of housing (Bernardin-Haldeman 1982). Mediocre living arrangements can lead to health, mobility, and functional problems, possibly creating a need for hospitalization or treatment of mental health problems. Older persons spend much of their days at home – a fact consistent with findings in the *Beyond Shelter* study (Canadian Council on Social Development 1973) and subsequent studies. Thus, older residents in apartment and row housing are more aware than younger residents of the physical limitations and shortcomings of their buildings and the management. Because of limited mobility, they are also more aware of the quality of services in the immediate neighbourhoods. Older persons in many ways may be our most experienced and critical housing consumers.

The Provision of Housing Stock for Seniors
The amount of stock and size of self-contained and hostel housing developments varies by region (Table 13.2). Not surprisingly, since 1946 the largest stock of self-contained dwellings was produced in the provinces with the largest populations of seniors: Quebec, Ontario, and British Columbia. The Atlantic and Prairie provinces along with Quebec have the largest stock of small and medium-size seniors' buildings. On the other hand, British Columbia and Ontario have most of the large-scale, high-density housing complexes. On average in Canada, self-contained apartment developments in the early 1970s were fairly small (up to forty residents), while hostel residences were almost twice as large. Since that time, the economics of housing finance and development has changed considerably, particularly in urban areas. There has been a shift to medium and high-rise buildings (100 or more units) compared to the early 1970s when only 17% of the housing stock was high-rise.

There has been a change in location and concentration of housing as well. In the early 1970s some seniors' housing developments were visibly segregated or concentrated, and in the Prairies, British Columbia, and Ontario, they were often located in the suburbs. In contemporary public and non-profit housing, some seniors' housing is located in more central locations and not exclusively in age-segregated projects. The inclusion of apartment units or separate buildings for senior citizens within family housing has become more common.

Styles and attitude toward housing development have changed. Features such as income and age integration, neighbourhood quality, access to public transit and central location are more important. The range of housing stock

Table 13.2
NHA loans for production of elderly persons' accommodation
and total production, by type of housing unit:
Canada, 1946-1981
(new and existing)

	Loans (#)	Loans (%)	Total production, 1946-1981 Self-contained (#)	Self-contained (%)	Hostel beds (#)	Hostel beds (%)	Total expenditure ($000)
Newfoundland	39	1	1,179	1	1,246	3	38,279
P.E.I.	62	1	750	1	272	1	13,569
Nova Scotia	289	8	6,289	4	2,324	5	129,914
N.B.	162	5	3,256	2	2,574	6	97,384
Quebec	638	18	23,453	16	14,910	32	643,787
Ontario	988	27	65,958	45	6,356	14	1,106,008
Manitoba	350	10	12,025	8	4,733	10	220,381
Saskatchewan	499	14	10,302	7	4,354	9	229,456
Alberta	106	3	4,849	3	1,718	4	91,261
B.C.	466	13	17,806	12	8,179	18	384,283
Yukon	3	0	32	0	–		994
N.W.T.	11	0	171	0	–		3,993
Canada	3,613		146,070		46,666		2,959,309
1946-1955	27	1	800	1	–		2,970
1956-1965	325	9	11,047	8	2,408	5	71,149
1966-1971	933	26	36,217	25	24,534	53	504,019
1972-1977	1,561	43	73,575	50	13,049	28	1,467,977
1978-1981	767	21	24,431	17	6,675	14	913,194
All Canada	3,613		146,070		46,666		2,959,309

SOURCE CHS (1981, Table 63). Production under NHA includes Loans to entrepreneurs and nonprofit corporation (Section 6, 5 & 15.1), Cooperative Housing (Sections 6 & 34.18), Public Housing (Section 43) and Federal/Provincial Rental Housing (Section 40).

provided has been broadened by the use of cooperative and non-profit housing (NHA, Section 56.1). New forms of tenure and living arrangements have also been accommodated and special income supplement and shelter allowance programs have been targeted to seniors in the provinces of British Columbia, Manitoba, Quebec, and New Brunswick.

The increased provision of self-contained apartments for seniors also eased the need for beds in homes for special care; 83 per thousand seniors in 1980 across Canada, down from 86 per thousand in 1975 (Table 13.1). Fewer older persons in younger age groups entered homes for special care and the average age for admission to facilities increases. The average age of admission to Ontario's Extended Care program has increased progressively (to 83 years in the mid 1980s) reflecting increased longevity, new health and social services, and greater availability of housing alternatives (Ontario, Ministry of Community and Social Services 1986).

DISABLED PERSONS

Before the mid 1970s only a few reports pinpointed the variety of housing needs of disabled persons. As Brown (1977) noted, "a decade ago, a section in a report dealing with accommodation for the disabled would have been directed almost entirely to institutional care." Disabled persons and community service organizations articulated the need for alternative living arrangements, accessible community services, and greater personal mobility through special transportation services. Above all, these groups advocated housing dedicated to an independent lifestyle.

The need for alternative living arrangements was acute for younger and middle-aged disabled persons. Many young disabled persons requiring only a minimum level of assistance were confined to extended care hospitals, private nursing homes for seniors, or rehabilitation units or acute care hospitals. Often, custodial care was the only service offered in nursing homes, a type of care that made it difficult for the disabled to achieve self-sufficiency (Brown 1977). Another kind of difficulty was faced by disabled persons who, after being wholly dependent on their families and/or guardians for care, find one day that these caregivers are unable to carry on. Similar concerns arise about the problem of dislocation from family and one's community because special kinds of care, treatment, and rehabilitation programs are not available in the local community.

Disability Status

Using World Health Organization classification, disability is a functional departure from the norm falling between "impairment" – which is any interference with the normal structure and functioning of the body and "handicap" – which reflects the value attached to an individual's status when it departs from the norm. Disability then becomes the loss or reduction of functional ability and activity resulting from impairment (Wood 1975). Health and Welfare Canada has participated with the OECD in extending the definition of disability status further to "the consequence of the effects of ill-health on activities essential to daily living" (McWhinnie 1982, 12-5). Many disabilities are short-term or episodic in nature, and thus do not always fall within the exact time-frame of data collection. Therefore, disability status reflects three dimensions: time, severity, and deviation from the usual level of functioning. As of the late 1980s, other terms such as "physically" or "mentally" challenged" have been used to replace disability.

Reporting on the Population with Disabilities in Canada
In 1983-4 the *Canada Health and Disability Survey* showed a little over 2.4 million disabled persons in Canada, the largest proportion of which is age 65 years and over. Approximately 39% of the population in this older age group were disabled (Statistics Canada 1985, Table 1). Given the larger number of older women in the population, there were also more women with disabilities. In the age group under 65 years, only about 9% of the population is disabled. In terms of type of disability, mobility problems are the most prevalent, followed by problems of agility and hearing (Statistics Canada 1985, Table 10). Of the total disabled population, only 8% reported using special exterior accessibility features. Similarly, only 8% reported using inside aids such as handrails or lift devices (Statistics Canada 1985, Tables 19 and 20).

The survey data must be put into perspective. These figures are drawn from self-reporting in a single time period and as such do not represent the total picture of the nature and degree of disability experienced by the survey population over time. Second, it is important to remember that many people have multiple problems so they may be counted more than once in the categories by nature of disability. Third, the classification of disability status must be considered as well as how assigned status allows access to federal and provincial income supplement as well as supportive housing and service programs. The survey data do not provide the linkage to these areas of concern.

Housing Policies Promoting Independent Living
In the early 1970s a philosophy of normalization (Wolfensberger *et al.* 1972) was adopted by many provincial and local governments and community service organizations. Normalization promotes the integration of disabled persons into the mainstream of social and economic activity. Normalization can be comprehensive, providing a home environment that satisfies all needs with accompanying support services and opportunities for training, education, and gainful employment. The aim of normalization is to eliminate existing unproductive activities and inefficient programs which overprotect and segregate disabled persons. The majority of disabled persons have a physical limitation that is both permanent and stable which they learn to accept. What the persons and organizations helping the disabled must learn to accept is that almost all disabled persons are willing and able to rise to the challenges and risks of a normal lifestyle

(Falta and Cayouette 1977). These challenges and risks are clearly articulated in the housing environment. Housing policies to promote independent living embrace three different areas of concern:

Adding Support Services: Integration of support services into housing often shifts responsibility from health and social service agencies to municipal and provincial housing corporations or to community non-profit providers. Integration broadens the possible combinations of community services and living arrangements. Provision of housing becomes a shared responsibility between housing providers and social service agencies that requires more collaboration among providers, disabled persons, and their families. Different disabilities have differing requirements for support services. As an example, not all accessible units need ground floor access or adherence to high-level fire codes and housing standards.

Small Group Living Arrangements: Cooperative and small-group living in a semi-integrated setting may be included in communal residences, group homes, halfway and transition houses, and residential treatment or care facilities. Such a living arrangement is an option for some physically and psychiatrically disabled persons. It is most appropriate for those who need a maximum level of supervision, constant companionship, or attendant care. It is also appropriate for those who want to live within a more communal arrangement for longer periods of time where some housekeeping and food services are provided.

Home and Community Adaptations: For many years, these adaptations were achieved through personal arrangements or charitable community projects. In home adaptation, physically disabled persons remain with families or friends in their own homes or apartments. Over the last fifteen years, CMHC and other agencies have produced design and planning guides to aid selection of options for home and community adaptations. Modification, renovation, or upgrading of the dwelling can be undertaken to make it more suitable for the disabled person's requirements. The renovations may be fairly extensive including an elevating device or an accessory apartment to accommodate a live-in companion or assistant. The delivery of support services must also be considered. Home-delivered community services are needed, but the adequacy of service provision may depend on location and distance. Hospitals, rehabilitation, and treatment centres are often concentrated in one area of a community, and few outreach services are provided for necessary maintenance, monitoring, or therapy programs that disabled persons may need in their own homes.

As the 1981 *Obstacles Report* suggests, independent living is more cost effective over the long run, and housing providers now have a mandate within the NHA to build integrated housing. The most suitable funding source for groups building for disabled consumers continues to be the non-profit housing programs and the cooperative housing programs. However, other programs such as residential rehabilitation assistance can be just as suitable.

For integrated housing to be successful, it must facilitate self-sufficiency and entry into the mainstream of social existence – recreation, employment, and education – the ability to interact and share with others. The key to this objective is mobility, accessibility, and sensitivity. Sensitivity to personal needs is emphasized in the Mayor's Task Force report in Toronto (1973): "The handicapped and the elderly, like the non-handicapped ... should be able to visit friends and relatives without running into embarrassing, sometimes insurmountable problems of entry, getting around inside, using the bathroom, etc." Sensitivity goes further to include promotion of greater understanding of physical and mental disabilities and support for community service programs from within housing development sectors, local governments, as well as among consumer groups.

Psychiatrically Disabled Persons
The needs of psychiatrically disabled persons are different from those of the physically disabled. More than any other segment of the disabled population, those with psychiatric problems have been part of large-scale social integration experiments in most provinces that have led to more outpatient treatment and deinstitutionalization. The impetus for deinstitutionalization came from parents and families, mental health professionals and care providers, and provincial governments convinced that homes for special care were not cost effective for this group. In British Columbia, for example, well-monitored community and small home environments were preferred because these could be seen as "an individual's home," and because persons with mental disabilities needed to be in a residential environment (Canadian Council on Social Development 1985, 13).

Attitude and preferences for accommodating the psychiatrically disabled changed quickly, starting in the late 1970s. There was little time to develop and promote integrated and semi-integrated housing stock with a more normal living environment and established links to support services. As a result, how well the psychiatrically disabled are housed is still an open question. In a 1979 position paper, the Toronto-based organization, Community Resources Consultants, stressed that between 1960 and 1970, "new methods" resulted in a 40% reduction in hospitalized patients. This led to a growing gap in the provision of housing and community care for a group of adults who needed transitional living arrangements and better aftercare services. As shown in a 1982 report on deinstitutionalization, the psychiatrically disabled have had to rely on boarding homes and rooming houses (Parker and Rosborough 1982). Boarding homes vary in size and focus. Homes for special care offer room and board, twenty-four-hour supervision and are licensed by the Ontario Ministry of Health. Only those discharged from a psychiatric hospital may live there. Domiciliary hostels also offer room and board and twenty-four-hour supervision, but are unlicensed. Commercial boarding homes are unlicensed and vary in price, usually offering only room and board. In addition, a few halfway houses and group homes are operated by non-profit organizations in smaller-scale, more individual-oriented settings. Some of these residences restrict length of stay from six to eight months. Privately-owned rooming houses vary in size, condition, and quality of living arrangements. Municipal non-profit housing corporations in Montreal, Ottawa, and Toronto, for example, have begun to develop and upgrade existing rooming houses, improving living conditions for disabled persons dependent on this housing stock.

The Special Needs of Disabled Women

The female population forms over 50% of the population with one or more disabilities in Canada. Younger women have more problems with mobility, agility, and sight. The lives of disabled women are also governed by life changes including marital, social, and economic influences. Younger women want to move out of protective institutional environments, and increasingly, disabled women want to live on their own. In the planning and development of supportive housing for disabled residents, a number of women's needs have not been adequately considered: parenting alone, location close to work, family, and friends, and the need for privacy (McClain and Doyle 1984; MacDonnell 1981). Because of the demands of employment and family life, disabled women also need to travel and to move to other communities for opportunities like everyone else. Supportive housing for disabled women must encompass a broad range of services with enough flexibility in management and organizational structure to accommodate changes in a disabled woman's life and lifestyle.

WOMEN AS VICTIMS OF DOMESTIC VIOLENCE

A 1982 brief by the Canadian Advisory Council on the Status of Women estimates that one in ten Canadian women will be abused or battered by her husband over her lifetime. The incidence of battering cuts across social, economic, cultural, and geographic lines. The degree of injury varies, but it is known that some women are battered while pregnant or after birth of a child, and many women suffer repeated assaults. Some women become disabled; others are already disabled. In one of the first major books published on domestic violence, Martin (1976) outlines a compelling argument for why some women suffer repeated abuse – they find the alternatives of no place to go with their children, or of living alone, more frightening.

Several conferences in the mid 1970s and local community efforts have resulted in the growth of specific services for abused women and in the development of a network of emergency shelters and transition houses across Canada. Shelters offer temporary housing usually up to a maximum of three to four weeks. Supportive services such as crisis intervention, legal counselling, and referral to social services and social welfare assistance programs are usually offered as part of shelter programs. Transition houses offer longer-term accommodation usually from two to six months. They are often better equipped to help care for children and to provide anonymity and safety, as well as training and education. Often, women remain in transition houses longer than expected because suitable permanent accommodation cannot be found. Shelters and transition houses usually have few paid staff and operate with a large contingent of dedicated volunteers for daily services as well as for management and fund-raising. The funding and development of transition houses, like sexual assault centres, has been difficult and unstable. (Norquay and Weiler 1981; Allen 1982)

The first women's shelters were provided by charitable organizations such as the Red Cross, YWCA, and religious groups primarily in Canada's largest cities – Vancouver, Toronto and Montreal. With the focus on women's needs provided by the 1970 Royal Commission on the Status of Women and 1975 International Women's Year, new attention was brought to women's legal rights and family violence. Out of these discussions came the initial federal and provincial funding for women's emergency shelters and support services. CAP provides for equal federal cost sharing of operating expenses with the provinces.

As of 1982 the National Clearinghouse on Family Violence estimated 145 shelters and transition houses in various stages of development across Canada. CMHC statistics on women's shelters and transition houses combined with figures from the Solicitor General show a slightly lower total figure of 113, with 6 of these shelters directly serving Aboriginal women. Ontario and Quebec have the largest number of shelters; Saskatchewan has the most facilities per capita.

Because shelters are still largely concentrated in urban centres, they face increasing demand for services. Social values about the benefits of reporting have changed; there is better follow-up of child abuse and assault victims whose cases may uncover a history of domestic violence; and these centres must accommodate forced migration because of a lack of services in smaller towns and rural areas and increasing cases of homelessness resulting from displacement. According to the Advisory Council on the Status of Women (1982), 45% of the Canadian population in 1980 lived in areas without access to transition houses or shelters. Women and children in rural or isolated areas have few alternative institutions available. Where shelters have been developed, they are frequently overcrowded – turning women and children away – depending on the time of year, availability of social services, and other factors. Allen (1982) notes that in 1981, 33 transition houses in Ontario accommodated 10,332 women and children and turned away another 20,000. The provincial organization for shelters and transition houses in Quebec estimated that only 12% of the women and children needing services are accommodated by member facilities (Canada 1982).

Part of the problem with lack of transition houses and shelters results from uneven demand. There is much variation from one night to the next. Occasional local publicity about domestic violence may serve to increase temporarily the number of women seeking assistance. Cost reimbursement is a second problem because it comes mostly from per diem payments for room and board to women and children eligible for emergency and social assistance. Amounts of reimbursement and the determination of levels of service vary by province and municipality, based on cost-sharing arrangements. Some women meet eligibility requirements for social assistance better than others. If women have travelled to seek shelter from outside a particular jurisdiction, then they may not meet the residency requirements. Another problem is the inability to relocate once women have used up their tenure in transition houses. It is expected many women need to stay up to three months in a transition house, but many more need additional time to search for employment, to participate in job training, and to sort out complicated legal problems and the status of their children.

Second-stage housing has been developed for women with longer-term housing needs. Cities such as Halifax, St. John's, Winnipeg, Regina, Calgary, and Vancouver have second-stage housing (Klodawsky and Spector 1985). Second stage housing provides a supportive living environment for displaced women and, given availability, it offers more extended residency depending on their needs. An independent living situation is provided in second-stage housing for women needing job and life skills training and an opportunity to rebuild connections with family and friends.

THE HOMELESS AND DISPLACED PERSONS

In a 1961 study of homeless transient men, the Canadian Welfare Council characterized the majority of the population as "on the move" both within provinces and across Canada. These men passed through a community only staying for a short period of time. Some men adopt the traditional transient lifestyle and are cut off from family and friends. Others are migratory or seasonal workers transplanted from their home community because of poverty, lack of work, or unsuitable living conditions. During periods of recession and economic boom, the numbers of migratory workers increase considerably in some Canadian regions. The same situation occurs in rural communities during peak harvest time. A third group consists of unattached residents in urban core areas who have made skid row their home. The 1961 study differentiates between mobility and transiency of this population. Essentially, persons with a certain amount of mobility usually have expectations, plans, and independent means to underwrite their plans. Thus, migrant and seasonal workers have mobility to a limited extent while transients move without plans or because they have little income, no security, or chronic problems.

These definitions symbolized the population for whom the early municipal shelters and charitable missions were established. Since this time, urban core areas have undergone significant redevelopment, and the nature of the population without secure housing has changed considerably. The number of people with no secure housing or with chronic housing problems has reached record levels comparable to the Depression era. Skid row can no longer be confined to specific neighbourhoods or represent a particular "way of life" (Ward 1985). Homelessness represents a multitude of problems today that have a cumulative effect on cities and the providers of emergency housing assistance. Canada does not have a federal policy directed at the needs of the homeless. Responsibility falls on municipalities, non-profit organizations and a network of volunteers to provide the bulk of housing assistance. In the 1980s municipal non-profit housing corporations have also become more involved in providing housing. If the homeless are eligible, then provincial social assistance and social service programs provide the major means of support, in addition to federal pensions, old age security, and vocational rehabilitation assistance programs. Some homeless persons work most often in short-term, low-paying jobs.

Single Men Still a Majority

In Metro Toronto's profile of the homeless population, *No Place To Go,* (1983) an estimated 3,400 homeless were documented with no fixed address through a survey of existing shelters, hostels, and social service agencies in Toronto. Single men accounted for 77% of all hostel residents. Persons aged 18 to 24 years made up the largest proportion of hostel residents and agency clients. Most of the hostel residents came from within the boundaries of Metropolitan Toronto, giving unemployment or transiency as their main reason for seeking hostel accommodation. Less than 10% of hostel residents reported they were post-psychiatric, alcohol or drug treatment clients. Given the young age of this group, it was surprising that only 8% originated from family homes. Most young homeless men had lost private rooming house accommodation due to eviction or rent increases. A 1985 survey of men in an Ottawa Salvation Army hostel confirmed these results. Approximately 47% of the users were under age 30, 54% of this group had some high school education, and 78% reported drinking problems (Ontario, Ministry of Housing 1986, 56).

A 1979 study by the Downtown Eastside Residents Association of their Vancouver neighbourhood noted a decline in the number of rooming houses from 1,200 in 1973 to 495 in 1979. Generally housing conditions were poor. The majority of rooms averaged less than 120 square feet (10.89 square metres) with no private bath and inadequate wiring. Over half the population was older men over age 50 years, and most had lived in the area for at least 13 years. At least 10% of the male population was difficult to house, even in older hotels or rooming houses, because they were chronic alcoholics. Women were a small minority of the population in this area. Most women were over age 35 years, and more than half lived alone. Despite the higher mortality rate of women in the area compared to men, they remained because of "the small town atmosphere and drop-in centres" which provided more of a feeling of home than other parts of the city. In addition, many of these women were too young for subsidized seniors housing and had a lifestyle or problems which prevented them from finding more secure accommodation (Hooper 1984).

Developing Permanent Accommodation

The 1983 brief by the Social Planning Council of Metropolitan Toronto states that the problems for the homeless are no longer temporary or emergency conditions; they are long-term and "no single type of permanent accommodation will meet all needs." A range of housing options is needed, including renovation and new construction of rooming houses and flop houses; low-cost subsidized apartment housing suitable for singles and families; and cooperative housing or communal residences which provide direct supportive services. As well, alternative housing must be provided that accommodates the longer-term, dependent population residing permanently or repeatedly in emergency shelters and hostels. These residents take up what little space is available for crisis housing.

Many homeless are housed by social service agencies in unsuitable accommodation that is sometimes too costly or in poor condition. Agencies may end up housing women with children on a temporary basis in hotels or motels not located in the safest of locations. Often, the homeless lack social and life skills which make them incompatible with other roomers or tenants. This lack of skills also presents difficulties when dealing with private landlords or public housing authorities. Trying to search for a job when your home is a hostel presents innumerable difficulties resulting in further discrimination against the homeless.

Where the gaps continue to exist is outside of Canada's largest cities for families, particularly women with children, and for those who face chronic unemployment. There are few housing services for the homeless who are confined to smaller towns and rural and remote areas. The difficulties of securing family rental housing in urban and rural areas have increased as well. Women with children face discrimination in tighter rental markets because of their children. The rental allowance limits under social assistance payments also make finding housing more difficult (Mellett 1983, 32). In rural areas, few organizations are able to develop low-cost alternative housing.

Unique Special Needs Housing Projects

What follows is a brief description of selected housing projects developed across Canada. These projects represent a range of housing types, and while this list is not exhaustive, it exemplifies projects with unique characteristics.

Cheshire Homes 1972-86: Canada. Semi-independent homes for young physically disabled persons modelled on small-scale homes, these are found in British Columbia, Saskatchewan, Ontario, Quebec, and Nova Scotia. Newer residences have separate units and several apartments to promote independent living. Tenants live alone or in a more communal setting and participate in the operation of the home.

Regina Native Women's Centre 1971: Regina. A network of 47 single detached dwellings dispersed throughout the City of Regina, most homes are rented to single Aboriginal women with children, but services are offered to all women. Housing support, employment, education, health, and social services are provided.

Kuanna Housing Cooperative 1977: Edmonton. A non-profit cooperative that provides housing and a supportive environment in an integrated setting for physically and mentally disabled persons. Care and assistance is provided by live-in non-disabled residents. All residents participate in management and serve on the board and committees which control the small-scale, linked dwellings.

Jack's Hotel 1979: Winnipeg. Jack's is a renovated single room occupancy hotel serving mostly older men. Currently, the non-profit residence is run in cooperation with a resident's association.

Constance Hamilton Cooperative 1982: Toronto. This cooperative housing project, a 31-unit townhouse complex near downtown Toronto, was designed primarily for women with young children. All units can accommodate families and have a large combined kitchen and dining-room area. The cooperative also has a six-bedroom second-stage housing component that allows women-in-transition to stay from six months to a year.

Homes First Society 1985-6: Toronto. Each floor of the 17-unit high-rise building has two large apartments housing from four to six residents. Some of the apartments house older and younger singles (both men and women), and one cooperative apartment is for women with children. While there is no specific programming in the building, several centres and services are nearby.

Résidence Esplanade II 1983: Montréal. A unique residence for severely disabled adults, twenty adapted apartments are located on two floors in a recycled school building of recent vintage. Care is provided on almost a one-to-one basis by jail inmates (both male and female) who work for community service points toward their parole.

Adsum House 1983: Halifax. Accommodating up to eighteen women daily, its clientele are primarily single women (over 16 years of age) who are homeless and transient – the only shelter primarily serving this clientele in Atlantic Canada. The house offers short-term accommodation and houses a few women in transition from the Halifax County Correctional Centre for up to three months. Services offered include counselling, referral, and assistance in finding permanent accommodation.

Conclusions and Recommendations

Only recently have governments in Canada recognized special needs consumers as targets of housing policy. Governments have moved towards adapting and developing housing to fit the special needs and requirements of these consumers. Complementing this, philosophies of care and service delivery have changed considerably over the last forty years. Governments now provide a spectrum of support in care and services: for example, re-established care facilities in existing higher density housing, integrated housing with on-site support services, and existing residences adapted to accommodate consumer needs. They have expanded their commitment to de-centralized, home-based care and service delivery, and they increasingly recognize that the circumstances of special needs consumers may change over the course of their lifetime.

Developing housing aimed at special needs consumers is a difficult task because the population is diverse. The primary groups are different in age, health condition, and mental health status; and the groups differ in access to employment and income support. Nonetheless, the problems of securing permanent housing with adequate levels of support services are much the same in these groups. The existing private-sector low-income rental housing stock continues to shrink; the public and non-profit sectors are unable to keep up with the demands for housing by displaced residents; and de-institutionalization has swollen the population of special needs consumers in some communities. While some progress in developing supportive housing has been made in Canada, much remains to be done to achieve access, security, and independence for special needs consumers.

To ensure that special needs consumers are considered in housing policy and program development, better research data and documentation are needed. Information about tenure in the stock of accessible housing and access to the delivery of supportive services is lacking. Also little is known about how special needs consumers access housing and support services. Information in the existing local community case studies and program evaluation reports needs to be collected systematically and shared across Canada. As well, much of the research on public and non-profit senior citizen housing is now dated. A new look needs to be taken at how senior housing has aged along with its residents. Similarly, congregate housing and small group living arrangements for older adults and younger people need to be the focus of research so that their effectiveness as long-term care facilities and distinctive types of housing services are better monitored.

The assessment of social housing policy and programs has to be recast to meet the challenges posed by community-based delivery of care and support services. Housing takes on new meaning and functions, and standards and features must be adapted accordingly. To point out just one difference, care deliverers are as much users of the housing as the residents. Also, each housing type has its strengths and the ability for individual consumers to choose is important. Beyond bricks and mortar, consideration in future program development must be given to financial arrangements; improved maintenance and repair; access to home adaptations in private as well as publicly-subsidized housing, access to childcare, food services, and other community services; and access to a system of social supports that evolves with changing needs. These are essential to housing's future performance as part of a human service network fostering independent living for all special needs consumers.

Notes

1 I would like to express my gratitude to the following people for their thoughtful comments: Peter S.K. Chi (Cornell University); and consultants Novia Carter (late of

Winnipeg), Ladia Patricia Falta (Montreal), Sylvia Goldblatt (Ottawa), and Myra Schiff (Toronto).

2 According to Priest (1985), the divergence in the choices of living arrangements among elderly men and women is "due largely to differences in their marital status." His analysis of the 1981 Census shows an increasing proportion of widows and widowers in the oldest age groups. Among older women, more live alone among those aged 55-59 years and age 75 years and over.

3 Schwenger and Gross (1980, 251) report that the corresponding rates were 5.1% for England and Wales (1970-1) and 6.3% for the US (1973-7). See also Schwenger (1977).

═══

CHAPTER FOURTEEN

Post-war Social and Economic Changes and Housing Adequacy

Damaris Rose and Martin Wexler[1]

THE HOUSING environment has been profoundly influenced by three changes in Canada's social and demographic structure since 1945: (1) the growing number of elderly; (2) increased labour force participation among married women with children at home; and (3) the growing number of mother-led families. A consideration of the scale, scope, and impact of these changes on the housing environment, in turn, raises questions about the extent to which housing and neighbourhood design can respond to changing needs, about the new aspirations of the groups affected by social and demographic change and how they have adapted to existing residential environments, and about what generally constitutes adequacy of housing and neighbourhood.

In Canadian cities, it has been commonplace for households in different life-cycle stages to live in distinct neighbourhoods. Moreover, it is often assumed that there is a best form of housing for each stage of the family life cycle; with transitions between stages, the household is expected to move to a "more suitable" location (Stapleton 1980). In contrast to the dramatic social and demographic changes, much of the new post-war housing and many new post-war neighbourhoods were designed and sited for the life-cycle stage in which one adult stays home full-time to take care of a family. However, the traditional nuclear family with a male breadwinner, young children, and a full-time housewife is today in the minority.

The increased diversity of household types and flux in ways of living necessitate a reconsideration of the appropriateness of current housing and neighbourhood design. What should and can be done to increase the flexibility of residential environments, to allow for a greater range of choice in living arrangements, and to provide social support for major life-cycle transitions? As their needs have changed, the aforementioned three groups have developed different strategies to adapt existing housing and communities or have acquired new housing aspirations. In discussing how the three groups have coped, it is assumed in this chapter that these strategies or aspirations reflect a degree of choice, whether individual or collective.[2]

Housing and the Elderly

Over the past half-century the population aged 65 years or older in Canada has burgeoned – up fourfold to more than 2.36 million in 1981 compared to 1931. The number aged 75 years or older rose even faster – up fivefold over the same period. This aging of Canadian society will become even more important over the next half century as the post-war baby boomers eventually move into this age group.

The number of households with an elderly maintainer or head also increased significantly: up 832,000 households in 1981 over 1951.[3] The importance of this increase does not simply lie in a need for additional units; many elderly already had homes and often wished to remain in them. As important is the use of existing environments or creation of new ones to meet the needs and lifestyles of this aging population.

The stage of the conventional life cycle that follows childrearing is often divided into two phases (largely resulting from the standardization of retirement age and from increasing longevity, especially among women). The first represents the time when both spouses are alive and generally in good health. The second is typified by the death of the male spouse and by onset of chronic health problems. This phase may, in some instances, be longer than the period of employment or the period that the couple lives together.

As the number, proportion, and typical age of the elderly increase, the residential strategies used to cope with aging and concomitant economic, social, and physical changes have heightened significance for the use and adequacy of existing housing and neighbourhoods. Possible strategies include aging in place, living alone, and living in institutional environments.

AGING IN PLACE

Although many elderly move to smaller, possibly more convenient dwellings, the elderly as a group are less mobile than the rest of the population (see Stone and Fletcher 1982). And while the phenomenon of aging in place is not limited to elderly home owners, greatest attention has, in the past, been paid to this group (see Wexler 1985). A majority of elderly households (63%) were home owners in 1981.[4] Even among elderly living alone and elderly aged 75 years or over, about one-half were owners in 1981 (*Census of Canada 1981*).

Why do so many elderly stay on in a family home for as long as possible? The elderly often have strong emotional attachments to their homes and are accustomed to familiar surroundings. For owners, the out-of-pocket costs of their housing may be low. Also important are the stress, effort, and cost of finding more suitable housing, of disposing of or moving possessions acquired over a lifetime, and of setting up the new home. Owners experience the additional stress of having to sell their homes. Finally, the options available may, unless subsidized, be no more attractive or inexpensive than their present homes.

Aging in place is, of course, just one alternative. Increasing numbers of elderly are choosing to purchase condominiums and to move to retirement communities. While these tendencies are important, especially among elderly living in major urban centres, aging in place will probably continue to represent the preferred choice for many elderly.

It is sometimes argued that elderly occupy more space than they need, want, or can afford. In so doing, it is claimed, they continue to live in units that are unsuitable for their needs, and thereby inhibit a more efficient allocation of housing, especially the use of larger dwellings by young families (Myers 1978). At the community level, it is thought that aging in place results in an inefficient use of community and service infrastructure (Lewinberg Consultants Ltd. 1984).

In terms of selling price, elderly owners consume only modest amounts of housing. Stone and Fletcher (1982) show, using the 1974 Survey of Housing Units microdata sample (SHU), that 28% of owners aged 75 years or older reported an expected selling price under $4,000 per room.[5] Part of the explanation for this is that the dwellings owned by the elderly are older. Brink (1985) reports, for example, using the 1976 *Survey of Consumer Finances*, that half of all elderly owners live in dwellings built before World War II.

However, available data indicate that the elderly use more space (in terms of quantity if not necessarily quality) than do the younger population. Stone and Fletcher (1982), for example, use square footage counts obtained from SHU to show that elderly individuals consume more space on a per person basis than do the younger population.[6] Using bedroom counts, Stone and Fletcher (1982) and Brink (1985) show again that elderly use more space. According to SHU, 78% of elderly households have at least one bedroom per person in contrast to only 45% of non-elderly households.[7] Further refining this measure, Brink suggests that 50% of elderly owners in Canada in 1976 were over-housed based on a standard of one bedroom per person.[8]

Rather than moving to cheaper or easier-to-maintain dwellings, elderly people may decrease out-of-pocket expenditures by reducing maintenance. This practice may reflect limited know-how (especially in the case of older widows) or merely the decreased saliency of building maintenance to the elderly. Based on US data, Struyk and Soldo (1980) suggest that elderly owners under-maintain their homes. Available Canadian data, however, do not indicate that elderly owners do less maintenance than the younger population. Based on SHU, identical proportions (92%) of elderly and non-elderly home owners living in urbanized cores of major census metropolitan areas report their dwellings to have "good" exterior conditions (Stone and Fletcher 1980). Similar results are observed in the 1982 Household Income, Facilities, and Equipment microdata sample HIFE] in which 12% of elderly and 13% of non-elderly households report their dwellings to be in need of major repairs (see CMHC 1986c).[9]

LIVING ARRANGEMENT AND AFFORDABILITY

Women predominate among the elderly, and many of these women live alone. The reasons for this have been well documented by Stone and Fletcher (1982). Our understanding of the "preference" for living alone or for privacy is, however, less well developed. Certain other living arrangements, such as boarding and lodging, have become less commonplace (Modell and Hareven 1973). The elderly also have more money available for housing-related expenditures as a result of increased income through not only direct sources such as improved pensions, property tax rebates, and housing allowances but also other, more indirect sources such as free medication and reduced public transportation fares. In addition, the need to save to pay for health and medical care has been reduced by medicare, while even nursing and home care is covered in certain provinces.

The public and non-profit housing programs, which have resulted in the construction of over 200,000 purpose-built units since 1953, have also contributed to the stock of highly affordable bachelor and one-bedroom units reserved for the elderly.[10] Numerous other federal and provincial programs have also been developed to make housing more affordable to the healthy elderly, including rent supplements and cooperative housing.

LIVING IN AN INSTITUTION

Although the vast majority of elderly continue to live in private dwellings, a growing number will spend at least some time in an institution even though a significant proportion of those living in institutions have been shown to be no more dependent than those in private dwellings. Institutional shelter in Canada consumes the largest part of provincial funds devoted to housing for the elderly (as shown by Corke 1986 and Renaud and Wexler 1986), although other countries have found ways of delaying or avoiding institutionalization through community support services and the availability of sheltered housing, that is, housing provided with supplementary services.

Although the proportion of institutional beds occupied by the elderly has kept pace with their growing numbers, or has even increased slightly, this increase is not dramatic, especially in view of the larger number of elderly 75 years of age and over and other factors, such as the reduced availability of daughters to assist their aged parents (Schwenger and Gross 1980).The numerical increase is nonetheless remarkable. During the period 1962-3 to 1976, 58,600 institutional beds for the elderly were added, representing a 53% increase; from 1976 to 1981-2 a further 53,400 beds were added representing a 32% increase. These figures represent significant capital expenditures (in which CMHC has played an important role) as well as important operating costs both now and in the future.

Again, however, the context has changed. There are more elderly living to a great age. Sometimes, their children are themselves aged; in other cases, their children are all in the paid workforce; and in such cases, professional care-providers typically assume much of the responsibility for the frail elderly.

FIGURE 14.1 :e by age of youngest
child, as a percentage of those with employed husband:
Canada, 1976-1985.

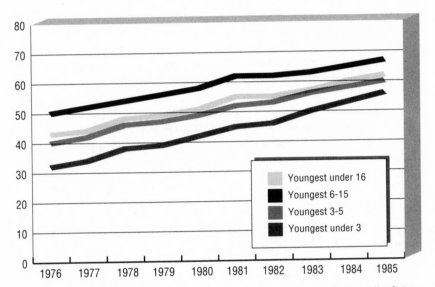

SOURCE: Statistics Canada, *Labour Force Survey*. Unpublished data. Micofiche.

Housing and Two-Earner Families with Dependent Children

In 1961 in only 18% of husband-wife families (with at least one child aged 15 years or under and with at least one parent in the labour force) [11] were both-spouses wage earners (*Census of Canada 1961*, Catalogue 93-520, Table 93). Yet as the 1960s progressed, "consumerism" spawned a need for higher household incomes to maintain the middle-class lifestyle, and expansion of the service economy created new jobs into which married women increasingly slotted (usually on a part-time basis, and once children were in school) (for example, see Armstrong 1984, Chapter 3). The decline in fertility and compression of the child-bearing years reinforced this trend. The result was an acceleration of the steady increase in married women's labour force participation rates that began in the early 1940s, when only about 5% of married women had paid employment (Eichler 1983, 44).

The increase in labour force participation of married women (with children under 16 years of age) became a social trend in the 1970s (Figure 14.1). *Labour Force Survey* data show that the participation rate of wives with at least one child under 16 years of age stood at 57% in 1984, up from 44% in 1977 (Statistics Canada 1977-84). [12] During the early 1970s most of the increase in married women's labour force participation was among by women without pre-school children, but since 1976 labour force participation has surged among married women with children under 3 years old (Figure 14.1). Well over half of this group are now in the labour force despite the shortage of daycare, and the difficulty of juggling job and family (Truelove 1986). The growth in importance of two-earner families with pre-school children may in part have been a coping strategy for the decline in real earnings of male spouses experienced since the late 1970s (Gingrich 1984; Pryor 1984), and the rising real costs of home ownership in the late 1970s and early 1980s (Langlois 1984; Social Planning Council of Metro Toronto 1980).[13]

SETTING UP HOUSEHOLD IN WELL-SERVICED NEIGHBOURHOODS

For the dual-earner family, an efficient organization and division of time between waged work, domestic labour, child care (possibly also the care of an aging parent) and other household maintenance activities is crucial. Given the spatial separation of homes and neighbourhood units from workplaces as well as commercial and community services in larger urban areas, this need to manage time carefully may well have shaped household residential location strategies (see for example, Michelson 1985).

Accessibility to good daycare and after school care can be important to women's decisions to enter the labour force and to whether they work full-time or part-time (Michelson 1983). In the absence of formal daycare in suburbs and smaller communities, another strategy entails the use of informal childcare networks and unlicensed daycare centres, which often exist in contravention of zoning regulations (Mackenzie 1987). Such informal childcare is dependent on the presence of non-employed women in a nearby location (as pointed out by Truelove 1986); an increasingly scarce resource in many suburban areas.

Given these requirements for well-serviced neighbourhoods with good public transportation, two-earner families tend to gravitate toward higher-density areas. In practice, this often translates into the older, inner housing stock in larger cities (Wekerle 1984), neighbourhoods whose relative location itself facilitates time management in a schedule built around multiple roles; although one American study points to similar advantages of nucleated "exurban" communities (Genovese 1981).

PHYSICAL CHANGES TO THE HOUSE: RENOVATING OLDER BUILDINGS

The selection of a housing unit whose design is more efficient for the carrying out of household tasks can be another way to manage time. Within the older stock, this may involve remodelling of kitchens to include an eat-in area and to make space for appliances such as a dishwasher and a microwave oven.[14] Gentrification is partly a response to the lack of flexibility in suburban neighbourhood design to accommodate the schedules and lifestyles of two-earner families (Rose 1984; Wekerle 1984) or, indeed, families with adolescent children not old enough to drive (Social Planning Council of Metro Toronto 1979). Unlike in the US, Canadian "gentrifiers" are more likely to be families with children (Rose 1986). New townhouses in the inner city – one kind of gentrification – tend to be built for two-earner, childless couples and are often located on redevelopment sites close to the CBD; these offer fewer facilities for the two-earner family with children than does a renovated older house (or large "plex" apartment) in an inner-city neighbourhood well-equipped with small parks and community services (Klodawsky and Spector 1984).

Housing and Lone-Parent Families Headed by Women [15]

In 1981 there were 397,000 mother-led families with at least one child under 18 years old at home in Canada. The age structure and marital status of lone mothers with minor children have changed, especially since reform of federal divorce legislation in 1968. In the Montreal CMA, for instance, the proportion of all single mothers who are under 35 years reached 41% in 1981, up from 34% in 1971. By 1981 two-thirds of lone mothers with minor children were either separated or divorced, whereas previously many had been widows (see Rose and LeBourdais 1986). [16] Although many women do eventually remarry, remarriage tends to occur after many years of lone parenthood (Klodawsky, Spector and Rose 1985, Chapter 4)

Mother-led families are poor, and lone mothers under 35 years are the poorest among these (Klodawsky, Spector and Rose 1985, Chapter 5). [17] A family with school-aged children that experiences the loss of the husband often must choose between staying in place (often in a suburban home) or moving. Whether the family continues in owner occupancy depends partly on the monthly payments given what is usually a reduced family income, and partly on whether the lone parent can physically handle the maintenance of their dwelling. In spite of the difficulties, the lone mother left with the family home may well simply accept "excessive" housing expenditures because no cheaper accommodation is available in the area. For lone parents who work downtown, the alternative may be to move to the inner city in the quest of affordable (if smaller) quarters and socially supportive environments.

Mother-led families are rarely able to buy their way out of the problems of inadequate community environments. Low incomes mean that they are less likely to own a car, or even a washing machine, the most time-saving of all household appliances. [18]. Low rates of ownership of those commodities underline the importance, for this group, of proximity of the home to public transportation and commercial services.

Younger, never-married or separated lone parents suffer the worst affordability problems: for instance, in Montreal in 1981, 50% of separated lone mothers paid over 35% of their income in rent, and one-third, over 50% (Rose and LeBourdais 1986). Doubling-up with parents or other relatives (often in single-family houses in the suburbs) is one strategy used by the recently-separated for economic and emotional support; interestingly, in the Montreal region, separated lone parents predominated among those living as "secondary families" (that is, living in a dwelling where others are responsible for major household payments), and one in ten separated lone mothers headed a secondary family.[19]

Among lone-parent family home owners under 35 years of age, 11% own condominiums, compared to only 3% for young husband-wife families (Statistics Canada 1984a). A sample study in a few "gentrifying" neighbourhoods in Montreal found that 10% of the purchasers of renovated condominium and undivided co-ownership units were middle-income female-headed families (Choko and Dansereau 1987).

Non-profit housing cooperatives provide modestly-priced housing to a socially and economically-mixed clientele.[20] This sector, though small, provides significant housing opportunities for lone parents. Cooperative housing can also offer a rent-geared-to-income subsidy for low-income lone parents. Yet many lone parents of modest incomes (in between those of lone mothers in condos and those in public housing) live in cooperatives without receiving an individual subsidy. This is especially the case in cooperatives converted from existing rental units, which are the most common form in the Maritimes and Quebec and especially in the Montreal area where, furthermore, more than one in five cooperative households was a female-headed lone-parent family in 1982 (Klodawsky, Spector and Rose 1985, Chapter 10). In renovated cooperatives, rents have typically been lower than in the new construction cooperatives which predominate elsewhere in Canada (Simon and Wekerle 1985).[21] Cooperatives have the great merit of almost absolute security of tenure and predictable housing expenditures, both of which are important for a lone-parent family.

Non-profit housing corporations, run by municipalities or private bodies, also form part of Canada's social housing sector and have also played a small but valuable role in offering housing alternatives for lone-parent families. Unlike cooperatives, these corporations are run by a management board without any requirement for resident involvement. They tend to be larger bodies than cooperatives and may thus be able to acquire land and negotiate with developers. In Halifax, for example, the municipal Non-profit Housing Society has been successful in providing low-cost housing for local residents on low incomes, about 40% of whom are lone parents (Klodawsky, Spector and Rose 1985, Chapter 9). Some lone mothers, especially those coming out of a crisis-ridden domestic situation, may prefer this type of non-profit housing to a cooperative in which they have to be involved in management and maintenance from the outset. Thus, non-profit housing corporations also have a particularly important part to play in second-stage housing for battered women needing to live in an environment providing multi-faceted support but permitting personal autonomy (Wekerle 1988b, Chapter 6).

Many lone-parent families remain dependent on the private housing market.

There is an increasing spatial dispersion of lone-parent families within metropolitan areas: the growth of lone-parent populations in suburban apartment areas is a striking phenomenon, for instance (Rose and LeBourdais 1986). Yet many still come from inner city neighbourhoods and want to stay in them for the services they offer; in contexts of gentrification and inflating rental housing markets, they are vulnerable to displacement. Strategies adopted include apartment sharing with people not related by kin, for instance, another lone-parent family. Such strategies are more difficult where family-sized apartments are scarce or where sharing is not permitted.

For female-headed families, all of these housing strategies have a social support element. Further research is needed on this topic, but it may be that, for those purchasing condominiums, price is not the only consideration, particularly in the case of relatively inexpensive undivided co-ownership property in well-serviced neighbourhoods (particularly favoured by women purchasers, according to Choko and Dansereau 1987); in Montreal, for instance, social networking and exchange of services are facilitated by the "plex" format (that is, superimposed flats with separate outside entrances) of much of the newly-converted co-ownership units (Dansereau and Beaudry 1985).

For other lone parents, the cooperative formula may fulfil similar needs and provide similar security, while in addition often enabling creation of improved facilities for children, such as shared outdoor play areas easily supervised by adults (Klodawsky, Spector and Rose 1985, Chapter 10; Rose and LeBourdais 1986). Farge (1986), Wekerle (1988a), and Simon (1986) discuss the active role of women in organizing cooperatives and the benefits of this form of self-management, especially to those with little experience in the world of employment. However, the frugal standards to which many new construction cooperatives are built and their often-peripheral locations do much to undermine these advantages, which can be seen much more strongly in renovation cooperatives in inner-city Montreal. Furthermore, although daycare is heavily used by lone parents, efforts to include daycare centres in cooperatives have usually been frustrated, mainly because of funding criteria which place tight restrictions on non-shelter components within a project (Simon 1986; Wekerle 1988b, 157).

In contrast, the present way that public housing is organized in Canada does not offer these same advantages of control over one's environment and a sense of empowerment, although social support networks do exist (Klodawsky, Spector and Hendrix 1983). Nor has there been much progress in establishing daycare and other community services in larger older projects in which large concentrations of lone-parent families live. Nonetheless, in 1981, over one-quarter of public housing tenants were female lone parents (unpublished CMHC data). However, women have been in the forefront of public housing tenants' campaigns for greater involvement in management of their housing and for services appropriate to families (FRAPRU 1984; Klodawsky, Spector and Rose 1985, Chapter 9; Sirard *et al.* 1986, 61-8).

Discussion

The growing numbers of elderly Canadians, of two-earner families with children, and of lone-parent families reflect and illustrate the diversification of family and household types. These three should not be considered narrowly as client groups as regards housing issues. The sheer numbers of elderly, for example, point to the importance of accepting that aging is not a special circumstance and that becoming physically frail is not abnormal.

Getting away from the concept of client groups brings into question the notion that the elderly or lone parents, for example, act irrationally in order to stay where they have been living, or because they have to pay a premium for a particular location that suits their combinations of roles. This mode of analysis shifts the focus and the onus to the environments that are built or that have been inherited. To what extent are housing and neighbourhoods inappropriately designed for a growing number of Canadians? Do zoning and other planning controls, set up to ensure rational development, prevent or inhibit adaptation within existing communities?

Important strides have been made in providing for privacy within the self-contained unit; yet we have inadequately provided for community at the scale of the building and the neighbourhood. Examples may be found in non-profit as well as public housing. Even though such housing is possibly the largest source of purpose-built housing for the elderly population, these units were built without the provision of support services and are only available on a limited individual basis in those provinces which have developed home-care services. Once tenants living in such housing become frail, they are expected and in some instances forced to leave. One of the reasons for this is that the definition of housing adequacy in the case of public housing was and continues to be limited to affordability. Criticism could also be made of public housing now used by many lone-parent families. Public housing, unlike cooperatives, lacks a tradition of encouraging mutual service exchanges and social support and rarely provides needed community services such as daycare.

Much of our existing housing and neighbourhood design and planning were initially predicated on a view of a rigid correspondence between life-cycle and an "appropriate" type of residential environment, and also, less explicitly, on a particular gender-based division of labour within the family between employment or "career" and the day-to-day running of the household, rearing children, and attending to aged parents. All of these themes have been reflected and reinforced in the physical structure of building types and community forms and in the functional and spatial separation of homes and neighbourhood units from the activities of work and commercial, and community services (Wekerle and Mackenzie 1985; Willson (n.d.). Zoning regulations determined by municipalities have often reinforced generational as well as socio-economic segregation and have prevented neighbourhoods adapting as new demands have been felt.

The support and backup needed by elderly living alone, by frail elderly or families caring for frail elderly, by two-earner families with children, or by lone-parent families exceed those typically provided in most buildings or groups of buildings and communities. Some of the alternatives such as institutionalization have been found inadequate and costly. Providing single-type, age-concentrated environments for the elderly or for lone-parents, despite the possibilities of group solidarity and cohesion, presents problems that can only be resolved through integrated environments. For example, the effect of aging in unison of elderly living in public and non-profit housing has reached near-crisis dimensions in certain provinces.

In spite of this criticism of ghettoization, powerful arguments can be made for the value of social solidarity and mutual caring among those with similar problems. Spatial concentration may enhance such networks. These points have been made by organizations of lone parents and public housing tenants' associations (Klodawsky, Spector and Hendrix 1983; Klodawsky, Spector and Rose 1985). Similarly, many elderly prefer age-concentrated environments to mixed-age communities.

One of the common themes linking the three groups considered here is the need for environments that reduce the time, cost, and effort often required to manage tasks of everyday living. For two-earner and lone-parent families, it is a priority to use time efficiently and, for the latter, economically; for the elderly who often have surplus time, priority is given to the ability to circulate easily and safely, which is especially difficult in winter, as well as to cost. Yet the way that residential environments have been organized and managed rarely reflects these concerns, with the exception of the elementary school (see, for example, Social Planning Council of Metropolitan Toronto 1979).

Conclusions: Common Principles

One of the certainties that can be drawn from this chapter is the difficulty in predicting and responding to social and economic changes that affect housing and community. This is reinforced by a document published in 1970 by Central Mortgage and Housing Corporation on its 25th anniversary which recognized that

> ... amongst the factors that have affected the design of housing and the content of what is built, surely nothing has been more important than changes in the role of women. The married woman still remains the central figure in the family home but now the single woman has gained independence to set up another kind of household, living alone or with a friend or group.... Women have become the principal tenants of the new volume of apartment houses (1970, 34).

Remarkably, the authors failed to recognize the growing importance of the participation of women with young children in the labour force and of elderly women living alone, both trends already well advanced in 1970.

The three groups discussed in this chapter were chosen because of their quantitative importance and as an illustration of the need to evaluate critically and, where needed, to reorient our thinking about housing and neighbourhood design to make them responsive to the changes and needs outlined in this chapter. The elderly, for example, continue to age. With increasing age, they experience various types of losses which affect their capacity to live independently. These losses can be compensated for by various physical, service, and emotional supports. While many elderly prefer to move to smaller units or to environments with more formal support services, the vast majority prefer to remain for as long as possible in their own home, in which they may have lived for many years. When they do move, it is generally within the same neighbourhood.

With the growth in importance of two-earner families, a restructuring of "family time" and "work time" is taking place. Families need more options for time management and task sharing between household members. They need accessible community services that can totally or partially replace work formerly done by full-time homemakers and nurturers. Where the physical environment imposes time-space constraints and limits such access, it is typically only the more affluent who can purchase solutions (such as two cars, a renovated single-family home in a centrally-located neighbourhood and with a designer kitchen, and individualized childcare services). Those of more modest resources juggle their time and struggle to adapt their living environments as best they can.

Mother-led families have similar accessibility needs but their low incomes give them less choice and flexibility. In the initial period after relationship breakdown, social support services are particularly important, and a sheltered environment is sometimes needed. After this transitional phase, the family is likely to be more stable but often still economically vulnerable for many years before remarriage occurs. A range of inexpensive residential choices appropriate to different phases of the lone-parent experience, and permitting the flexibility to stay or to move, is therefore desirable.

All of these groups have adopted strategies that permit them to satisfy their needs and ambitions, often despite physical, regulatory, or administrative constraints.

We consider flexibility a principle rather than a response to particular "client groups." During the period discussed, these client groups have changed, while the houses and communities that have been built, and are still being built, continue. Moreover, one aspect of consumer adequacy must be the choice between continued residence either in the same dwelling or within the neighbourhood, and a number of reasonably attractive mobility alternatives.

Although in practice it is often impossible to distinguish between these two options, in this section we make certain recommendations about them. For instance, addition of an accessory unit to a single-family home for a divorced daughter and her child represents a type of physical flexibility (and flexibility in terms of municipal regulations), as well as providing for a more supportive living arrangement for all three generations. Our recommendations are far from exhaustive. They are intended to stimulate further debate, research, and experimentation at all levels of government, within the private sector and from the "grass roots."

Encourage new housing prototypes that facilitate more supportive living arrangements: Sheltered housing, group homes, housing complexes designed by and for lone parents to live in proximity to one another, and units to be shared by two unrelated adults (for instance, two divorcees each with a master bedroom but sharing kitchen and living room), represent special housing types which permit the creation of "families by choice." Not only do such arrangements facilitate the delivery of formal support services, they make possible informal services such as sharing of facilities and exchange of household services.

Encourage innovative ways of designing for kinship without forcing relatives to live together: There exists a richer range of kin groups and ways of housing them – superimposed plexes, accessory apartments, granny flats, and bi-family units – than has been considered in the past. New options will have to be considered as the role of women as kin-keepers changes and as the number of children available to care for aging parents declines.

Allow and encourage communities of mixed age and household type: There are numerous reasons for favouring such mix. Service needs can be "averaged out" and a wide range of community facilities furnished. If a choice of unit types is provided within a single community, people can remain despite changes in life cycle, income, and health status. Intergenerational contact and service exchange may be promoted, while social polarization may be reduced. The apparent contradiction between mixed and homogeneous residential environments can be resolved, so to say, "on the ground." Parts of or entire buildings or groups of buildings of moderate size can be geared toward a single group, while the community as a whole is heterogeneous. Land-use zoning is, however, a crude tool for obtaining such a mix, although certain innovative techniques are being tried such as performance or impact zoning and incentive, inclusionary, and age-restrictive zoning. Other land-use and architectural approaches that would provide mixed neighbourhoods also need to be adopted. An example of an architectural approach is the St. Clair-O'Connor Community in Metro Toronto which combines housing for different age groups and has different levels of support services within a single complex.

Provide better physical integration of housing and community services: Integration of housing and community services is especially important in suburban areas, both as they age and as families change. For example, planning requirements for new subdivisions should include provision for daycare, and subdivisions should be designed and located with access to public transportation in mind. Furthermore, services should be tailored to the evolving needs of the local population. For instance, locally-based health care, home care, and improved snow removal need to be made available as the population ages.

Encourage physical adaptations to housing and to the use of housing by people who want to remain in their homes, for example, the elderly and the lone mothers in the suburbs with teenage children: While the professional "ideal" has been to presume that households move when there is a change in their situation, many households modify or alter their homes or occupancy in order to remain (Teasdale and Wexler 1986). Public policy makers and planners must not only begin to acknowledge and tolerate such modification; they must facilitate and encourage physical adaptations to private owner-occupied and rental housing, favouring flexible housing forms that can accommodate shifts in life-cycle and income. Home sharing could thus, for instance, become a more widespread option.

Modify zoning and building occupancy regulations to permit dwelling modification and diversification of uses: Zoning and building occupancy regulations frequently constrain the use and modification of single-family homes: for example, in prohibiting or making extremely costly the addition of an accessory apartment or the use of one's home as a place of self-employment, community enterprise, or service such as a daycare facility. Households frequently make such adaptations – and now do so illegally – to accommodate various changes in the life-cycle and important ruptures in the family. Such adaptations represent often successful strategies for obtaining additional income or social support. In some cases they may permit households to remain in their homes where they would not otherwise have been able to do so.

Existing public housing needs to be restructured: As the largest source of purpose-built housing for the elderly and as currently used by lone-parent families, public housing should be restructured to respond to the evolving needs of these populations. It is important, for instance, to offer secure tenure to lone parents whose children have left home; presently, such tenants often suddenly lose their right to public housing and have no satisfactory alternative options. The French experience in improving public spaces, developing a full range of services and democratizing the management of "les grands ensembles" – vast suburban complexes of public housing, usually constructed in large slab or tower blocks – has shown that thoughtful renovation and adaptation of existing public housing projects can result in supportive environments with cost-effective on-site service delivery for low-income lone-parent families (Klodawsky, Spector and Rose 1985). Current experimental programs in public housing in Ontario, through the creation and support of group living arrangements, offer promising alternatives (Corke and Wexler 1986).

Encourage experimentation by cooperatives and non-profit housing corporations to provide housing that includes a service element, especially in inner-city areas undergoing gentrification and other neighbourhoods close to jobs and services: Given adequate funds, and within the limits of community resources, cooperatives, and other non-profit housing corporations have often provided innovative and supportive housing solutions for certain groups within the lone-parent and elderly populations, with respect both to housing design and the delivery of social services and health care, in addition to an environment that fosters informal social networks (Hayden 1984, Chapter 9). As well, non-profit housing in inner-city areas can be a means of allowing such groups to stay in place in spite of gentrification. We do not see "third sector" housing as a replacement for fully subsidized housing for these groups but rather as a complementary mode of provision deserving of strong financial and institutional support by all levels of government.

How successfully has post-war housing adapted to changing social needs? A succinct response to this question would be, "very unevenly." The actual physical structure of single family and plex housing, at least, is adaptable (Teasdale and Wexler 1986). Yet the capacities of the public and private sectors are restricted by perceived, although not necessarily real, market limitations. Probably more important still are the effects of zoning and building occupancy regulations. In the case of social housing, innovation has been limited by NHA provisions and by an orientation that emphasizes only physical adequacy, affordability, and suitability. The new federal-provincial housing agreements, that since 1986 have replaced previous modes of provision of social housing, may lead to a more flexible approach adapted to local conditions and particular clienteles. However, continued financial restraints do not give cause for optimism that adequate resources will be channelled into supportive services as well as into modestly-priced housing. With respect to the other sectors, zoning and public attitudes toward zoning, at least in major urban centres and in particular in areas dominated by single-family housing, seem to constrain adaptation and neighbourhood mix. Changing these attitudes – as has been attempted, for instance, by the Vancouver Planning Department (Vancouver 1986) – seems to require innovative physical and social design, a greater concern with coordinating housing and other services, and a rethinking of how neighbourhoods are viewed and managed.

Notes

1 The authors are equally responsible for this chapter. We thank Michael Ellis for research assistance on some statistical sources, and the editor and anonymous referees for comments on an earlier draft.

2 Significantly lower proportions of elderly in countries such as England and Wales, Sweden, and the Netherlands are housed in institutions compared to Canada, despite their generally older populations. See Schwenger and Gross (1980) and Brink (1985). In these countries, subsidized home-care services are more broadly available to elderly living in the community. As noted by Schwenger and Gross (1980, 253), "we have asked older people to pay for community-care services, while we have made it easy and cheap for them to go into hospitals. The incentive has been almost irresistible." Despite this incentive, institutionalization remains an undesirable situation for most households, to be accepted only when informal supports have been exhausted or are unavailable.

3 While the overall proportion of households with an elderly head remained about the same, the proportion of elderly varied over time and among provinces. In 1981, for example, the highest proportions (about 22%) were in Saskatchewan and Prince Edward Island, the lowest (15%) in Quebec. Somewhat surprisingly, the incidence of home ownership among elderly households decreased from 77% in 1951 to 63% in 1981 (calculated from *Census of Canada 1951*, 2: Table 100 and *Census of Canada 1981*, Catalogue 92-933, Table 9). This decline is probably linked to the increasing number of elderly aged 75 years or over, to urbanization of the elderly, and to migration toward more preferred climates within Canada.

4 See *Census of Canada 1981*, Catalogue 92-933, Table 9. More rural provinces, such as Newfoundland, Saskatchewan, and Nova Scotia, had high incidences of elderly home ownership in 1981, while Quebec was the lowest. In 1981 the incidence of condominium ownership was still low everywhere but in British Columbia. While only 3% of all elderly home owners owned a condominium in 1981, 8% of elderly British Columbia home owners lived in condominiums.

5 In addition to the reliability of estimated selling prices, especially those reported by elderly owners who had purchased their homes many years ago, there is some question about the use of a per room measure. A large, old house with, for example, four or five bedrooms may represent a less "desirable" commodity given today's smaller family sizes than a house with just two or three bedrooms. The use of the SHU is further complicated because it only considered households living in census metropolitan areas. Nonetheless, the literature reinforces the general conclusion that, at an aggregate level, elderly owners live in less valuable properties than younger owners.

6 In fact, elderly households were three times more likely (27% *versus* 9%) to occupy dwellings having 152 square metres or more per person than do non-elderly households.

7 These data may well underestimate space consumption. Many households with extra bedrooms convert them to other uses such as a den, study, or sewing room which are then not counted as bedrooms. See Wexler (1985). On the other hand, because of illness or different sleep patterns, elderly couples often need separate bedrooms or may want a spare bedroom to receive family overnight. See Howell (1980). As well, square footage data may exaggerate overconsumption because of the high proportion of elderly living alone. A certain space minimum is required for basic services such as kitchen, bathroom, and laundry, regardless of household size.

8 More generally, the concern about overconsumption, except where unwanted or involuntary, needs to be questioned. Should the elderly be penalized for overconsuming housing, while other individuals in society are not generally penalized for overconsumption, as in the case of owning a vacation home or a luxury car? Some types of overconsumption are even encouraged, if not rewarded, as in the case of non-taxation of capital gains on a principal residence.

9 Certain data from the RRAP evaluation suggest that elderly are not less likely to make repairs in their homes that affect their safety. In the RRAP Physical Inspections, carried out by professional building inspectors, of buildings that received RRAP funds, dwellings occupied by elderly households were less likely to have a safety failure than those

occupied by younger households (significant at the 0.001 level). Those elements considered included the condition of attached structures, the furnace, the electrical system, the interior stairs, fire hazards, and the presence and location of smoke detectors.

10 AGRFP and PMSH computer systems, Canada Mortgage and Housing Corporation.

11 We use the total number of husband-wife families with at least one parent in the labour force (or employed – depending on data availability) as the denominator in all calculations pertaining to "two-earner" families.

12 While the rate remained highest in Ontario and Alberta (respectively 64% and 62% in 1984) and lowest in Newfoundland (48% in 1984 – a figure which in part reflects poor economic conditions in that province), in only two provinces (Newfoundland and New Brunswick) were fewer than 50% of wives with children under 16 years of age in the labour force. Nor is this trend limited to the more urbanized provinces: even in Saskatchewan and Prince Edward Island, 61% of married mothers with children under 16 years of age were in the labour force in 1984.

13 Yet provincial variations suggest a more complex picture. If we examine the increase in labour force participation among married women with children under 6 years old, national figures rise from 37% in 1977 to 48% in 1981 and 52% in 1984. In 1984 Ontario (at 59%) and Alberta and Quebec (at 53%) are well above the Canadian average, while British Columbia (48%) remains well below. Although many women may have been impelled into the labour force by increased costs of access to home ownership, these provincial variations are not correlated positively with house prices. In any case, the two-earner family with children at home is now an essential part of the reality of home ownership.

14 This achievement of greater efficiency through private purchase and consumption of appliances contrasts with early 20th-century designs and experiments for collective provision of cleaning and food services. See Hayden (1981). Ironically, the drive toward greater efficiency resulting from the shortage of servants after World War I generated the first wave of consumer demand and technological innovations for household appliances. See Luxton (1980).

15 The percentage of all single parents who are female is around 85% and has changed little since 1951. The average income of female lone parents was 59% of that of male lone parents in 1981 (*Census of Canada 1981*, Catalogue 92-935, Table 20). We have therefore chosen to limit our discussion to mother-led families; although father-led families face some of the same problems in the residential environment, they are able to use financial resources to overcome them to a much greater extent.

16 This is different from the United States situation where the teenage never-married are the fastest growing type of single parent (Holcomb 1986).

17 In the 1981 Census, the average annual family income among lone-parent families headed by a woman under 35 years of age was just $7,600. Lone-parent families headed by a woman 35-44 years of age averaged $13,200. Those headed by a woman 45-54 years of age averaged $17,100. Among those over 55 years of age, the average was $19,300.

18 Statistics Canada, in *Household Facilities by Income and Other Characteristics 1985,* reports that lone-parent families had an average income of $12,401 compared with

$36,431 for husband-wife families. Among lone-parent families, 29% had no washing machine, and 42% had no automobile, compared to just 7% and 11% respectively for husband-wife families.

19 Statistics Canada, Census of 1981, special tabulations compiled for INRS-Urbanisation and funded by Fonds FCAR (Quebec).

20 Fuller details on cooperative housing programs in Canada, including the recently-abolished Section 56.1 and the new Conservative government program, can be found in Bourne (1986).

21 Under the cooperative housing program (NHA 56.1) operating from 1978 to 1985, individual cooperatives managed a subsidy pool to provide rent-geared-to-income assistance for members with low incomes. The extent of these subsidies thus depended on the circumstances of the individual co-op. Since 1986 section 56.1 has been replaced by financing of co-ops under and index-linked mortgage program in which 30% of units funded are eligible for a rent supplement to serve households in core need. It is still too early to evaluate the impact of these changes on the availability and affordability of co-op housing for single parents. The federal government cancelled the cooperative housing program in 1992.

CHAPTER FIFTEEN

The Affordability of Housing in Post-war Canada

Patricia A. Streich

SINCE PUBLIC housing was first initiated in Canada, analysts have wrestled with the meaning of "affordability." Though volumes have been written, affordability appears to defy objective measurement. [1] The central argument of this chapter is that the magnitude of the housing affordability problem is largely conditioned by the perspective from which it is defined and measured. Affordability is more than a concept of positivist economics; it encompasses issues of social standards, the notion of reasonable payments for housing in attainment of a level of social well-being, and questions about social equity and equality of opportunity.

The chapter demonstrates how conventional approaches to the definition and measurement of affordability reinforce a dichotomy between housing as a consumer good for the more affluent and housing as a social necessity for the less affluent. An alternative concept of the affordability problem is outlined, a life-time or longitudinal approach, that considers the opportunities for households to move towards their housing goals and to adjust their consumption with changing needs and resources over their life-cycles. A longitudinal approach is helpful in considering the extent to which policies have become more effective in addressing affordability issues. In this sense, progress is defined by the improvement in housing opportunities rather than static counts of households currently paying an excessive proportion of their incomes on housing.

The two themes – dichotomization of the affordability problem and the need to develop an opportunities approach to affordability – are developed through discussion of three questions. In what sense is there a housing affordability problem? Who has the affordability problem? What have been the impacts of post-war policies and programs on affordability problems?

The Affordability Problem: Post-War Trends

Miron (1988) discusses the apparent paradox that rising prosperity (as measured by real growth in incomes) since 1945 has been matched by corresponding increases in consumer expenditures on housing. Although housing should have become more affordable, consumers did not spend a decreasing proportion of their budgets for housing.

From 1946 to 1981 the Statistics Canada housing price index increased about five times, approximately matching the overall Consumer Price Index (CPI). In the same period, per capita personal disposable incomes surged ahead by about twelve times. By these indicators, affordability should have improved.

However, shelter as a proportion of total consumer expenditure remained fairly stable from 1949 to 1967 (31% to 32% of consumer expenditures); it increased after that, reaching 35% by 1978. The 1937-38 Family Income and Expenditure Survey showed that families who rented spent 20% of their total living expenditure for shelter and home owners spent 19% (Carver 1948, 74). In 1982, according to results from the Household Income, Facilities, and Equipment microdata sample (HIFE), home owners with mortgages spent 24%, home owners without mortgages 17%, and renters 23% (Canada 1985, 10). After four decades Canadians were still spending about one-fifth of their household incomes on housing. Why housing expenditure kept pace with income when shelter prices were increasing much less quickly is not clear. What is clear is that, overall, Canadians did not find their shelter costs declining in importance as real incomes rose.

While most Canadians live in affordable housing, an affordability problem exists for some households because their monthly shelter costs exceed a "reasonable" proportion of their incomes. What constitutes a reasonable proportion is a value judgment. Views vary among countries, and they have changed over time; but as a 1981 CMHC report concludes: "in North America, a range of 20% to 30% has been historically regarded as a fair rate of expenditure for housing" (CMHC 1981, 7). Comparisons over time are difficult from published sources. However, available data indicate that proportions of renters and owners with mortgages have experienced affordability problems.

The Low-income Housing Task Force, using data from the 1969 Family Expenditure Survey microdata sample (FAMEX) found:

> There are 1,831,000 Canadian households spending in excess of 20% of their incomes for shelter. Two thirds of them are low income. There are 1,076,000 spending in excess of 25% of income for shelter. Four-fifths of them are low income. One Canadian household in three spends in excess of 20% of income for shelter, one in five in excess of 25% ... 400,00 households (spend) more than 40% of income for shelter (Dennis and Fish 1972, 59).

Miron's (1984) analysis of the 1978 FAMEX data shows that 28% of renters and 17% of all owners paid more than 25% of their incomes for shelter. Using the 1982 HIFE and the 1982 FAMEX, CMHC reported that 18% of all households or 1,512,000 households were paying more than 30% of income for shelter. The data showed that 23% of renters and 19% of owners with mortgages spent over 30% of income for shelter.

Although different benchmarks have been used, trends are evident. Not only have the proportions of Canadian households with shelter-to-income ratios

Table 15.1
Affordability problems among renter households in metropolitan areas,
by income category: Ontario, 1972-1983
(as % of all renter households in income quintile)

Household income quintile	Renters spending more than 25% of income on shelter			Renters in core need		
	1972	1976	1983	1972	1976	1983
1 (lowest income)	87	93	93	86	93	90
2	53	53	46	15	40	19
3	17	16	15	1	1	1
4	1	3	5	0	0	0
5 (highest income)	0	1	2	0	0	0
Mean	32	32	32	20	25	22

SOURCE Arnold (1986, 100).
Study based on the Statistics Canada Household Income Facilities and Equipment (HIFE) microdata sets for 1972, 1976 and 1983. The data in this Table refer to non farm, unsubsidized renters in Ontario metropolitan areas larger than 100,000 population; i.e. Ottawa, Toronto, Hamilton-Burlington, St. Catharines-Niagara, London, Windsor, Oshawa, Kitchener-Waterloo, Sudbury, and Thunder Bay.

above the benchmarks increased but also the numbers of households with affordability problems have increased with population growth. Whereas in 1969 just over a million Canadian households were spending in excess of 25% of income for shelter, by 1982 over one and a half million households were spending over 30% of income for shelter (Dennis and Fish 1972, 60; CMHC 1984b).

Available data also reveal that the majority of households with affordability problems have low incomes. Disaggregation of the affordability problem by income level is possible from HIFEs that have been available since 1972. Table 15.1 summarizes the incidence of affordability problems among renter households in Ontario metropolitan areas by income category using two definitions of "affordability." Arnold (1986) uses Table 15.1 to demonstrate that the incidence of affordability problems in the lowest income quintile of renters increased from 1972 to 1983 under both definitions. The data are inconclusive with respect to the second lowest income quintile.

Miron's analysis of the 1978 FAMEX shows that the incidence of affordability problems is above average among the elderly and for single-parent families (Miron 1984). The highest incidence of affordability problems is among lone-parent families with young children (under five) who rent their housing; 72% of lone parents with young children spent over one-quarter of their incomes on rent compared with 28% for all renters. The pattern repeats itself for owners;

46% of lone parents with children, who own, spent over one-quarter of their income on shelter compared with 17% for all owners. The incidence of affordability problems is also above average for elderly households; 54% of renters 65 years of age and over and 30% of owners 65 years and over spent more than one-quarter of their income for shelter in 1978. These data suggest that the proportions of Canadian households spending above the standard proportions of income on shelter has not declined and the problem is concentrated among the poor, the elderly and single parents.

Another widely used indicator of accessibility to home ownership is the proportion of households in general (and of families living in rented accommodation in particular) able to carry (or afford) an average-priced dwelling within a specific proportion of their incomes. Accessibility to home ownership has declined since the 1950s. In 1951 over half of all Canadian families would have found that carrying the average new single-detached NHA-financed house was within 30% of their incomes. By 1983, based on data from *Canadian Housing Statistics, (CHS)* fewer than 15% of households would have found carrying an average-priced house was within 30% of their incomes. CMHC (1984b, 16) data which consider only renters in the prime home-buying age groups (from 25 to 44 years of age) show that 50% of this group could have afforded the average priced house in 1971 compared with 28% of the group in 1983 (again using a maximum carrying cost of 30% of income).

Approaches to the Definition of Affordability

Until the mid 1970s the shelter-cost-to-income ratio approach was used to measure the affordability of housing expenditures linked to household incomes. This approach used some benchmark of affordability such as 25% or 30% of income for shelter as an acceptable, affordable level. Families and individuals who pay larger percentages of their incomes for shelter are defined as having an affordability problem. Several conceptual problems are readily apparent with this approach.

One is the problem of estimating the housing expense of home owners. It is commonly thought that some part of home owners' shelter expenses are actually a form of obligatory savings which is recouped as owners' equity on the sale of the dwelling. Furthermore, home owners can benefit from capital gains on the resale of a dwelling and, where it is their principal residence, such gains are not taxable in Canada. Numerous attempts have been made to develop a common shelter-cost-to-income ratio criterion for both renters and owners. One approach is to impute the rental value of owned homes and compare this with the amounts paid by renters with similar incomes. However, people do not pay imputed rents; such an abstract measure is not meaningful for determining the ownership shelter cost burden on purchasers.

A second limitation is the failure to account for variations in preferences in relation to the life-cycle of housing consumption. Some households may voluntarily consume more housing than their budgets might comfortably bear. Often this is done in the short run in expectation of longer term improvement; although a household may currently be experiencing an affordability problem, the consumption pattern might be affordable when viewed over the household life-cycle. Some analysts advocate use of a permanent income rather than current income definition to overcome this problem, particularly among young home buyers. Since 1945, families have appeared willing to spend large amounts of their current incomes on a house purchase, presumably expecting that fixed mortgage payments and rising incomes would alleviate the shelter cost burden after a few years. The shelter cost-to-income approach in contrast is a static, cross-sectional view of affordability problems.

The application of the rent-to-income ratio approach to pricing in publicly-assisted housing poses another paradox. Whereas housing analysts tend to use flat rate scales to measure affordability problems, Canada has a long history of graduated rent-to-income scales for clients in subsidized housing. Rent-to-income scales in public housing since 1944 have been based on the premise that low-income households cannot afford as large percentages of their family budgets for rent as middle-income households. Hence, the public housing rental scale (known as the graduated rent scale) required rental payments of as little as 16.7% for the low-income clients up to a maximum of 25%. Furthermore, over the 1960s and 1970s, different rent scales emerged as provincial housing authorities developed variations of the federal GRS (see Archer 1979). An additional variation in the affordability criteria applied in public programs was introduced with the provincial housing allowance schemes for senior citizens.[2] Whatever the rationale, the standards applied to affordability among the population at large were different from the program standards for pricing in assisted-housing programs.

The shelter-cost-to-income approach does not deal with over and under consumption of housing. The affordability problem is understated to the extent that households make housing affordable by living at higher densities (more than one family sharing overcrowded accommodations) or living in poor quality homes. These consumers appear to have no affordability problem because they have reduced their expenditures to fit their budget. At the same time, some households with an affordability problem may have voluntarily chosen to consume "more" housing. To the extent that housing preferences are above the normative standard, affordability problems would be overestimated with the shelter cost-to-income approach.

Concerns about the ratio approach led to the development of the core need approach in the mid 1970s. "The core need approach seeks to identify those households currently experiencing housing problems who would be unable to obtain minimum standard housing without paying an excessive proportion of their income on shelter" (CMHC 1981, 4). Core need incorporates the concepts of physical adequacy, crowding, and housing affordability into one measure, a more comprehensive definition than the simple shelter cost-to-income ratio approach. Furthermore, by relating the shelter cost measure to local rents, the

262 Patricia A. Streich

approach provides the potential for flexibility to local market conditions. The core need approach defines some normative housing requirement and relates affordability to norm rents in the local market. Under the core need definition, the affordability problem includes only those households that, within their household incomes, would not be able to afford an adequate unit in their area without spending in excess of some maximum amount of their income on shelter.

Core need and the shelter cost-to-income approaches yield different measures of the affordability problem. An analysis of 1974 Survey of Housing Units (SHU) data found that 589,000 households in metropolitan areas were in core need compared with 702,000 households in need under the shelter-cost-to-income ratio approach; the incidence of "need" was reduced to 17% from 24% of all households (CMHC 1981, Table 2). The same study notes that "the core need definition has eliminated the so-called voluntary over-consumers of housing" (CMHC 1981, 22).

As a comprehensive measure that combines adequacy, crowding, and affordability in one indicator of housing problems, the core need concept is a conceptual advance over previous definitions. However, application of the concept to the analysis of housing programs poses some challenges. First, any definition of a normative level of housing consumption is arbitrary and subjective; choice of the norm inevitably affects the magnitude of the problem. Second, assuming a consensus exists on a normative standard for consumption, measurement of the affordability of that standard requires detailed data on the rents of norm units in each locality. Third, the approach implies that households could occupy the corresponding norm units. Such a perfect matching of units to households may rarely (if ever) occur. Even if there are enough norm units in a market, there is no assurance that these will be allocated to the appropriate households. [3]

Since the core need concept was first developed, its application to housing programs has evolved: particularly in the 1986-8 implementation of federal-provincial (F/P) agreements for the non-profit program (NHA 56.1). The normative level of housing consumption is embodied in a National Occupancy Standard (NOS) introduced in 1987. NOS is sensitive to household size and household composition and is used to determine the number of bedrooms a household should have. For example, parents are eligible for a bedroom separate from their children; children aged 5 years or more of the opposite sex do not share a bedroom; and household members aged 18 years or more are eligible for a separate bedroom unless married or cohabiting as spouses.

Under the terms of the F/P agreements, CMHC requires that the NOS be applied to determine eligibility by virtue of overcrowding or affordability problems. Core need income thresholds are compiled for units by bedroom size, and households with insufficient income to afford the requisite sized unit are defined as in core need and eligible for subsidized housing. The F/P agreements do not require that provincial housing bodies apply the NOS in placement of households in specific dwelling units. Therefore, eligibility for a single elderly person or elderly couple would be based on the income required to afford a one-bedroom unit, and for a single parent with one child on the income required for a two-bedroom unit. Depending on the urgency of the housing need and the types of units available, a single senior may be placed in a bachelor unit and a single parent and child in a one-bedroom unit. Placement policies are the responsibility of the provincial housing agency delivering the program.

In addition to using core need to determine program eligibility, the concept has been applied at an aggregate, provincial level to estimate the numbers of households in need for planning and budget allocation purposes. As of 1986, housing budget allotments to the provinces were based on an agreed formula – known as the 1984 Regina Accord – based on counts of households in core need. The allocation formula was further revised after 1989. In addition, core need estimates are developed for planning areas within provinces as part of a three-year planning cycle to guide decisions about the location of housing projects among local market areas (see CMHC 1986b).

Thus, the core need concept has come to guide housing planning, budget allocation and program eligibility for subsidized housing across Canada. The concept does not purport to take account of all housing factors such as locational considerations, many of which are taken into account in the detailed project planning and placement practices of provincial and local housing authorities.

A Consumer Life-Time Opportunities Approach

Current affordability is only one issue affecting a household's choice of housing. Important too are factors affecting the decisions to search for alternative housing and the decision to move.[4] As Steele and Miron (1984) point out, transaction costs (psychic and monetary costs) of moving are impediments to adjustment of housing consumption and housing costs. Experience with the Canadian housing allowance programs has been similar to that of the United States; in both countries, households eligible for additional subsidies to enable them to afford better housing tend not to increase their housing consumption as much as possible.[5]

A more dynamic, longitudinal approach focuses on the opportunities for households to adjust consumption to changing needs. Provided that the opportunities exist, that consumers have the means to learn of the opportunities, and that they are able to access the opportunities without systematic barriers or constraints, the affordability situation may improve through individual choice. Progress in affordability is defined in terms of individuals having the opportunity to move towards more affordable housing during their housing careers, related to individual goals and trade-offs across a wide range of expenditure needs (Myers 1980).

Individual choice and opportunities are more difficult to measure than simple shelter-cost-to-income ratios, particularly when they span a household's life-cycle. Measurement requires longitudinal data which are costly to compile and analyze. Nevertheless, a more dynamic concept of affordability may become even more valuable in coming decades as such demographic changes as the aging of the Canadian population raise serious issues about the matching of housing demand and supply and the use of the nation's housing stock.

Disaggregating the Housing Affordability Problem

In the 1960s it became customary to define "problem" groups and from these to develop targeted assistance programs. Canada never went as far as US housing legislation which defined low-income families by their inability to afford adequate housing.[6] However, Canadian public housing was perceived to be serving the lowest two quintiles of the income distribution. Compartmentalizing the population into groups for discussion of the housing problem had the effects of relating the problem to the persons experiencing the problem rather than to the forces creating the problem and of deflecting attention from broader issues of housing needs and equity; specifically, it reinforced a basic dichotomy in housing policy between housing for the poor and the affordability of home ownership for the more affluent.

HOUSING AND POVERTY

In 1941 Toronto families with incomes below $1,000 spent 40% of their incomes on housing compared with 21% for families with incomes from $1,500 to $2,000 (Carver 1948, 75). HIFE data on Ontario renters in metropolitan areas suggest that the affordability picture had not changed by 1983. Renters with incomes below $5,329 in 1972 and below $12,454 in 1983 spent 46% of their incomes on rent compared with average rent-to-incomes ratios of 24% in both years (Arnold 1986, 87).[7] The poor who own their own homes, mainly the elderly in cities and rural areas, are in no better situation than those who rent; as many as two-thirds spend over 30% of their pensions on heating, taxes, and maintenance.

The shelter cost problems of the poor in cities were exacerbated during the 1960s and 1970s. The private-sector, low-rent housing stock was whittled away by conversions and demolitions. In some inner-city neighbourhoods, gentrification has contributed to the losses. Despite concerted efforts, the public sector was unable to compensate for these losses in most Canadian cities.

Private market responses to the demand for low-rent units by low-income persons and families have been seen by local municipalities and residents as undesirable or below acceptable community standards. For example, in Vancouver, illegal basement suites were a response to the demand for low-rent units; in Toronto, the Parkdale area was identified as a problem because of the creation of bachelorette units. For consumers who could afford only the most minimal housing, boarding and rooming houses and group homes have all been subject to increased municipal control and limitation. Partly a consequence of the constrained supply of housing opportunities, Canadian cities in the 1980s faced a growing problem of homelessness. The spectre of men and women living on the street, in parks, and subways has become a visible sign of the lack of suitable housing alternatives for those of low-income.

As well as the absolute shortage of accessible dwelling space, Canadian cities have undergone some demographic shifts. Inner cities had tended to house the older and younger population, while families with children sought housing in the suburbs. In the late 1970s concerns emerged about the graying of post-war suburbia. As families age in suburbia, first teenagers, and later their now-elderly parents, find themselves in residential environments that were created for raising small children. Is the single-family suburb flexible enough to adjust to changing needs as has the older inner-city housing stock in some cities?

Support services are generally more available and more accessible to users in central cities than in suburban areas. Furthermore, meeting service needs may be more efficient among a concentrated user population. However, central city municipalities face the problem of an increasing demand for services from a dwindling tax base, and their populations are becoming increasingly polarized between the lowest incomes and the highest incomes. The missing middle-income households are located in outer suburban areas.

With this broader view of the problem of housing poverty in the city, the period since 1945 has seen an evolution in thinking about the nature of the housing poverty problem. Whereas in the 1950s, and much of the 1960s, the problem appeared to be the clearance of "slum" housing, by the 1980s the problem has become the ghettoization of the inner municipalities into homes for the poor and the rich. The simple slum removal strategy is unlikely to provide a meaningful framework for dealing with the contemporary low-income housing problem. Housing poverty has become a function of urban development and urban society.

AFFORDABILITY OF HOME OWNERSHIP

In a nation which is still a country of home owners, an obvious yardstick of housing well-being has been the ability of households to afford a home of their own. Accessibility to home ownership has been one factor in shaping the map of housing poverty in our cities since 1945, for the affordability of housing governs where a family could live. Lenders assess the affordability of, hence access to, home-owner mortgages using the gross debt service (GDS) ratio. Over the years, the GDS ratio permitted on NHA loans has been increased, and the definition of eligible family income has been liberalized. However, even with liberalized lending practices, access to home ownership for low-income families is constrained by the amount of owner equity available for a downpayment.

Low-income home owners also face serious affordability problems according to a special Statistics Canada publication based on the 1981 Census. Among mortgagor households with incomes below $10,000 in 1980, 90% were spending over 30% of their incomes for owners' major payments (principal, interest, taxes, and utilities); three-quarters of mortgagor households with incomes below $15,000 were spending over 30% of income for housing (Che-Alford 1985,

60). Low-income home owners are more likely to be living in dwellings in need of repair than are high-income owners. In 1980, for example, roughly one in three mortgagors with an income below $15,000 reported that their dwellings needed repairs, while the comparable figure for all mortgagors was 22% (Che-Alford 1985, 56).

Home ownership has not provided a hedge against the increasing cost of housing. Mortgage rollover makes home purchasers vulnerable to the volatility of interest rates. Over the longer term, many home owners eventually come to hold their homes mortgage-free, but the rising costs of taxes, heat and other utilities, and repairs and maintenance place financial burdens on the fixed incomes of old age.

Low-income owners, like other owners, have equity in their homes which, if realized, would allow them to increase their current consumption. Home equity can be accessed without sale of the dwelling and displacement through home equity dissavings or reverse annuity mortgage (RAM) schemes. Currently, RAMs are available only in British Columbia. Generally, few options remain for home owners in financial difficulties. Sale of the family home and the search for affordable rental accommodation involve psychic and financial costs which are especially severe for elderly households. Delaying necessary repairs (that is, disinvesting in the dwelling) is a short-term consumer strategy to deal with unaffordable housing expenses.

As the cost of home ownership rose during the 1960s, government introduced home ownership assistance programs. In 1970 an experimental Innovative Housing Program attempted to assist poorer families toward home ownership. Some low-income families were able to buy homes, but the program was a special demonstration project and was not continued. AHOP was originally conceived as an alternative to public housing for low-income families. As the amount of federal assistance was reduced and as provincial matching subsidies failed to materialize in most provinces, AHOP eventually became another moderate-income home ownership scheme.[8] Outside of rural, Aboriginal, and northern areas, housing programs have provided little assistance to the poor to buy homes.

In the 1970s CMHC explicitly distinguished between two broad categories of housing programs, namely, "social housing" and "market housing." These two categories reflected the dichotomization in thinking about housing problems – subsidized housing for the poor and assisting the market to supply housing for moderate-income Canadians. Problem group targeting has tended to reinforce the dichotomy and has done little to stem the tide of diminishing housing choices and opportunities.

Impacts of Government Policies and Programs

While the proportions of Canadian households spending above the standard percentage of income on shelter have not declined substantially since 1945, there may be some grounds for optimism if the policies and programs used now are more effective than the means adopted previously to address housing problems. An important question to ask about policy impact is: to what extent have housing programs and policies become more effective in dealing with affordability problems since 1945?

Overall, Canada has made little progress in devising more effective tools to address affordability problems since 1945. Indeed, on some bases, the tools introduced during the 1970s seem less attuned to solving the core of the affordability problem. In 1971, the time when public housing was reaching a peak in terms of production, the overwhelming majority (over 90%) of assisted social housing units produced were allocated to low-income households. By 1980 federal and provincial housing programs were yielding roughly one low-income unit for one moderate-income unit of social housing assistance. The principal reason for this shift relates to the application of the income-mixing policy under the federal nonprofit housing program which replaced public housing during the 1970s.[9]

The non-profit and cooperative housing programs initiated in 1973 embodied the concept of a mix whereby a portion of the units in each project were to be allocated to low-income families and senior citizens, the balance of the units to be provided at "market rents" to moderate-income households. Housing projects that mixed higher-income households and low-income households were thought to be more socially viable and to generate less local (community and municipal) resistance than did low-income housing projects. Had the federal government provided sufficient funding, the output of low-income units could have been maintained at the levels of the early 1970s under the public housing program. To reach the low-income unit production level of 1971, it would have been necessary to produce over 65,000 units of non-profit housing annually (compared to only 20,000 non-profit units actually authorized in 1979).[10] In terms of addressing the affordability of housing for low-income Canadians, the non-profit approach was less effective than previous housing programs.

One notable attempt to initiate a more effective tool to address affordability problems has been the housing allowance approach. Provincial housing allowance measures in British Columbia, New Brunswick, Manitoba, and Quebec have targeted assistance to the needy low-income population (mainly senior citizens in rental accommodation). In part, the rationale for such programs stems from a realization that traditional supply-side housing programs are incapable of providing relief of the volume of needy households with affordability problems. Housing allowance programs are, however, not without their own limitations. As noted above, eligible households do not necessarily adjust their housing consumption to the extent possible with allowances. Furthermore, demand-side strategies alone may be ineffective without measures to address the supply of suitable, adequate dwellings.

In addition to the choice of policy instruments, the difficulty of addressing affordability problems in Canada may be related to the institutional context within which housing has been situated. For much of the post-war period, the primary impetus for housing policy in Canada derived from the federal level of government even though constitutionally the provincial governments were ascribed the major responsibility in the field of housing. The roles and relationships between the federal and provincial housing agencies have evolved from the 1960s when Ottawa adopted the lead role, to a position by the 1980s where parallel federal and provincial housing initiatives were adopted and impacted on housing problems. Although certain provincial governments (notably Quebec, Alberta, and British Columbia) have repeatedly reaffirmed their claim to provincial sovereignty in the housing field, the majority of provinces have enjoyed the continued involvement of Ottawa in financing housing programs.

Experience suggests that the priorities of the two levels of government are sometimes at odds, and the capacities of the two levels to adequately address the defined problems have varied considerably. During the 1960s there was debate about the distorting effects of cost-shared programs; yet with the adoption of unilateral financial mechanisms in the 1970s, provincial governments have sought to maximize the impact of available federal subsidies through adding provincial financial assistance voluntarily onto federal program subsidies. Therefore, although there has been greater separation of federal and provincial housing measures during the 1970s, inter-dependency of federal and provincial housing measures continued.

While the policy instruments adopted have addressed affordability problems, solutions seem elusive. Constitutional arrangements for housing in Canada may have impeded the search for solutions. Fiscal restraint among governments after the mid 1970s has increased the difficulties of dealing effectively with housing affordability problems. Canada may have fewer grounds for optimism about solving this housing problem in the 1980s than it had in the 1940s when it embarked on its first subsidized public housing project.

Past Experience and Future Prospects

While unequivocally some people are unable to afford some housing, and hence said to have an affordability problem, the dimensions of the problem are more difficult to specify. Not only are there conceptual difficulties in determining the normative and behavioural yardsticks for specifying the "ability to pay" criterion, but also the precise measurement of shelter expenses and incomes pose many problems (see, for example, Miron 1984). Definitions of affordability have been modified from a simple cross-sectional measure of what people actually spend on housing in relation to their income to a more composite measure of householders' ability to afford suitable, adequate shelter within a specific proportion of their income. Posing questions such as the ability of people to achieve their desired consumption of housing over their own housing careers creates the need for longitudinal housing data.

Canadian governments have made investments through housing programs and the tax system to increase the supply of decent housing and to subsidize housing costs for consumers. An affordability problem still exists for many renters faced with high rent-to-income burdens and low-income households face problems whether they rent or own. Canadian housing policy has relied heavily on supply-side, new construction strategies for a variety of reasons. Although public, non-profit, cooperative and other housing programs have created a stock of affordable housing, the supply is insufficient to meet the needs. In 1971 the Dennis Task Force recommended a shift from supply-side strategies to demand-side subsidies and some steps have been taken in that direction. However, the policies and programs adopted have not achieved the goal of affordable housing for all Canadians.

What then are the implications for the future? May we expect to achieve more progress in making housing affordable in the decades ahead? This chapter has suggested two promising avenues for thinking about housing policies. Cross-sectional, static, and problem group approaches to defining affordability limit the vision and the prospects of solution. The chapter suggests a longitudinal, life-cycle of opportunities approach which considers the ability of households to achieve their housing goals. The life-cycle perspective is a promising avenue that is consistent with a private market environment for provision of housing services. In addition, the criterion for evaluating public policies should be the extent to which they represent more effective approaches to affordability than past policies. Do policies and programs improve the opportunities for households to achieve their housing goals, or remove barriers that prevent households from reaching their goals?

Notes

1 This Chapter does not review the growing literature on the subject as several reviews are available. See, for example, Miron (1984).

2 In most of these programs, subsidies were applied to only a portion of the seniors' rent above some preset maximum; in effect, seniors receiving housing allowances continue to spend more than 30% of their incomes on rents after receipt of the allowances. In part, housing allowance schemes provided less generous subsidies because seniors remained in the private housing market and were expected to pay a larger proportion of their incomes for greater freedom of choice this entailed.

3 The US housing allowance program, Section 8 (existing housing) demonstrated the problem of many housing allowance households being eligible to receive benefits but being unable to find units that met minimum housing standards and, therefore, being unable to benefit from the program.

4 See, for example, the review by Clark (1982).

5 Bradbury and Downs (1981) reports on the American EHAP experience and Steele (1983) discusses the Canadian experience.

6 In the US, affordability criteria were used to define low-income families in the 1937 Housing Act:

   Families of low-income are defined as those who cannot afford to pay enough to cause private enterprise in their locality or metropolitan area to build an adequate supply of decent, safe and sanitary dwellings for their use (quoted in Carver 1948, 71).

7 The use of per capita incomes to measure income change is not ideal. For the working poor and moderate income families, a more meaningful measure may be the change in average wages and salaries. Prior to 1976, the Statistics Canada index of average wages showed real growth in earnings; wages were increasing faster than prices. Since 1978, however, wages have stagnated; from 1978 to 1981 wages increased by 12%, the CPI by 11.5%, and the shelter index by 13.5%. Stagnation of wages coupled with uncertainty about employment, lay-offs, and long periods of unemployment may more accurately reflect the income situation of working families. These data raise doubts about the apparent growth in personal incomes and about the ability to afford house price increases: especially when housing already consumes a large proportion of household budgets.

8 Some provinces did initially provide matching subsidies to increase the income penetration of the AHOP subsidies; Ontario was operating its Home Ownership Made Easy Program (HOME); Nova Scotia developed an innovative use of AHOP with its building cooperatives program.

9 The income-mix policy may illustrate an example of a shift in social goals in housing policy. Whereas earlier policy sought to produce the maximum numbers of subsidized units (frequently in high density projects), the creation of more income-integrated projects seemed to offer the potential to produce more socially-balanced and stable residential communities.

10 Streich (1985, 125) notes that, of the 20,000 non-profit units produced in 1979, only 6,000 were affordable to low-income households, effectively reducing support for low-income housing in Canada.

CHAPTER SIXTEEN

The Changing Settlement Environment of Housing

L.S. Bourne

ALL HOUSING is tied to land and location, and thus to an external environment. Aside from the isolated cabin or farmstead, all dwellings form part of a local community. Where we live, moreover, adds social meaning and economic utility to housing quality that are essential in discussing housing progress. It affects the quality and form of houses we occupy, what we may do with it, the kinds of physical and social services provided to the dwelling, and the amenities and services to be found nearby. This chapter, then, assesses post-war housing progress in terms of changes in the nation's settlement system.

Evolution of the Settlement System

At the turn of the century, just over 37% of the nation's population of 5.37 million lived in urban settlements.[1] That proportion increased to 50% in 1921, then to 63% by 1951, before stabilizing at around 77% in 1986. In 1941, 46 out of every 100 Canadians lived in rural areas, and 65% of those lived on farms (Table 16.1). A further 11% lived in small towns and cities with under 10,000 population. At that time, 54 Canadians in 100 lived in urban areas of all sizes, but of those, 40 lived in metropolitan environments (that is, urban areas over 100,000 population), and just 22 lived in the three largest metropolises (Table 16.2). By 1986, in contrast, only 3 in 100 Canadians lived on farms, and 20 in 100 in non-farm rural communities. Of the 77 who lived in urban areas, 59 lived in metropolitan areas, and 31 of these people resided in the three largest metropolises (Table 16.2). Between 1941 and 1986, Canada added as many urban residents as it had total population at the beginning of the period. The nation's urban population rose from 6.3 million to over 19.4 million in those forty-five years.

The nation's settlement system was reshaped accordingly (Table 16.2).[2] In 1941 Canada had only 63 urban places with over 10,000 population and only 8 with over 100,000 population. By 1986 there were 27 urban centres over 100,000 and 139 in total with over 10,000 population (Table 16.2).[3] The latter figures do not refer to individual municipalities as defined by political boundaries, but to functionally-defined urban areas; most of these represent aggregations of city and suburb on the basis of proximity and the degree of economic and social

Table 16.1
The urban transformation – population by place of residence:
Canada 1941-1986

	1941 (000s)	(%)	1961 (000s)	(%)	1986 (000s)	(%)	Change 1941-61 (%)	1961-86 (%)
Rural	5,254	46	5,266	29	5,962	23	0	13
Farm	3,117	27	2,237	12	895	3	-28	-60
Non-farm	2,137	19	3,028	17	5,067	20	42	67
Urban	6,252	54	12,972	71	19,392	77	108	42
Under 10,000	1,259	11	2,188	12	1,521	6	74	-30
10-100,000	1,506	13	2,860	15	3,042	12	90	6
Over 100,000	3,487	30	7,924	44	14,829	59	127	89
Canada	11,507	100	18,238	100	25,354	100	58	39

SOURCE Census of Canada, various years.

integration. They include census metropolitan areas (CMAs) and smaller census agglomerations (CAs), as well as any other free-standing municipalities that are not included within the boundaries of CMAs and CAs. [4]

The era since 1945 can be characterized by three distinct periods, corresponding roughly to the years 1945-64, 1965-80, and 1981-90. The first, a period of urban boom, was characterized by: (1) a rapid growth in population (and, for the most part, in economic production) in most regions of the country, including the addition of entirely new settlements, principally resource towns outside of the settled ecumene and new suburban communities around established urban centres; and (2) the growing concentration of population in larger metropolises. In this period the first large suburban developments also emerged.

During the second period, typically one of national decentralization, the overall rate of population growth slowed as fertility and immigration declined. As part of a structural adjustment process in the economy during the 1970s, growth shifted from the older urban areas in the industrial heartland of central Canada to resource-based regions and newer urban settlements, particularly in the west. As the latter settlements were on average smaller, it first appeared that the largest metropolitan areas were losing population (and jobs) to smaller towns and cities – a process described elsewhere as de-urbanization or counter-urbanization. Indeed, during the latter part of the period (1976-81) the entire set of census metropolitan areas (CMAs) did lose population on balance to the rest of the country. For the most part, however, this was not due to a process of de-urbanization, or to the renewed growth of rural settlements in general. Rather, it reflected the continued spread of urban populations into urban fringe

Table 16.2

Growth of the Canadian settlement system, 1921-1986.

	1921	1941	1961	1986
Size of settlement	*Number of settlements*			
100,000 and over	7	8	18	27
30,000-99,999	11	19	25	54
10,000-29,999	25	36	60	58
5,000-9,999	45	49	87	105
All settlements of 5,000 or more	88	112	190	240
Size of settlement	*% of all settlements of 5,000 or more*			
100,000 and over	8	7	10	11
30,000-99,999	12	17	13	22
10,000-29,999	28	32	32	24
5,000-9,999	51	44	46	43
All settlements of 5,000 or more	100	100	100	100
Indicators of urbanization and concentration†	*% of Canada's population*			
Population urbanized	47	55	70	77
Population in 25 CMAs	35	40	48	59
Population in 3 largest CMAs	19	22	25	31
Total population in 3 largest CMAs (000s)	1,651	2,551	4,725	7,730

SOURCE *Census of Canada,* various years; Stone 1967; Simmons and Bourne 1989. Census metropolitan areas (CMAs) as defined at each census.
† See footnotes 1 and 2 in text.

areas outside the current boundaries of the census-defined metropolitan areas, but still within more broadly defined metropolitan regions.

During the latter part of this period, a widespread decline of smaller urban settlements became apparent, particularly in peripheral regions. Although Canada has always had declining communities, the 1970s witnessed decline on a much broader scale. Fully 20% of small settlements in Canada declined in size during this decade. For the first time in this century, two smaller census metropolitan areas, Windsor and Sudbury, also declined: victims of sharp downturns in their specialized economies. Slower national growth and industrial restructuring had created a new era of urban change and settlement reorganization.

At the local or urban scale, the predominant direction of change was one of metropolitan growth and concentration accompanied by continued intraurban decentralization and suburbanization. Rapid population growth and a pent-up demand for new housing after fifteen years of depression and war, fuelled unprecedented suburban development after 1945. Easier access to transportation, especially the automobile, and easier access to credit, notably for mortgage loans, facilitated this growth. Local governments opened vast areas of land for development and set in motion the policies and practices necessary to regulate suburban growth and to provide services. Higher real incomes, in turn, increased the amount of space consumed and the quality of building and service standards demanded.

Canada, as a result, has become a predominantly suburban society. In 1941, only 24% of metropolitan residents lived in suburbs (that is, outside the politically-defined central city). By 1961, nearly 45% lived in suburbs and by 1986 over 60%. This process of spatial population dispersal, combined with the ongoing demographic transition, reshaped local residential patterns and living environments through the late 1960s and 1970s. Declining fertility rates and shrinking household sizes led to an older and socially more polarized population, particularly in small towns and in older central cities. The population in many of these communities underwent a remarkable "thinning out," even in neighbourhoods with a stable or increasing housing stock. Typically, the population housed in a fixed stock of dwellings, holding constant all other types of neighbourhood transition, would have gone down by 30% in about twenty years simply through the effects of smaller household sizes and increased housing consumption.

The period from 1981 to 1990 presents yet another era in Canada's evolving settlement pattern, and thus different signposts to the future distribution of housing demands and needs. Beginning with the most severe economic recession of the post-war period, a lower level of population growth, and sharp drops in resource and commodity prices, the pendulum of national settlement growth swung back toward central Canada and toward those communities within or near major metropolitan regions. Large metropolitan areas again grew most rapidly and smaller cities and rural settlements on average declined.[5] At the local level, rapid decentralization has continued apace, as population (and employment) and housing are dispersed over larger areas at decreasing densities. Within the older parts of many urban areas, especially the inner cities of the large metropolises, however, population decline was finally arrested. The primary reason for this latest reversal is that the combined impacts of new residential construction (often condominiums) and in-fill housing were sufficient by then to compensate for the losses of population attributable to declining average household size and residential conversion. The zone of intraurban population decline has now spread into the older suburbs, many built in the early post-war period.

Whether these trends will continue through the 1990s obviously remains to be determined. Slower urban growth and a stable percentage of the population classified as urban should not, however, lead to the conclusion that urban growth has ceased. Rather, it has changed form. Urban populations have now spread over a massive territory, sometimes called exurbia, or the urban field.[6] This has blurred the traditional boundaries between rural and urban and between city and suburb. Moreover, the majority of small towns and rural areas in Canada that are growing in population are also located within the broadly-defined regions surrounding the large metropolitan areas and owe their growth to their functional linkages with, and particularly commuting to, the metropolitan core.

New Suburban Forms: The Corporate Suburb

Although Canadian cities and towns have always had suburbs, some of which have been both planned and large, the community development process since 1945 brought not only rapid suburbanization but also a new settlement form: the much larger, integrated, and self-contained suburban community. It also produced a new industry, the corporate development sector and a new type of suburb, the corporate suburb.[7]

This suburban form did not arise by accident. It was aided by federal government policies designed to establish a more efficient building industry and by an increasingly elaborate system for regulating land use and development at the local level. It was also facilitated by the practices of mortgage lending institutions that emphasized loans on new rather than existing houses and was encouraged by economies of scale in new construction and service provision. As a result, large corporate developers quickly assumed a role in producing not only new dwellings and subdivisions but entirely new suburban communities with a range of land uses. Some developments, of which Don Mills in Toronto is perhaps the best-known example, were in effect planned new towns. Similar suburban developments have since appeared around other large cities, and their size often exceeds the population of many free-standing Canadian cities[8].

Two inferences of relevance to this chapter can be drawn from this trend. One is the parallel rise of comprehensive urban planning and the perceived need to create integrated and self-supporting communities that balance housing and employment opportunities. These designs were as much in vogue in post-war Canada as they were internationally.[9] The obvious objectives were to minimize commuting, to encourage a sense of community, and to facilitate rational development overall. The second implication is the commercialization or commodification of housing, in which housing has come to be more widely treated as a near-liquid asset to be easily exchanged in a market. The emergence of the corporate suburb also suggests a growing commodification of entire communities and the living environments they provide.

The social consequences of the new corporate suburb have been mixed. To what extent do these new suburban forms represent progress? No doubt higher volumes of construction have created efficiencies in building and scale economies in the provision of services. Whether this has translated into lower housing costs for consumers or higher profits for the developers is unclear. On the other hand, a more balanced mix of residential, commercial, and industrial uses has reduced some types of land-use conflict, but not the level of commuting. Furthermore, examples of more plentiful and well-integrated services, in a relatively safe, pleasant, and planned (but perhaps boring) environment, are numerous. Traffic flows were rationalized, separating local from through traffic and pedestrians from vehicles, and local amenities were added. The quality and standards of building, site design, and infrastructure were also improved through planning and regulation.

Yet these same principles also often produced new problems. Post-war suburbs were designed for only a narrow range of household types and social or income classes. This specialization meant a reduced range of housing opportunities as well as more limited social experiences. These suburbs, given their housing designs and low densities, were also less adaptable to subsequent changes in housing demand. They may have contributed to feelings of social alienation and physical isolation, particularly among married women and children, and were often poorly prepared to accommodate diverse social needs.[10] Later suburbs were an improvement in terms of the range of services, but also tended to be homogeneous and frequently more expensive. Only the newer (post-1985) suburbs have begun to achieve a greater mix of income and social groups.

An additional element in these compositional changes in the provision of housing is the appearance in a number of urban areas, but especially in Toronto, of a new public-sector variation of the corporate suburb theme: the dispersion of socially-assisted housing from central areas to the suburbs and their geographical re-concentration within those suburbs, usually in the form of homogeneous townhouse and high-rise apartment developments. In these localized concentrations of the socially disadvantaged in the newer suburbs, we may have created one of the next generation's major settlement problems.

Inner City Forms: Revitalization and Decline

At the other end of the intra-settlement scale, a marked residential restructuring of many older inner cities has given rise to another new community form. As in other western countries, a resurgence of capital investment in many older neighbourhoods within Canadian cities has revitalized housing and the social milieu in those areas. The complex process of social change and residential restructuring involved, often referred to as gentrification, has now been widely documented although its long term consequences are uncertain.[11]

Evidence indicates that certain types of urban areas and neighbourhoods are more likely to undergo revitalization and gentrification than are others (see Ley 1985, 1988). Those communities tend to be either: (1) large metropolitan centres that have service-based economies, historic and culturally-rich business cores, and other urban and environmental amenities which attract middle-income professional households; or (2) small towns and cities with similar amenities and that are located near metropolitan areas or in recreational or retirement areas. The specific inner city neighbourhoods selected for revitalization tend to be those with older but architecturally substantial housing, a pleasant physical environment and proximity to office and institutional employment.

The consequences of this new urban form are also mixed. On the one hand, inner city revitalization has improved the quality of housing, community services, and living environments generally, at least for the new residents, and in the neighbourhoods directly involved. Although the geographical extent of gentrification is still quantitatively small, the effects on the central city's economy, fiscal base, and ambience have been generally positive. On the other hand, this trend has added to the polarization of social classes and residential conditions within the inner city. Many low-income renter households have also been displaced. Some find that they are also denied entry to other traditional lower-cost inner city neighbourhoods because of rising housing prices and rents and the conversion of rental housing to ownership. Still others have ended up relocating in the new corporate suburb, but often in the social housing sector.

The Role of Public Policy

What has been the role of public policy in directing these varied settlement changes? In responding to that question, it is necessary to differentiate between national, regional, and local scales and between the intended and unintended impacts of policy. At the national level, explicit or targeted policies (that is, those relating to housing and urban development) have not had much effect on the overall structure of the national settlement system. This should not be surprising since the federal government has not had either a national settlement policy or an urban policy *per se*.[12] Some provinces have had or still have such policies, and the local impacts may be considerable. Their aggregate effects, however, have been small, and at times have conflicted with other national policies. Similarly, housing policies on the whole generally have not acted to redistribute growth among regions or units in the settlement hierarchy at a national scale. The impacts of regional development policies and programs, on the other hand, have been more substantial but uneven.[13]

In contrast, the indirect and largely unintentional impacts on settlement of aspatial policies deriving from outside the housing sector have been immense.[14] Policies relating, for instance, to tariffs and trade, transportation, resource pricing, welfare provision, taxation, and equalization payments have profoundly influenced how the settlement system in Canada has evolved and at what rate.[15] For example, federal transfer payments and equalization programs have served as an implicit settlement strategy by encouraging the stay option for residents of depressed regions. Indeed, the settlement system has been created and, in broad outline, designed by such national policies, even though these policies have been developed with little or no regard for housing needs or community goals.

At the intra-settlement or intraurban scale, however, the direct impacts of public policies have been greater. The location, design, and density of new suburban developments, as well as the relative pace of renewal and revitalization as described above, have been conditioned, if not accelerated, by a variety of policy decisions. These include policies and regulatory practices relating to local government administration, mortgage lending, taxation, housing supply subsidies, land-use regulation, user servicing charges, transportation funding, subdivision controls, building codes, environmental assessments, heritage preservation, and agricultural land preservation. Combined, these have shaped the spatial configuration of land uses and of places of work and residence, the provision of public goals and services, and thus the nature and quality of living environments provided by the settlement system.

The absence of national urban policies and the strength of provincial and local controls on urban development, however, may well have contributed to the apparent divergence of settlement forms between Canada and the United States. On the whole, Canadian urban areas exhibit higher residential densities and less suburban sprawl, lower levels of inner city disinvestment and housing abandonment, less political fragmentation and higher levels of public services than do their US counterparts.[16] Undoubtedly, the widespread introduction of regional governments has contributed to this equalization of services and tax rates within Canadian urban areas.

Of course, such broad generalizations are open to debate, and they ignore wide regional variations in urban forms and settlement trends within both the US and Canada. Nevertheless, US cities, on average, are becoming even more decentralized and politically fragmented. In contrast, Canadian cities, despite extensive suburbanization, have retained proportionately more population, family households and jobs in their central areas. This, in turn, translates into higher levels of local demand for public services and housing, and thus into a higher quality inner city housing stock.

The Housing Impacts of Alternative Settlement Forms

At the macro scale, the settlement system influences housing conditions primarily through changes in the economic base and the social attributes of the individual communities in question. As the Canadian settlement system is markedly differentiated in terms of economic base, and relatively open to external influence, any shifts in the national economy or in the international economic order translate into highly variable rates of urban and regional growth or decline.[17] These sectoral shifts impact directly on local labour markets, on occupational structures and income levels, and thus on the aggregate demand for housing. These, in turn, influence the level of investment both in new housing construction and in maintenance of the existing stock. Thus, at the national level, aggregate measures of housing progress are continually redefined by the shifting economic fortunes or misfortunes of entire regions and communities as places to earn a living.

Canadian society, as a result, continually writes off housing resources simply because they are in the wrong location. Surplus housing abounds in some communities and regions, notably those with a declining economic base, while at the same time, severe shortages persist in other growing regions. Examples of such losses include the decline in rural farm housing, abandonment in resource towns, demolitions in declining cities and neighbourhoods, and the removal of housing located in the path of commercial or transportation improvements. Who benefits from and who pays for this depreciation of the stock depends on the situation, but typically it is the most vulnerable members of society.

At the local scale, within individual settlements and communities, the focus of the relationships shift. Housing quality then hinges not only on exogenously determined economic and demographic variables but also on the collective fortunes or misfortunes contingent upon location within the community. The principal mechanism at this latter scale is that of spatial externalities, the spill-over effects that link the viability of all properties within a given local community or neighbourhood. These relationships are articulated through the operation of competitive land and housing markets and are supported by the actions of the public sector through zoning and the provision of neighbourhood services. In effect, the criteria that we are looking for here relate primarily to local conditions and to the attributes of communities as *places to live.*

Evaluating Alternative Settlement Forms

The settlement environment of many Canadians has not changed dramatically since World War II. Most of the older communities and settlements dating from the 1900-45 period are still here, still intact, and still viable. Yet for the majority, the environment in which they live and work is different, either because it is new or because it has been transformed almost beyond recognition. How might we classify those living environments and how might these be linked to measures of housing progress?

A SIMPLE TYPOLOGY

No formal and widely accepted typology of settlement types presently exists. Such a typology, preferably, should be sufficiently robust to encompass all common settlement types and simple enough to facilitate its widespread application. For present purposes, the concise classification outlined in Table 16.3 should suffice to identify the varied settlement and community environments in which Canadians are housed. In total, twelve distinctive living environments are suggested, classified on the basis of their size, age, and internal diversity on the one hand, and the form, composition and density of neighbourhoods within those settlements on the other.

The actual classification depends on five basic attributes of settlements: location, size, density, homogeneity (in built form), and age. These criteria suggest that housing environments are influenced first and foremost by where one lives within the rural-to-metropolis spectrum. Location within this continuum implies differences in community size that, in turn, are a surrogate for other

Table 16.3
A simple typology of settlement environments:
density, built form, community type

Settlement type: by location, size and diversity	Low-density single-family	Medium-density low-rise apts multi-family	High-density high-rise apts. other multi-family
Rural areas, hamlets, villages	1	–	–
Towns & small cities	2	3	–
Larger cities & metropolitan regions			
Inner city	4	5	6
Older suburbs	7	8	9
Outer suburbs/exurbs	10	11	12

attributes: for example, the density of land uses and population, the degree of social diversity (or heterogeneity), the range of choice in housing types and employment, as well as levels of congestion, land costs, and the range of public services available. Larger settlements inherently provide different opportunities and constraints in housing choice.

Over time, the distribution of households and housing resources among these varied environments has shifted. Prior to World War II most of the nation's population lived in settlement types 1, 2, 3, and 4, while in the post-war period almost all new population growth has been accommodated in types 8 through 12. In absolute terms, types 7 and 10, the typical low-density suburban prototype, and type 8, the medium density older suburbs, represent the dominant settlement form.[18] At the same time, types 6 and 9, the large high-rise apartment concentrations in both inner city and suburb are a product entirely unique to the post-war period, particularly the 1960s and early 1970s. Combined, these high-rise living environments now house over two million Canadians. The newest forms of community development are obviously types 10 through 12, and these can be expected to accommodate most of the population growth in the future.

HIGH-RISE LIVING

Unlike the single-family dwelling, the high-rise apartment building is unique to the post-war era. In 1945 almost no one lived in multi-unit buildings over five storeys in height, although shared accommodation and multi-family occupancy were common. In 1990 over one million households, or over 12% of the Canadian total, lived in such buildings. As a built form, and in aggregate as a community, the high-rise, multi-unit apartment building, has altered the urban landscape and housing inventory of most cities as well as the living environments of many Canadians. Typically, the high-rise rental building was seen as a short-term holding facility for the young, the childless, and the mobile – while they waited to move on or out into single-family housing – and more recently for the elderly. For those who might not previously have been able to afford to establish their own household, this was progress; for others who were trapped in housing unsuited to their needs, it was not.

One perhaps extreme example of the mid 1960s high-rise rental housing community is the St. James Town development. Post-war Canada has produced many such communities, but usually on a much smaller scale. This development, located just east of downtown Toronto, contains over 11,000 people (larger than many small urban settlements) and some 6,000 apartments in 15 buildings, all on 14.2 hectares of land. It is the most concentrated living environment in Canada. Some Canadians would be appalled at the density, homogeneity, and blandness of St. James Town. Others, in contrast, including many of the residents, appreciate the central location, the accessibility, and the relative affordability of the accommodation it provides. Trade-offs between quality and rent, convenience and living space, density and accessibility are being made that further complicate any overall assessment of housing quality.

It is also difficult if not impossible to paint a scenario of life in such a large residential community. Nevertheless, it is feasible to outline some of the attributes of, and problems facing, the residents of this development and similar communities elsewhere. Many of the residents are young singles or small households, usually childless. Over time, however, as in many other similar developments, the composition of the residents has changed. An increasing number are now single-parent, lower-income, or socially-dependent households, as well as recent immigrants; and many are elderly. Local on-site services have proven to be inadequate for such a diverse population. As a result, the quality of the housing and of the external environment have both tended to decline over time.

Quality of the Settlement Environment

The concept of housing quality clearly incorporates components of the environment external to the stock, including both the local community and the broader settlement context. While improvements in the quality and diversity of living environments help define housing progress, measuring these components is another matter. The difficulties inherent in developing consistent and unambiguous indicators of environmental quality, and more generally of the quality of urban life and social progress, are well documented. The limited existing literature testifies both to the complex conceptual and measurement problems involved and to the immense number and variety of indices that are necessary to capture the quality dimension of our changing settlement fabric. Early efforts by the former Ministry of State for Urban Affairs, for example, to develop a set of urban indicators,[19] and by others to define quality of life measures,[20] have been largely unconvincing and difficult to replicate elsewhere, precisely because of their level of generality and subjectivity. Research has also shown a divergence

Table 16.4
Criteria for measuring the quality of the settlement environment of housing.

General criteria	Examples of quality indices
Size, form and level of development	Overall degree of diversity, richness; employment and social opportunities; potential for social interaction.
Physical infrastructure	Quality of sewers, water supply; roads, waste disposal; other utilities.
Public social services	Quality of schools, libraries, community centres, health clinics and hospitals; social support services, recreational, cultural facilities, parks.
Natural environment	Levels of air and water pollution; vegetation, landscape preservation; scenic amenities.
Built environment	Building material and quality; architectural style and ambiance; physical layout and design.
Transportation, accessibility	Ease of mobility, job choice; density of road and transit networks; congestion, safety, noise.
Private services	Retail facilities and services; cable TV and telephone; entertainment facilities.
Public participation and regulation	Voluntary organizations; openness and responsiveness of public agencies and institutions; efficiency and equity of regulatory mechanisms.
Social environment	Richness, diversity and density; socially-supportive and enhancing; sense of place and belonging; absence of social tensions and civil strife.
Personal autonomy	Degree of control over one's housing and local environment.

between subjective and objective measures of environmental and contextual quality.[21] The residential desirability of Canadian urban areas depends not only on who one is but where one lives.

With these considerations in mind, this chapter can now turn to an enumeration of the attributes of places as urban settlements that impinge on the quality of housing. Table 16.4 provides a selective checklist of factors that are relevant to assessing housing conditions and residential satisfaction in different types of settlement. This checklist is not, and could not be, applied here to even a representative sample of the settlement environments defined above. Instead, it is meant to emphasize the diversity of services that flow from the occupancy of housing in particular places at particular times as well as the complexity of measuring housing progress.

The indices themselves are largely self-explanatory. They define, in ten broad categories, the quality of living environments in general and the external components of housing quality in particular. These include: the dimensions of urban form and levels of social development; the quantity and quality of public infrastructure, social services, and the natural and man-made or built environment; levels of accessibility and mobility; private sector services; the efficiency and equity of the public regulation of urban development; the diversity and richness of the social environment (including a sense of place); and the degree of influence or control that individuals and households feel that they have over their residential situation.

Applying these indices to the varied living environments of Canadians over the post-war period leads to a number of general observations. Evidence indicates that the quality of our settlements – rural and urban, large and small, old and new – has on average increased since 1945. [22] For almost every index in Table 16.4, building and servicing standards and environmental quality have risen and the range of social choice has been extended. [23] Canadians, again on average, now have greater accessibility to distinctively different social milieus (although perhaps internally more homogeneous) and community types, as well as to a wider range of social services and work locations, especially within or near large urban areas.

This demonstrable improvement in environmental quality and housing choice in effect represents a net positive benefit to society – meaning simply that there are more winners than losers. But there are losers, both individual households and entire communities, which have been left behind by the changes in the settlement system documented earlier. Many of these households were simply in the wrong place at the wrong time or had housing needs not met by the market. These people should be the primary concern of housing policy.

Even for the winners, defining housing progress involves making explicit trade-offs. This is perhaps most apparent in the contrast between current living conditions in large metropolitan areas and those in small towns. Selecting a residence in the former generally affords access to higher incomes, more choice in jobs, services, cultural facilities, and of course access to more people. It also brings increased living and housing costs, higher levels of congestion, and some environmental disamenities. Small towns, on the other hand, are seen to offer the advantages of lower housing and land costs, easier access to recreation, relatively low levels of crime and pollution, and greater social stability and community cohesion, but at the cost of fewer services and limited employment opportunities and choice. [24]

Within urban areas, these trade-offs assume a more immediate expression. Households can choose, for example, between outer suburban areas, which offer the attractions of lower costs, newer construction, more space, and less pollution, and older neighbourhoods in the central city [25] that offer easier accessibility to the downtown core, a wider range of services, and often more varied and interesting social environments.

Thus, progress in terms of the external environment of housing is not only subjectively defined but also conditioned by an individual household's residential history, current place of residence, and long-term expectations. Over time, the criteria used to measure progress will shift subject to the reorganization of settlements as places to live, on the one hand, and the changing aspirations, incomes, and lifestyles of consumers on the other hand. In many instances these conditions and preferences are contradictory. There is also a tendency to confuse the quantity of choices available in urban environments with the quality of choice. They are not the same.

Emerging Settlement Trends and Housing Policy Issues

Forecasting how many Canadians there will be in the decades ahead is difficult, but it is easier than predicting where and how they will live. The settlement geography of Canada, and the configuration of the housing stock, have undergone a continuing reorganization in the past that will persist in the future. Each period superimposes a new layer of urban development on the old, adding new settlement forms and selectively altering and adding to what has gone before. Each period in turn offers a new set of challenges for both research and public policy. [26]

What do we know about our settlement future? We can anticipate additions to and adjustments within established settlement patterns over the next decade that will affect future housing progress. We know, for example, that overall population growth rates will be lower, but will likely be more volatile and unpredictable over time and from place to place (see Brown 1983; Simmons and Bourne 1989). Restructuring of the economy – in response to international competition and technological change – will have uneven locational consequences. At the national and regional levels, we might expect two distinct trends: a further concentration of urban development in the larger metropolitan regions, combined with continued decentralization of population and employment within those regions. Fully 80% of new population growth and housing construction will occur in the extensive urban fields surrounding the major urban centres.

As a result, the redistribution of growth within the national settlement system will likely accelerate. The contrasts between winners and losers will also become more apparent in an environment of slower population growth and continued economic uncertainty. More communities will register zero growth or an absolute decline in population (and subsequently in the number of households) and perhaps in employment as well. Other settlements, both large and small, will witness renewed if not accelerated growth, depending on their economic base, demographic structure, local environmental amenities, and their location and linkages to the metropolitan centres.

At the regional scale, a continued concentration of new growth is expected. Most of this growth will not occur within the metropolitan areas themselves, but in the surrounding suburbs, small cities, and non-farm settlements. In most parts of the country, these metropolitan regions, and particularly their fringe areas, will continue to expand at the expense of peripheral regions.

Despite a reduction in total new housing requirements over the next decade, housing issues may well intensify as a result of these settlement trends. Policy analysts and private market investors will have to be forward-looking and more responsive to fluctuations in demand and social needs over time and among communities. The aging of the baby-boom generation, for example, given their large numbers and changing life styles, will alter the demand for different types of dwellings and the locations at which that demand is expressed. Reduced rates of aggregate population growth may also increase the level of inequalities in the provision of housing and in housing quality. Localized housing shortages and price escalations in growing regions will be juxtaposed with surpluses of housing in other regions. This, in turn, will lead to serious problems of under-maintenance and declining quality in both housing and infrastructure. In more extreme cases, a mounting problem may well be one of downsizing, that is, devising strategies for the removal of surplus housing or the closure of entire communities.

Similar problems will manifest themselves at the local or intra-settlement level, particularly within depressed regions and the older metropolitan areas. At this scale, the polarization between declining and prospering neighbourhoods will be sharper and more visible, adding to established patterns of social segregation. A number of older suburban neighbourhoods are also vulnerable to decline, pinched between the attractions provided by newer and expanding outer suburbs and more trendy central city neighbourhoods. It will be a continuing challenge to adapt and upgrade these older suburban areas to meet changing demands. Despite low vacancy rates at present, a declining aggregate demand for housing in the future might, in fact, pull the blanket of high prices from many older and less attractive inner-city neighbourhoods as well as from the most poorly designed or serviced suburban communities. The rate and scale of decline that follows will depend in part on the success of recent initiatives in community renewal and in reforming the local planning and regulatory system, and in part on the rate of new suburban construction. [27]

The potential problems of declining investment and future reductions in housing demand are likely to be most serious in the high-rise sector. In those communities, both new and old, containing concentrations of high-rise rental and low-end-of-market condominiums, concerns about future quality levels are now widely appreciated. Given the choice of alternative accommodation, at or nearer to ground level and at comparable cost, a substantial proportion of existing high-rise renter households would almost certainly move. Urban renewal, in the next decade or two, could return with a vengeance and in the form of a high-rise crane for renovation or demolition.

Canadian urban areas have not to date approached the levels of decline and under-investment that afflict many older American inner cities, particularly in that country's depressed industrial belt (see Berry 1982). In part, this is attributable to lower levels of industrial specialization among urban areas in Canada. Nor are Canadian cities and suburbs likely to be in this state in the near future, in part because of continuing government efforts at equalizing regional economic opportunity and at maintaining investment in social overhead capital and social welfare across regions and among urban municipalities. Governments, particularly at the federal and provincial levels, must maintain a similar role in the future to ensure that regional differences in living conditions and housing quality do not increase to unacceptable levels.

Conclusion

Location and the changing settlement environment clearly exert a continuing impact on the flow of services from the housing stock. That environment has changed in many ways and for many reasons since 1945, but two processes stand out: the rapid urbanization of population and jobs; and the immense suburban and exurban landscapes created around the older central cities. These two processes have contributed directly to housing progress: first, by upgrading the quality of new dwellings while improving the physical and social services provided to households in those dwellings; and second, by facilitating access to more varied living environments. Such progress has not been attained, however, without costs, and these costs have not been evenly distributed among regions, neighbourhoods, or social groups.

At the same time, changes in settlement forms in general and suburban design in particular do not by themselves solve social problems or result in housing progress. There is no single urban form that is good by definition.[28] Instead, we have learned to plan for uncertainty and flexibility to accommodate increasingly diverse social and housing needs. Areas of tension will persist as the nation's settlement fabric is altered to accommodate changing demands, and in light of increasing constraints on public action and investment. Future housing conditions in urban Canada and continued housing progress will ride in large part on those changes and our ability to influence and to adapt to them.

Notes

1 The traditional definition of an urban settlement in the Census of Canada is a population concentration of 1,000 or more living at a density of at least 1,000 per square kilometre. The prairies did not become 50% urbanized until the mid 1950s, and the Atlantic region only in 1960.

2 The number of distinct urban centres in any size category can increase or decrease in either of two ways: (1) through growth, in which smaller centres pass the minimum size specified for a larger size category, or through population decline in which a centre drops below that minimum threshold; (2) by incorporation, in which smaller centres are merged into larger places through annexation or (for measurement purposes) through their statistical amalgamation into one or other of the extended urban area definitions

employed in the Census (for example, census metropolitan areas). In an increasingly urbanized society, the impact of the latter has become as important as the former.

3 This inventory of 139 urban centres, based on the 1986 Census, includes 25 CMAs and 114 CAs (2 of which had over 100,000 population). In the 1986 Census all urban areas over 10,000 population were defined as CAs. The CAs, like the CMAs, are defined largely on labour market criteria and include both a central city and those surrounding suburban municipalities closely linked to the central city through commuting to work and related criteria of interaction and interdependence.

4 The practice of defining extended urban areas, combining incorporated urban centres and their surrounding municipalities, was initiated by the Dominion Bureau of Statistics (DBS) in 1956. Stone (1967) applied and extended this concept, under the title "urban complexes," in his extensive study of urban development. Over a relatively long period of time, such as the period covered in this study, however, the use of a uniform set of definitions for defining urban areas is problematic.

5 Data for the 1981-86 Census period show an average growth rate for all urban places of less than 50,000 population of −1.1%.

6 The concept of the urban field defines an extensive region stretching 100 kilometres or more beyond the built-up urban area. It represents the expanded living space of urban residents, serving as a place of recreation, residence, and subsequently of employment.

7 For a discussion of the emergence of a corporate land development industry and of corporate suburbs in Toronto, see Lorimer and Ross (1976). The term is used here to describe the production of entire suburbs, including housing, commercial uses, and infrastructure, by large integrated development firms.

8 Other examples of large-scale developments include Bramalea and Erin Mills near Toronto and Mill Woods in Edmonton.

9 For reviews of the evolution and practice of community planning in Canada see Hodge (1986) and Cullingworth (1987).

10 An excellent case study of the social transformation of the suburbs in the 1970s is contained in a report of the Social Planning Council of Metro Toronto (1979).

11 Smith and Williams (1986) provide an overview of the international experience in inner-city gentrification.

12 Examples of explicit and direct settlement policies include Newfoundland's village consolidation and resettlement program, federal programs to concentrate population in selected northern settlements, and various resettlement programs associated with major transportation, resource, and hydro-electric developments.

13 The long chronology of federal regional development policies, including DRIE (Department of Regional Industrial Expansion), ERDA (Economic and Regional Development Agreements), and IRDP (Industrial and Regional Development Program), have undoubtedly had significant impacts on the settlement system in particular regions, but the overall effect has been uncertain. See Savoie (1992). These policies may, for example, have reduced the rate of metropolitan population concentration at the national level but increased it at the provincial level.

14 Some of the impacts of government activities on the settlement system are outlined in Simmons (1986).

15 Examples include the differentiated impacts on regional and urban growth of policies relating to oil pricing, the Auto-pact, transportation, government employment, and tariffs.

16 Many of these assertions are documented in a comparative analysis of US and Canadian cities by Goldberg and Mercer (1986). Of particular importance here are the destructive effects of intra-urban expressway construction on inner cities in the US, which are largely absent from Canadian cities.

17 Empirical evidence on the variability of urban and regional growth rates in Canada over time is provided in Preston and Russwurm (1980), Robinson (1981), and Simmons and Bourne (1989).

18 Although precise estimates of the total population living in each of these settlement types are not available, settlement types 1, 2, and 3 would include no more than 20% of the population; types 4, 5, and 6 about 25%; types 7, 8, and 9 about 45% and types 10, 11, and 12 perhaps 10%.

19 The results of these studies are summarized in MSUA (1975) and Gertler and Crowley (1977).

20 Of particular interest here are the results of the work of an interdisciplinary group at York University, as reported in Atkinson (1982), Greer-Wootten and Velidis (1983), and Lotscher (1985).

21 Roberts (1974), for instance, provided the initial comparative evaluation of the images that Canadians hold of their cities as places to live.

22 For example, in 1941 only 8% of farm dwellings had a flush toilet, and 20% had electric lighting, compared to over 80% and 90% respectively for dwellings in urban areas. The latter proportions also increased directly with the size of the urban place.

23 A simple index would be the proportion of suburban dwellings linked to municipal sewer and water systems in 1946 compared to the present.

24 Hodge and Qadeer (1983) provide an extended discussion of the relative attractions of small towns and large cities.

25 It is useful at this point to clarify the terms "central city" and "inner city." Central city refers to the principal (usually the largest) municipality within a census metropolitan area, and its boundaries are the limits of its political jurisdiction. The relative size of the central city typically depends on its age and when it ceased to annex surrounding territories. In Canada, the central city may include as little as 17% of the total metropolitan area population (Toronto) or as much as 95% (Calgary). The inner city refers to older residential areas near the downtown core and is commonly defined to include those areas in which the housing stock dates predominantly from before World War II. The boundaries of the inner city are therefore somewhat arbitrary.

26 Discussions of these research challenges are given in Lithwick (1983), Coffey and Polèse (1987), and Savoie (1992).

27 Continued high rates of new suburban construction in slow-growth metropolitan areas, as in the US, might further erode the central city housing stock.

28 Lynch (1981) offers an explicit set of criteria for evaluating urban form in general.

CHAPTER SEVENTEEN

Neighbourhood Differentiation and Social Change

Francine Dansereau

THAT NEIGHBOURHOODS develop in varied ways reflects their complex character. [1] A neighbourhood is a physical setting, defined through density, land-use or building form, and access to urban services and amenities. In terms of population composition, behaviour patterns, and activity spaces, it is a social reality which takes on a symbolic dimension in the images and expectations held by residents and outsiders. Finally, a neighbourhood is a local economy, in which the real estate market sorts social groups on the basis of lifestyle preferences and ability to pay, in accordance with general views on the area's economic prospects and viability. These dimensions interact to create diverse housing opportunities and constraints for people living in different locales.

An urban area reflects and responds to changes in society as a whole, according to its own character and internal dynamics. Among such changes, three types are of interest in this chapter. One concerns the ways in which people form households; that is, socio-demographic and life-style patterns. A second relates to shifts in the distribution of economic activities, occupations, and incomes which have accompanied the gradual move to a post-industrial economy. A final set of changes is linked to the post-war immigration into larger centres that has transformed the ethnic fabric of many neighbourhoods. The manner in which these broad changes influence neighbourhood differentiation is shaped by the social and economic histories and cultural patterns of each city.

One mediating influence concerns the different paths taken by urban regions in housing provision and community planning. Notable differences are found across Canada in neighbourhood mix of building and tenure types, the form and location of new housing, especially social housing, and the extent of public involvement in planning and development. To illustrate, consider how the size and location of the low-income housing stock affected the evolution of central-suburban disparities in Montreal and Toronto. In Toronto, with its strong metropolitan government, public housing – including some high-rise projects – was largely decentralized to the outer suburbs, while Montreal has favoured the infilling of small-scale projects within the central city. [2]

FIGURE 17.1 Age composition in central city and CMA: Selected cities, 1951-1981.

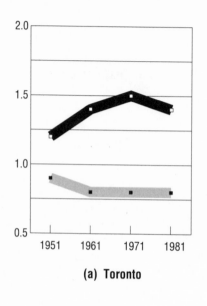

(a) Toronto

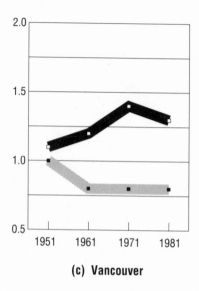

(c) Vancouver

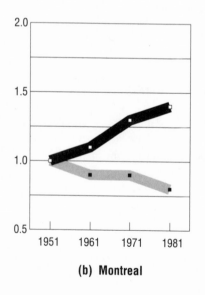

(b) Montreal

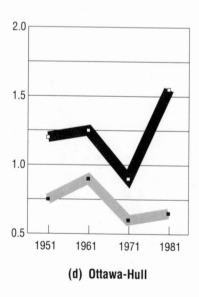

(d) Ottawa-Hull

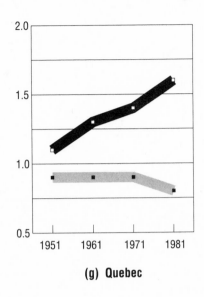

(e) Edmonton

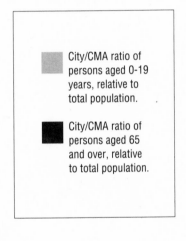

(g) Quebec

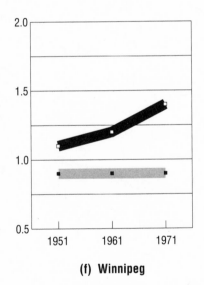

(f) Winnipeg

City/CMA ratio of persons aged 0-19 years, relative to total population.

City/CMA ratio of persons aged 65 and over, relative to total population.

SOURCE: Published reports of the Census of Canada, various years.
NOTE: 1981 data for Winnipeg omitted because of near-coincidence of city and CMA boundary.

FIGURE 17.2 Household composition in central city and CMA: Selected cities, 1951-1981.

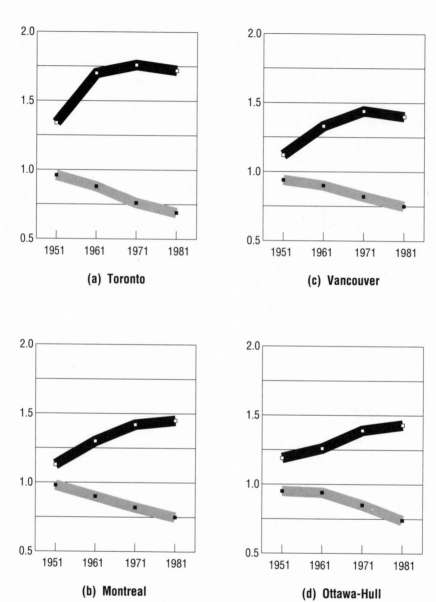

(a) Toronto

(c) Vancouver

(b) Montreal

(d) Ottawa-Hull

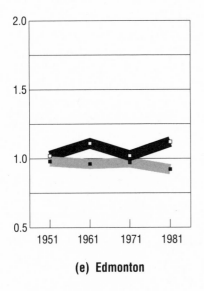

(e) Edmonton

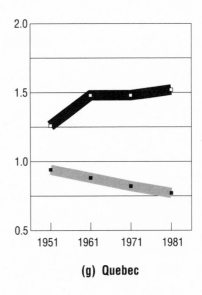

(g) Quebec

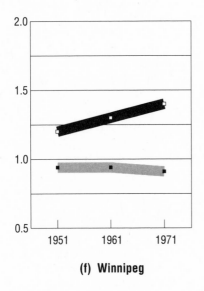

(f) Winnipeg

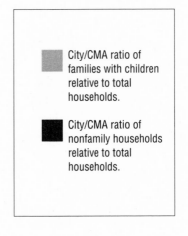

City/CMA ratio of families with children relative to total households.

City/CMA ratio of nonfamily households relative to total households.

SOURCE: Published reports of the Census of Canada, various years.
NOTE: 1981 data for Winnipeg omitted because of near-coincidence of city and CMA boundary.

Neighbourhood Differentiation: Continuity and Change

From the earliest studies of internal differentiation in Canadian cities, the spatial patterns of three dimensions of population characteristics have been useful for exploring social change: life-cycle, socio-economic status, and ethnicity. These are examined below. At the same time, there is an overlap among these dimensions: falling into a group – be it the elderly, newly-formed households, lone-parent families, or a particular category of immigrant – is likely to imply a distinct socio-economic position.

LIFE-CYCLE

The spatial distribution of "life-cycle" attributes – for example, family size, family composition, and age – has usually been found to follow a concentric pattern in post-war Canadian as well as other large North American cities. Early post-war suburbanization resulted in a concentration of families and children in low-density suburbs and in a rising proportion of non-family, elderly and newly-formed households in higher-density central areas.

Figures 17.1 and 17.2, showing Central City/CMA differentials for selected ages and household types, exemplify these patterns.[3] For most CMAs, the relative surpluses of non-family households in central cities increased rapidly in the 1950s, grew less quickly in the 1960s, and eventually stabilized (or, in Toronto and Vancouver, even decreased) in the 1970s.

The over-representation of the elderly and under-representation of the young in central cities have followed a similar trend. Again, only Vancouver and Toronto showed a halt in the relative aging of central city populations in the 1970s. Nonetheless, a few figures demonstrate that the elderly are now an integral part of suburbia. In Montreal in 1976, 51% of the CMA population aged 65 years or over lived in areas built-up after 1961. For the same year the Social Planning Council of Metropolitan Toronto (1979) pointed out that 44% of the Metro population aged 75 years or over were to be found in the outer municipalities.

SOCIO-ECONOMIC STATUS

While the persistence of an inner-city ring of economically-deprived households has been noted, affluent households have not systematically exhibited a corresponding tendency to cluster in an outer ring. Instead, income, occupation, and education have been found to vary by geographic sectors rather than rings.[4] Such sectors are typically delineated by topographic features, by socio-physical barriers such as the "right and wrong side of the track" or of a main artery (for example, Bloor Street in Toronto, Sherbrooke Street and Blvd. St. Laurent in Montreal), or by the initial elite or working class character of a neighbourhood onto which later developments have been grafted. These divisions have also, in many cases, been entrenched in municipal boundaries.

Table 17.1
Canada, 1951-1981, selected cities:
Percent of male labour force by selected occupations,
and female participation rate, central city and CMA

*(a) Selected occupations as percent of male labour force**

		Managers & professionals				Blue collar				White collar			
		1951	1961	1971	1981	1951	1961	1971	1981	1951	1961	1971	1981
Toronto	CMA	21	25	22	29	47	45	44	39	31	28	34	30
	City	21	18	21	32	58	52	43	35	20	30	35	32
Montreal	CMA	19	23	22	28	51	48	41	39	30	28	36	32
	City	17	19	17	24	51	51	44	40	31	30	38	35
Vancouver	CMA	18	24	18	27	47	46	44	41	30	26	33	29
	City	18	23	19	27	45	45	42	37	33	29	36	33
Ottawa-Hull	CMA	22	28	31	37	38	35	30	26	39	35	37	35
	City	26	31	34	40	32	30	26	22	42	38	40	36
Edmonton	CMA	19	24	20	27	47	45	43	44	31	28	33	26
	City	20	25	21	27	45	44	43	45	32	28	34	26
Winnipeg†	CMA	18	21	20	26	49	47	43	41	32	30	35	30
	City	17	19	18		48	49	46	32	30	34		
Quebec	CMA	17	22	22	32	48	45	36	32	31	31	40	34
	City	20	20	19	29	45	43	36	31	34	36	44	39

(b) Female labour force participation rate‡

		1951	1961	1971	1981
Toronto	CMA	35	39	49	61
	City	38	45	53	61
Montreal	CMA	31	32	38	51
	City	32	36	41	51
Vancouver	CMA	28	32	43	56
	City	31	37	48	57
Ottawa-Hull	CMA	33	37	46	58
	City	36	40	49	58
Edmonton	CMA	30	37	48	62
	City	31	38	49	63
Winnipeg†	CMA	33	38	47	57
	City	36	42	48	
Quebec	CMA	29	31	36	50
	City	32	36	38	47

SOURCE Published reports of the *Census of Canada*, various years.
* See text for definition of categories. Direct time changes cannot be inferred from these figures because of changes in grouping of occupational categories between census years. Percentages are relative to total employed male labour force.
† City percentage omitted for 1981 because of near-coincidence of city and CMA limits.
‡ Females ages 15-64 in labour force/total females ages 15-64.

A brief look at the evolution of Central City/CMA differences in occupational[5] structure (Table 17.1) and incomes (Table 17.2) illustrates how the redressing of the central-peripheral balance has been reinforced, especially in cities whose functions as national or regional service centres have grown most rapidly (as signalled by Balakrishnan and Jarvis (1979) for 1971). Toronto stands out as the exceptional case. In 1981 male workers residing in the central city were more likely to be managers or professionals, or white collar workers, than were men living elsewhere in the CMA; indices related to income have also increased, contrary to the changes observed in other urban regions (Table 17.2). Such shifts in occupational structure reverse a tendency, initiated in the 1950s, for such workers to flee the central city; indeed, the opposite trend is evidenced as early as 1971. By comparison, the slower development of office and high-level management activities in Montreal over the past decade is reflected in a modest narrowing of the gap between the central city and its periphery – the latter still retaining a higher occupational status.

What stands out from this overview is the inappropriateness of traditional centre-periphery dichotomies as frameworks for analyzing neighbourhood differentiation. The notions of middle-class suburbia and a working-class (or "underclass") inner city have been rendered obsolete by the movement of industry to the periphery – especially in fast-growing cities – and by broad changes in employment structure and in the relative benefits associated with different occupations. Diversification is not new, but it is taking on a heightened significance, creating new lines of demarcation within cities.

ETHNICITY

The ethnic make-up of Canada's large urban centres has changed since 1945. In 1951 barely one-quarter of the population in Toronto and Vancouver was of ethnic origin other than British or French (Table 17.3). By 1981 this population was in the majority in both cases (in Prairie cities, ethnicity other than British or French also characterizes a majority of the population, but this is the result of a history which goes back much further). In Montreal, the increase has been more gradual (from 13% to 24% between 1951 and 1981), while the populations of Quebec and of Atlantic Canada's major centres have remained homogeneous, with the exception of Halifax.[6]

Since 1945, and more so over the last two decades, immigration has been funnelled into the large cities, thereby contributing to their vitality. Immigrants are young and tend to have higher fertility rates than do the majority. They are hard workers and invest heavily in residential property, which for them is both a means of economic advancement and a symbol of success.[7] As new arrivals, often forced to start at the bottom of the ladder, immigrants are more likely than others to be concentrated in the inner cities; there is an overlap between areas of low socio-economic rank and those that are strongly ethnic – especially in the case of new arrivals employed in unskilled jobs (see maps in Hill 1976). The

Table 17.2

Central City/CMA income ratios*: Selected CMAs, 1951-1981

	Average family income			Average male employment income			
	1961	1971	1981	1951†	1961	1971	1981
Toronto	0.85	0.89	0.96	0.95	0.83	0.86	0.94
Montreal	0.93	0.88	0.85	0.98	0.91	0.86	0.80
Vancouver	0.98	0.98	0.96	0.99	0.94	0.93	0.90
Ottawa-Hull	1.06	1.03	1.02	1.03	1.05	1.00	0.97
Edmonton	0.99	1.00	1.00	1.01	1.00	1.00	0.97
Winnipeg‡	0.96	0.90		0.98	0.94	0.89	
Quebec	0.96	0.91	0.86	1.01	0.93	0.88	0.84

SOURCE *Census of Canada,* various years.
* Indices apply to family income, rather than household income, and male employment income, instead of total employment income, because these are the least influenced by changes in household and male/female labour force compositions.
† Calculations based on median instead of average income for 1951.
‡ 1981 data omitted for Winnipeg because of almost total coincidence of city and CMA limits.

Table 17.3

Central City/CMA differences: ethnicity and immigration indicators: Selected CMA, 1951-1981

		% of population of ethnic origins other than British or French				% of population born outside Canada		
		1951	1961	1971	1981	1961	1971	1981
Toronto	CMA	24	36	40	51	33	34	38
	City	28	40	50	51	42	44	43
Montreal	CMA	13	18	20	24	15	15	16
	City	15	21	25	30	17	19	23
Vancouver	CMA	26	34	37	49	29	26	30
	City	26	37	44	68	35	34	40
Ottawa-Hull	CMA	9	15	16	26	12	12	14
	City	11	19	20	32	16	16	20
Edmonton	CMA	38	48	48	55	23	18	21
	City	38	48	49	57	24	19	22
Winnipeg	CMA	41	47	48	56	24	20	19
	City†	45	52	54		29	25	
Quebec	CMA	1	2	2	4	2	2	2
	City	1	2	2	5	2	2	2

SOURCE Published reports of the *Census of Canada,* various years.
† See note to Table 17.1.

Francine Dansereau

Table 17.4
Overall ethnic diversity and residential segregation
indices for selected ethnic groups:
9 largest CMAS, 1961-1981

	Toronto			Montreal		Vancouver		Ottawa-Hull	
	1961	1971	1981	1971	1981	1971	1981	1971	1981
British	30	32	27	50	46	15	17	45	37
French	17	19	20	50	48	18	21	56	57
German	11	16	19	37	41	16	16	25	27
Italian	51	56	50	57	56	46	45	48	48
Jewish†	72	73	74	83	83	52	56	49	50
Ukrainian	37	34	34	44	48	18	16	31	33
Netherlands	23	30	32	+	57	22	25	33	35
Scandinavian	16	24	34	+	67	12	16	31	37
Polish	38	38	39	42	44	18	21	25	34
Aboriginal peoples		+	45	+	45	42	39	+	41
Ethnic diversity index	60	65	78	55	56	63	76	64	73

SOURCES For 1981: Bourne *et al.* (1986); for 1971: Hill (1976); for 1961: Richmond (1972) for Toronto and Driedger and Church (1974) for Winnipeg. See Bourne et al (1986, 58-59) for description of indices. Larger index value means greater segregation or ethnic diversity.
† Index calculated on the basis of religious affiliation.
+ Index not calculated because of insufficient numbers in ethnic group.

groups predominating in the classical "reception areas," that is, typically multi-ethnic enclaves located in older districts, have changed over time. But the fundamental role of these areas remains – helping new arrivals to make the transition and channelling successive ethnic groups into distinct tracks of employment and entrepreneurship. Housing conditions in such enclaves are often mediocre; physical deficiencies, overcrowding, rooming-houses, and doubling-up are more common than elsewhere.

When ethnic groups eventually relocate, some of those that are of "critical mass" tend to re-concentrate in new neighbourhoods. Such is the case, for example, for Jewish and Italian populations in cities like Montreal and Toronto.[8] Ethnic segregation is, then, an enduring feature of most of Canada's large cities, and some studies have illustrated that this ethnic segregation can be explained only partly by differences in socio-economic level or by the elapsed time since a group arrived.[9] The Jewish or Italian municipalities best exemplify this.[10] Even though they have been in the country for a long time, some visible minorities have not succeeded in breaking the double barriers against social

Table 17.4 continued

	Edmonton		Calgary		Winnipeg‡			Quebec		Hamilton	
	1971	1981	1971	1981	1961	1971	1981	1971	1981	1971	1981
British	14	12	10	8		24	21	29	21	16	15
French	17	15	11	15	57	39	39	28	23	18	18
German	15	15	11	10	48	19	20	31	39	14	17
Italian	45	41	35	33		39	34	+	39	38	37
Jewish†	59	65	48	49	62	69	72	+	+	59	68
Ukrainian	22	20	10	11	56	31	28	+	+	21	22
Netherlands	20	22	17	17		22	25	+	+	34	34
Scandinavian	10	14	10	14	47	15	18	+	+	23	32
Polish	19	20	15	17	59	25	28	+	+	29	28
Aboriginal peoples	37	37	44	36		49	49	+	61	+	52
Ethnic diversity index	75	85	66	77	76	77	86	13	12	60	69

+ Index not calculated because of insufficient numbers in ethnic group.
† Index calculated on the basis of religious affiliation.
‡ Winnipeg's indices for 1961 calculated in reference to British origin population rather than to total population (used for tabulation of all the other indices shown). Comparisons of the values of the indices between 1961 and 1971 are therefore precluded for Winnipeg.

mobility and residential integration; this is especially true of Aboriginal peoples in the western provinces and of blacks in Halifax, who remain "captive" populations in declining inner-city areas (Krauter and Davis 1978).

Over the past decade, the relative over-representation of people born outside Canada in central cities has generally continued or increased, except in Toronto (Table 17.3). The over-representation of ethnic minorities has diminished in Toronto. This may reflect the fact that Toronto's central core has been extensively gentrified, forcing immigrant workers and especially the new arrivals to seek out more affordable accommodation in the suburbs.[11] Immigrants are also drawn toward the suburbs by the extensive decentralization of manufacturing in the Toronto region and by its well-developed public transportation system.

Thus, for various reasons – be it deliberate choice, economic constraints, a lack of linguistic skills, or discrimination by the dominant groups – ethnic neighbourhoods have grown and multiplied over the last few decades. For many groups, segregation in large cities, far from decreasing, actually remained the same or even increased from 1971 to 1981 (Table 17.4). These indications do not necessarily reflect barriers to social or residential mobility, but they do indicate that ethnicity continues to be important in neighbourhood differentiation.

Inner-City Revitalization and Decline

Up to the mid 1970s most discussions about the evolution of the inner city revolved around population decline, the exodus of families, and growing obsolescence and disinvestment in housing and public facilities. Since then, the revitalization of inner-city neighbourhoods has captured the interest of the media, public authorities, and urban affairs professionals. What has taken place, and why the shift in perspective? Revitalization implies neither the repopulation of central neighbourhoods nor a return of large numbers of suburbanites. Rather, this term refers to private re-investment in housing and commercial activities and to the subsequent transformation of neighbourhood social composition. Revitalization is often equated with the rehabilitation of older housing. However, redevelopment operations and new infill projects have also contributed to urban revitalization in some cities, such as Vancouver and Edmonton.

One may contrast current forms of urban revitalization to the "bulldozer" urban renewal efforts of the 1960s, citing differences in methods employed and in the scale and visibility of public investment. However, there are also connections between them; urban renewal programs laid the groundwork for urban revitalization by eliminating slums and by reinforcing the commercial and cultural attributes of the central core. Halifax exemplifies this connection. Similarly, public assistance for housing rehabilitation and neighbourhood improvement influenced re-investment in central neighbourhoods by private developers and middle-class households.

Who are these re-investors or "whitepainters" as they were first termed in Toronto? The current image is that of the "yuppie," living alone or without children and attracted by the diversity and excitement of city life. This stereotype, however, overlooks the variety and, at times, marginal forms of re-investment. Family households [12] (including single-parent families headed by women) and professionals of modest means or insecure economic situation have been reported in gentrifying areas of some Canadian cities (Rose 1984). In the stage theory of gentrification (Pattison 1977), such anomalies represent only the first stirrings of a process initiated by "risk oblivious" innovators, but inescapably leading to a new homogenization of neighbourhoods favouring "risk averse" respectable professionals. This conceptualization is too narrow; re-investment attracts participants from a variety of social levels, including both newcomers and long-time residents. [13]

What causes urban revitalization? Economic factors related to the housing market, such as the rise in new house prices or in the cost of energy, are not as important as socio-demographic and lifestyle changes: for example, the increases in one-person households and in double-income or childless couples. The repercussions of these latter changes have been particularly strong in cities that have also seen an intensification of their service sector. Such cities attract many young professionals and non-traditional households. Conversely, slow-growth cities whose economy is essentially based upon manufacturing (such as Windsor) have seen little revitalization activity, and decline there remains the principal problem of the inner city.

However, post-industrialization has been accompanied by a dualization of the labour market. In high-growth sectors such as finance, insurance, and real estate, low-level jobs are increasing more rapidly than are planning and super-visory functions. [14] This dualization adds to the existing division of the labour market evidenced in Montreal and Toronto, where traditional light industries, drawing heavily from among poorly-paid immigrants, continue to occupy an important place. It is in the interest of both workers and employers within such industries to keep jobs and homes in the central city, in contrast to heavy indus-try where economic forces have dictated a relocation to the suburbs.

At the moment, in certain revitalizing neighbourhoods, this duality and divi-sion are being translated into a residential co-existence of groups from different segments of the labour market. [15] The waiter, the free-lance artist, the journalist, and the university professor live side-by-side with the hospital maintenance worker, the immigrant shopkeeper, and garment worker. In other words, segre-gation may well have been reduced in many cases. The main problem is the pre-cariousness of such delicate balances and the threat of eviction as a result of demolition or repossession. Vancouver has had high demolition rates (Ley 1986). The conversion of rental housing into condominium units has been banned or severely controlled by most provinces since the mid 1970s, but it is still a threat in some cities. In Montreal, for example, there has, until the end of 1988, been no control over the use of undivided joint ownership to repossess a dwelling in buildings of less than five units (approximately 50% of the Montreal housing stock). [16] Finally, conversion also operates in roundabout ways in buildings emptied of their inhabitants for the purposes of renovation. [17]

The rise in home ownership, which has accompanied revitalization of the inner city, constitutes a direct threat to renters, especially where the stock of moderately priced single-family housing is scarce. Tenants are exposed to repeated evictions or run the risk of finding themselves in suburban apartments where, in exchange for a possible improvement in the quality of their dwelling, they suffer in terms of affordability or accessibility (see Hodge 1981; Saint-Pierre, Chau et Choko 1985). The most underprivileged eventually swell the waiting lists for subsidized housing. A second threat involves gradual rent and house price increases that nevertheless lead to the displacement of households with limited means. The transformation of the commercial infrastructure and of the social fabric of revitalizing neighbourhoods also encourages departures because of the erosion of feelings of belonging and due to a growing "cultural disso-nance."

Rather than exacerbating private market tendencies by grants directed towards the housing stock, public policy should moderate and control central city re-investment. Public support could be withdrawn from already sought-after neighbourhoods and redirected to zones where the market is less active. [18] This would eventually run the risk of producing new areas of gentrification, but their proliferation might well reduce pressures on existing areas. Policies designed to retain industrial employment within the inner city can similarly serve as a counterweight; this has been done with some success, in conjunction with neighbourhood improvement initiatives, by the City of Toronto. It is also necessary to strengthen renters' security of tenure; and, finally, rather than continuing to give assistance indiscriminately, certain groups with a direct interest in remaining in central neighbourhoods should be targeted – students or young labour market entrants, elderly persons, and so forth. In the same spirit, projects run by cooperatives or non-profit organizations, and which serve to stabilize inner-city neighbourhoods, should also be given priority.

Yet while revitalization proceeds apace, the fabric of particular areas, or even the entire central core, of certain cities continues to be eroded. The inner city of Winnipeg, for example, continues to deteriorate, except for the few neighbourhoods that have been redeveloped or designated as historic preservation areas (McKee, Clatsworthy and Frenette 1979; Lyon and Fenton 1984). In many medium-sized cities, NIP and RRAP were the only support for any form of private re-investment whatsoever, and some municipal and provincial governments have not stepped in to fill the void upon termination of those programs.

A full range of inner-city neighbourhoods – declining, stable, revitalizing, undergoing massive redevelopment – still exists even in cities where revitalization has been extensive. For example, Strathcona in Vancouver or Kensington in Toronto are still declining zones or, at best, non-improving ones. In these latter cities the number and size of revitalizing areas have considerably increased, and several neighbourhoods have changed position, moving from the "stable" category in 1971 to "revitalizing" in 1981. In Montreal, increases in social status have been visible mostly in white collar or upper-middle class neighbourhoods, while several working-class neighbourhoods that were stable or declining in 1971 have lost ground.

The disparities between and even within older neighbourhoods are considerable and reflect a polarization of neighbourhoods in central areas. Solutions must be fine-tuned to each case. Local housing stock characteristics (tenure, size, and feasibility of conversions into smaller or larger units) require strategies and tools which can be developed and adjusted at a decentralized level to minimize the adverse effects of revitalization on low-income households.

Diversification of Suburbs

While the revival of inner-city areas has attracted much attention, almost nothing (with the exception of the 1979 study by the Social Planning Council of Metropolitan Toronto [1979b]) has been written on its hypothetical counterpoint:

disenchantment with, or deterioration of, living conditions in the suburbs. In general, post-war suburbs have aged rather well, contrary to the situation in France, for example, where large social housing projects, built in the suburbs from 1945 to 1960, have now become the primary target of national housing rehabilitation policies.

Except in the case of single-industry (or resource) towns suffering from economic difficulties, suburban neighbourhoods with depressed real estate markets and in the midst of a downward spiral are rarely seen in Canada. Problems related to the renewing of infrastructures or the renovation of commercial arteries are not infrequent, but the overall physical condition of suburban neighbourhoods presents few causes for alarm. Quite the contrary, trees have grown, hiding the often-criticized mediocrity of standardized houses, and households have invested much money and energy in improving their dwellings and in adapting them to their changing needs (see Teasdale et Wexler 1986). Services, too, have been developed, so that the present resident of a typical Montreal suburb, built twenty-five to thirty years ago, is likely to find all the necessary services within a ten minute walk from home (Mathews 1986).

SINGLE-FAMILY SUBURBS

This reassuring diagnosis applies primarily to suburbs composed of owner-occupied single family dwellings. The principal problems are to be found in certain suburbs established in the period 1945-60 whose housing stock has remained unchanged. Such communities exhibit a distinct aging syndrome (by comparison to CMA-level data): an over-representation of households nearing retirement age, with children over 18 years of age still living at home, and an under-representation of pre-school children. A series of mismatches between the new needs of this population and the available neighbourhood services may be cited: inadequate transit facilities and a lack of places for socializing among the retired or those in the later years of adolescence, under-utilized schools, and a lack of after-school child care in relation to the number of working mothers.

The housing stock, too, would benefit from adaptation: for instance, the transformation of existing houses (subdivision, accessory apartments, and granny flats); construction of smaller dwelling units in townhouse or low-rise apartment clusters for households wishing to age in place, while freeing themselves from the burden of maintenance; and construction of low-rise condominium or cooperative projects, with common facilities, to suit the needs of single-parent families. These are examples of solutions designed to increase the flexibility of the traditional suburban model.

PERIPHERAL HIGH-DENSITY LIVING

The picture is bleaker when one considers the case of apartment developments hastily built during World War II and up until the 1970s as a response to programs designed to increase the stock of rental housing. The signs of physical deterioration among the oldest buildings are evident. Significant too are certain areas of low-rent housing built during the 1960s and 1970s on what was then the periphery of Montreal and Toronto, and which have already begun to be identified as the new poverty pockets of our metropolitan areas (Social Planning Council of Metropolitan Toronto 1979; Dansereau et Beaudry 1986).

In both cases, the result has been the concentration of households in a precarious economic condition and an increased local demand for social services. The concentration of families reliant on social assistance, the high rate of unemployment, and the characteristically unskilled labour force are the principal signs of precariousness. In addition, some of these concentrations have become reception areas for recent immigrants. These situations are worrisome in that they sometimes involve the spatial segregation of dependent groups into locations of minimal accessibility, which reduces the possibility for entry or re-integration into the economic fabric and the general life of cities.

These conditions are certainly not the general rule and the urbanization of the suburbs also has its success stories. In many respects, suburbs have become self-sufficient in terms of services and job provision. The diversification of suburbs in terms of social composition and housing form is a problem only insofar as it leads to an uneven distribution between neighbourhoods or municipalities. If less affluent neighbourhoods continue to receive a disproportionate share of the economically marginalized, while the more affluent keep their privileged position, inequality and social tensions will inevitably be heightened. At present, little is known about the exact terms and consequences of the diversification of the suburbs. It is important to scrutinize the reality of the "suburban dream" and to assess the effect of suburban entry into the sphere of problems of social integration and inequality which, until recently, has been the domain of the older central city.

Conclusion

The diversity of neighbourhoods and of processes of neighbourhood change is obvious. Social and physical settings do not necessarily follow the same path nor change at the same rhythm. Neighbourhoods in different cities are subject to different pressures, depending on such factors as the rate of economic growth, the direction and composition of migration or prevailing attitudes towards social heterogeneity. Moreover, the influence of these factors may vary according to a city's economic structure or a zone's physical and social make-up: for example, an older city or newer urban centre; a single-family or apartment zone; the proportions of owners and renters.

This diversity makes it difficult to render a general verdict on the progress accomplished over the last forty years. Overcrowding and unsanitary conditions have diminished. Slums have practically disappeared from the large cities as have instances of uncontrolled fringe development devoid of the basic infrastructure. Significant improvements have been made in the design of new communities and in the integration of social and physical planning. Whole communities are no longer displaced in order to make room for skyscrapers. Still, in various respects, some neighbourhoods have become dysfunctional or less secure for some residents. The revitalization of inner-city areas is a good case in point: the physical environment has certainly improved, but the gains made by some households, in terms of accessibility or environmental quality, have created affordability and security of tenure problems for others. As a result, opportunities for residential choice and mobility have been seriously reduced for the latter.

All things considered, it is difficult to establish whether everyone has participated equally in the general improvement of neighbourhood conditions since 1945. To answer this question, one might try to specify the environmental gains made, for instance, by the 75-year-old widow, the unemployed youth, and the lone mother with young children, in relation to those made by an Asian immigrant, a construction worker, or a doctor. Even so, the results of such an exercise would remain of doubtful validity, since some household types or occupational categories existing today had no equivalent forty years ago. Others, such as large families, have virtually disappeared from the scene. The material conditions of some groups, such as retired people, have improved markedly in recent years, but their increasing weight in the population has brought them to the top of the policy agenda. In brief, a result of the profound changes in household and employment structure has been to produce new groups in precarious socio-economic situations, who bear little resemblance to the disadvantaged of past times.

Another concluding remark concerns the desirability or feasibility of national neighbourhood policies. Should the federal government review the decision made in 1982 to pull out of intervention at the local level on the basis of equalization principles? Any program to assist neighbourhoods should provide maximum flexibility, given the variety of conditions and the need to monitor changes at a highly decentralized level. Furthermore, most of the problems raised in this chapter relate to the interaction between the built environment and social policy. This raises the complex issue of intergovernmental relations as well as that of concerted action with nongovernmental organizations (NGOs) and citizens' groups.

A useful direction is to develop guidelines or instruments for a continuous assessment of residential environments, which would be administered by municipalities. In addition to data concerning households and dwelling conditions (similar to those provided by the US Annual Housing Survey) such an instrument could provide indicators of: (1) neighbourhood amenities or nuisances (such as open space, community services, transportation facilities, traffic, and sources of pollution); (2) housing market conditions (vacancy rates, rents and sales prices, investment in new construction and repairs); and (3) subjective evaluations by residents. A few questions might also be incorporated into the

Census questionnaire, in order to provide more insight into the fit between the characteristics of households and the environments in which they live.

Notes

1 For concise recent reviews of neighbourhood definitions, see Hallman (1984, 6); Grigsby, Baratz and MacLennen (1984, 15-18).
2 The long absence of metropolitan government from the Montreal scene, and Quebec and Montreal's later entry into the field of social housing (thus benefitting from lessons learned elsewhere), might explain the latter orientation.
3 In the 1971 and 1981 Censuses, a CMA is the main labour market area of an urbanized core (or continuously built-up area) having 100,000 or more population. CMAs are created by Statistics Canada and are usually known by the name(s) of municipal jurisdiction(s) forming their urbanized core. That municipal jurisdiction is referred to here as the Central City. In the 1961 Census, CMAs were delineated around municipal jurisdictions of 50,000 or more population meeting certain population density and labour force composition criteria; in total, CMAs still had to contain at least 100,000 persons. In the 1951 Census, a CMA is a group of communities which are in close economic, geographic, and social relationship.
4 This "sectoral" model is not specifically Canadian, dating back to Hoyt's (1939) discussion of the spatial structure of 142 American cities from 1900 to 1936. See Hamm (1982) for a recent review of research testing the sectoral and concentric hypotheses in various cities around the world.
5 The categories are as follows:
  • *managers and professionals:* managerial and administrative, teaching, medicine and health, technological, social, religious, artistic, and related occupations;
  • *white collar:* clerical, sales, and service occupations;
  • *blue collar:* processing, machining, product fabricating, assembling and repairing, construction trades, transport equipment operating, and "other" (including "not elsewhere classified") occupations;
  • *primary:* farming, fishing, forestry, mining, and quarrying, including oil and gas field occupations. (Table 17.1 does not show primary occupations, but these are included in totals.)
6 Halifax has a long history of ethnic (including racial) diversity, as Canada's main port of entry on the Atlantic; however, the flow of migrants has decreased compared to Toronto and the western provinces in the last two decades.
7 Unemployment rates are systematically lower and female labour force participation higher among immigrant groups, compared to the rest of the population. Holding two jobs is not uncommon. This explains the relatively high average family or household income of many groups in spite of low levels of qualification. Also, many immigrant groups have shown a higher propensity to become home owners and to engage in home improvements, compared to native-born Canadian of similar income levels in central

districts (Richmond 1972; Dansereau, research in progress on the Portuguese community in Montreal).

8 In contrast, other categories of immigrants (for example, from Northern European origin) are more likely to disperse.

9 Darroch and Marston (1971), in particular, have shown that the internal social stratification of a number of large groups corresponded to differentiated areas of concentration, including suburban areas with high levels of segregation and "institutional completeness." This notion, spelled out by Breton (1964), refers to the degree to which various ethnic communities tend to create and maintain distinctive organizations or institutions to satisfy a whole array of needs (such as education, jobs, medical care, food, and clothing), to mobilize the community's resources, and to express its specific character and interests through media or pressure groups. Institutional completeness tends to counter assimilation and, therefore, to foster residential segregation.

10 The Francophone/Anglophone linguistic barrier also remains a powerful basis for isolation and for institutional completeness wherever both groups are of sufficient numbers. Montreal's "two solitudes" is a well established reality which needs no further elaboration here. There are, however, less obvious examples such as Winnipeg, where the role of "institutional completeness" appeared crucial in explaining the residential segregation of the French minority, compared to the Scandinavians or Germans. See Driedger and Church (1974) and Matwijiw (1979).

11 In 1976 one-half of all immigrants to Canada from 1971 to 1976 (aged 5 years or over) had settled in Metropolitan Toronto's suburban municipalities: some of the "rapid growth districts" ranked among the highest reception areas (Social Planning Council of Metropolitan Toronto 1979, 175-93).

12 See Ley (1986) for a synthetic review of various Canadian studies. This should not be a surprise in the Canadian context in light of the data provided by Goldberg and Mercer (1986), showing that Canadian cities tend to retain higher proportions of families with children than do their American counterparts.

13 This variety has been stressed by numerous studies, including Phipps (1982), Bunting (1984), GIUM (1984).

14 For a more detailed discussion of the polarization thesis, see Sassen-Koob (1984); Berry (1985) also argues the same theme.

15 Villeneuve and Rose (1985) and Dansereau and Beaudry (1986) provide illustrations of these contrasts in Montreal's older districts.

16 The gradual conversion of rental units to owned ones as renters move out (either spontaneously or as a result of a combination of threats and of "offers that cannot be refused") has also spread to larger buildings. See Dansereau, Collin et Godbout (1981).

17 Renovations with tenant eviction, conversions to condominium or co-ownership, and conversions of standard to short-term rental (pseudo hotels) have multiplied, affecting in total close to 9,000 units, from 1978 to 1985 in the city of Toronto (CMHC 1985c).

18 This type of strategy has been put forward by McGrath (1982) who mentions that it has met some success in Boston, where the local administration has launched promotion campaigns to change the image of certain "neglected" districts.

CHAPTER EIGHTEEN

Social Mix, Housing Tenure, and Community Development

Richard Harris

CANADIAN PLANNERS and policy makers have only recently come to view social mix as a desirable goal. The idea that mixed neighbourhoods might benefit the working class, and indeed the community at large, was first applied by George Cadbury at Bournville (England) in the late 19th century. His project gave inspiration to the Garden City Movement and attracted attention in Canada. However, only in the late 1960s did social mix become a policy concern in Canada and a criterion by which housing progress might be measured.

Mix became an issue because of perceived failures of segregated low-income housing. Problems arose almost immediately after the start of Canada's postwar public housing program. Proposals for low-income housing typically were opposed by local residents, especially home owners concerned about effects on local property values. Residents of these projects were often stigmatized. Geographical segregation seemed to worsen the social isolation of the poor. To overcome the problem of social isolation, a new policy approach stressed social mix in the provision of subsidized housing for the poor.

At the federal level, the main instruments of this policy approach were programs for cooperative and non-profit housing, complemented by rent supplements in some provinces. Much of the discussion of social mix in Canada is set in the context of these programs (CMHC 1983a, 162-79; Vischer Skaburskis, Planners 1979a). The main purpose of the subsequent cooperative and non-profit programs (established in 1974 and substantially amended in 1978) was to provide subsidized housing to low and moderate-income households, and social mix was seen as a means to that end.[1] Politically, mix was to make subsidized housing more acceptable to recipient neighbourhoods; financially, it was to generate internal subsidies that would keep down project (and program) costs.[2]

In this context, social mix depends upon the availability of a range of housing, including a mix of housing tenures. In Canada today, a significant minority of households are unable to buy their own homes. Where the local housing stock is exclusively owner-occupied, low and moderate-income households tend to be under-represented. As a result, planners must encourage both rental and owner-occupied housing in mixed communities (Heraud 1968; United Nations 1978). Similarly, to preserve social mix, planners often must regulate changes in tenure composition. Such regulations are currently exerted in most Canadian cities, for example in regard to condominium conversions.

The Definition of Social Mix

To define social mix, it is necessary to identify the social groups for which segregation or mix is of concern. In the United States, most policies of social mix have focused upon ethnicity or race, and notably the location of blacks in relation to whites. In Canada, policy makers – and the directors of cooperative and nonprofit projects – have been primarily concerned with the mix of household incomes. Target percentages of low and moderate-income households are usually established, these groups often being defined as the bottom (fourth) and third income quartiles respectively.

Have concerns over ethnic and racial mix become more prominent in Canadian planning? To check this impression and others related to the subjects of social mix and housing tenure, in the spring of 1986, a mail questionnaire survey was conducted of the directors of 69 planning departments across the country.[3] Of the 44 who replied, 12 thought that ethnic mix was an issue in their community. Seven of these worked for municipalities within the metropolitan areas of Toronto and Vancouver. In general, ethnic and/or racial mix was perceived to be an issue in cities that contained a significant number of visible minorities. Elsewhere, and in the majority of cases reported, ethnic or racial mix seemed not to be a current policy issue. In Canada, the "social" in "social mix," with important local exceptions, still refers primarily to socio-economic position.

SOCIAL MIX AND SOCIAL CLASS

Although income determines the household's capacity to occupy or purchase decent housing, many social scientists do not view income *per se* as a basis of social stratification. Instead, they emphasize the importance of occupation, status, or class. There is disagreement as to the relative importance of these criteria, and how they should be defined. Some see social class as an economic phenomenon, while others view socio-economic status as being rooted in prestige (see Hunter 1982). Nevertheless, they share the view that income is significant mainly as a reflection of, or as a means of emphasizing, social differences.

If a policy of mix is concerned to improve the situation of, or the relations between, social groups, it should be defined in terms of class or status rather than income. This is easier said than done. Class is difficult to distinguish, and ambiguities make it hard to implement a policy of mix. In dual wage-earner households, a further difficulty arises because many households are socially mixed. In 1974, for example, considering white and blue collar workers separately, three-fifths of all working wives were in a different class position from that of their spouse. [4] To take account of the position of both, however, makes implementation cumbersome; hence, household income may be the preferred indicator in an operational definition of mix. If so, however, its imperfections should be recognized. The relationship between income and class is not simple. Not all members of the same class have the same income: in urban Canada in 1982, for example, 14% of households headed by blue collar workers received incomes that put them into the bottom income quartile, while 25% were in the top quartile. [5] In this context, income mix will not necessarily produce social mix. A mixed-income project, for example, could be made up of exclusively middle-class people, and indeed a few projects have shown a marked middle-class bias.

BENCHMARKS, EVIDENCE, AND THE PROBLEM OF SCALE

To measure social mix, it is necessary to identify the relevant social groups, benchmarks, and scales. An empirical benchmark, such as the "social composition of the metropolitan area," has been used in some Canadian projects. The most notable of these were Vancouver's False Creek, containing about 1,800 units built from 1975 to 1984, and Toronto's St. Lawrence project of 3,500 units built from 1977 to 1981 (Hulchanski 1984). In both, a compromise was struck between the desire to build a neighbourhood that was a microcosm of the city and the goal of housing a specific minimum number of low and moderate-income households.

It has sometimes been implied that a metropolitan benchmark is the best. Such a view, for example, was implicit in the idea that St. Lawrence be a "typical" Toronto neighbourhood. In Toronto, however, and indeed in all cities, few neighbourhoods come close to being a social microcosm of the metropolitan area as a whole (Ng 1984). The typical neighbourhood is often homogeneous. Homogeneity, of course, might be viewed as a problem to be overcome. Even then, is the metropolitan area a useful reference point? Arguably the ideal mix in an inner city neighbourhood is different from that in a suburb. In demographic terms, for example, the latter might contain more families with children.

Also important is the scale of analysis. Residential areas may be defined at many geographic scales, ranging from immediate neighbours – through the block, census tract and neighbourhood – to the municipality as a whole. Cities that are mixed overall can be composed of homogeneous neighbourhoods and blocks. Indeed, such a situation has probably typified Canadian settlements since at least the late 19th century (Sanford 1985). [6]

Typically, social scientists have examined the extent of segregation rather than of mix. Fortunately, segregation has usually been defined as the opposite of mix. The most commonly used statistic has been the Duncan's "index of segregation" (Duncan and Duncan 1956). Varying from 0 to 100, it is usually interpreted as the proportion of households in a social group that would have to move in order for the group's residential distribution to be the

Table 18.1

Income segregation (%) using Duncan's index of segregation*:
Selected Canadian cities, 1971 and 1981

| | Low income† | | High income‡ | | All groups | |
	1971	1981	1971	1981	1971	1981
St. John's	26	19	35	20	23	15
Halifax	20	28	33	21	20	18
Quebec	24	21	31	19	21	16
Montreal	24	25	36	23	22	17
Toronto	27	28	33	19	23	19
Hamilton	25	29	28	20	21	18
London	25	24	32	23	21	17
Kingston	–	24	–	22	–	16
Winnipeg	31	30	35	22	23	18
Regina	23	29	38	19	22	19
Calgary	30	27	33	18	23	19
Vancouver	20	22	30	16	22	16
Average	25	26	33	20	20	17

SOURCE Computed from *Census of Canada* (tract data).
* See text for definition. Six income groups in 1971; seven groups in 1981.
† 1971: $1-$1,999; 1981: $1-$4,999.
‡ 1971: $15,000 and over; 1981: $35,000 and over.

Table 18.2

Home owners as percentage of all households in class:
Urban and rural Canada, 1931 and 1979

| | Urban | | Rural | |
Class of household head	1931	1979	1931	1979
Owners and managers	58	72	–	91
Middle class	41	64	–	85
Working class	38	50	–	84
Self-employed	56	64	–	94
Other	60	47	–	81
All classes	46	55	79	87

SOURCES Calculated from 1931 *Census*; tabulations from the 1979 *Survey of Social Change in Canada*. See Harris (1986a, Table 5).

same as that of the rest of the metropolitan population. An index value of 0 implies the complete inter-mixing of the group in question, while values of 0 for all groups would imply that all areas in the city were mixed. The index of segregation provides insights into changes in social mix.

A serious limitation of existing research on segregation, however, is its limited coverage. Analyses of segregation have concentrated on the period since 1961. Larger urban centres have been given more attention than smaller settlements. Moreover, most studies focus exclusively upon the census tract scale of analysis, making it difficult to compare the amount of mix at various scales.

Taking these limitations into account, social segregation characterizes all Canadian settlements at the tract scale. The most complete evidence is available for income groups. In 1971 the average index value across six income groups ranged from 0.14 to 0.25 among the twenty-one largest urban centres (Ray *et al.* 1976, 44-5). Typically, within each city, the most segregated groups were those at each end of the income spectrum. In Calgary, for example, index values for the top and bottom income groups were 0.33 and 0.30, respectively, with values for intermediate groups falling as low as 0.15. These figures indicate a low level of income mix at the tract scale. Since then, judging from the index values calculated for 1981, the rich have become less segregated, while the poor have become a little more (Table 18.1).[7] In general, however, there was not much change in the overall level of segregation between 1971 and 1981.

Indeed, the amount of segregation does not appear to have changed much since 1981. This, at any rate, was the view of local planners who responded to the question: "is the amount of (income) mix increasing or decreasing in your community?" Five planners thought that mix was decreasing at the neighbourhood scale, while eight thought that it was increasing. At the block scale, four thought that mix was on the decline, but only one thought it was on the rise. The majority were unable to identify any trend. In recent years, certain inner city neighbourhoods have become more mixed. As the income distributions of neighbourhoods such as Toronto's Don Vale reveal, gentrification has brought affluent managers and professionals into close proximity with the working and welfare poor (City of Toronto 1984). Such dramatic examples of mix, however, are unusual and unstable. Overall, the level of income mix at the tract scale has changed little since 1945.

The same is likely true for social mix, defined in terms of class or status. Indices of socio-economic segregation can be calculated from occupational data on household heads in twelve metropolitan areas in 1981 (Table 18.2). These data pertain to occupation, rather than class, and to household heads rather than to all employed adults. In most cities the segregation index values were lower than those for income, ranging from about 0.06 to 0.20.[8] Overall, the pattern of variation in terms of class is similar to that of income: the most segregated are those at each end of the class "spectrum": owners and managers on the one hand, and blue collar workers on the other. This pattern is broadly typical of every city, regardless of size or regional setting. This accords not only with the Canadian indexes of income segregation for 1971 and 1981 but also with the evidence for socio-economic segregation in US cities, which dates back to 1950 (see Duncan and Duncan 1956; Marrett 1973).

The Case for Social Mix

A policy of social mix might be rationalized on democratic and paternalistic grounds. In democratic terms, governments might promote mix as a response to popular pressure, presumably from a group (or groups) that is being prevented from living in mixed neighbourhoods. In the US, this was arguably the impetus behind programs that promoted the desegregation of blacks from whites. Additionally, a paternalistic policy might be justified if policy makers could argue that, even in the absence of popular demand, mix would in fact benefit specific groups or the community at large.

THE DEMOCRATIC CASE AND HOUSEHOLD CHOICE

The democratic case depends on the existence of an unmet demand for mix. Little demand of this kind is apparent in Canada today. Only two planning directors considered that income mix was a "serious" local policy concern at either the block or neighbourhood scale, while a majority considered that it was not even a minor issue, and again at either scale. Social surveys confirm that popular support for mix is weak. The majority of middle-income people prefer to live in middle-income areas (Michelson 1977). There are good reasons for this. Similar people are more likely to share interests, views, methods of child-rearing, and patterns of public behaviour. Dissimilar people would have to cope with different, and perhaps mutually offensive, behaviour. This might account for the fact that those who are at "opposite ends" of the class spectrum are more segregated from one another than from intermediate social groups.

People seem to care most about social homogeneity at the micro-scale, corresponding roughly to a typical city block, and to care about it least at the scale of the municipality as a whole. This is generally acknowledged. Within the False Creek and St. Lawrence neighbourhood projects, for example, distinct social (and tenure) enclaves were created. In the first phase of False Creek, market condominiums for owner-occupancy were clustered at one end of the project (Hulchanski 1984, 157). In one of the more mixed enclaves, the early occupants were least satisfied, though this dissatisfaction may not have lasted (Vischer Skaburskis, Planners 1979b, 109). The point should not be overstated. In private apartment buildings containing only a few rent-subsidized units, the assisted families have not been stigmatized (Ontario Housing Corporation 1983). But where such families are at all numerous, a fine mix is distinctly unpopular.

There is disagreement, however, about the significance of mix at the neighbourhood scale. Even in Toronto, which prides itself as a city of neighbourhoods, such areas are not now widely used by most people as arenas for social life

(Wellman 1971). The social character of the neighbourhood is a matter of comparative indifference to most households. In mixed neighbourhood-scale projects, the existence of mix is widely acknowledged but not cited as a significant issue by project residents (Vischer Skaburskis, Planners 1979b, 98).

The unimportance of mix at the neighbourhood scale can, however, be overstated. Home owners, as opposed to tenants, remain sensitive to those changes at the neighbourhood level which might adversely affect property values (Michelson 1977; Vischer Skaburskis, Planners 1979a). This sensitivity has been demonstrated many times through community opposition not only to low-income housing but also to cooperative and market housing which threatens to bring strangers into the neighbourhood (Vancouver 1986). Moreover, families with school-age children are sensitive to the social composition of their neighbourhood, and specifically to the family background of their children's peers. This is consistent with evidence that social mix does affect the formation of social networks among children (Andrews 1986). Although the evidence is inconclusive, it seems that few Canadian families would choose to live in a neighbourhood (and especially on a block) that was highly mixed in social terms. Patterns of segregation reflect in a sensitive way class divisions within society as a whole, and most people would not wish it otherwise.

One possible exception should be noted. Low-income households have little choice where to live. They are also segregated. It is not clear whether, given the choice, they would live in more-mixed residential areas, although they do express dissatisfaction with the highly segregated environment of public housing (Ontario Housing Corporation 1983, 55). This may be especially true of large projects in the larger cities. The absence of political pressure from this group does not necessarily mean that they are content with their present situation; it might simply reflect a belief that mounting such pressure would be a waste of time. Governments, of course, respond more readily to active than to latent political demands, but the latter should be a concern in any democracy. Whether low-income people would prefer to live in more-mixed neighbourhoods is an important topic for research and a potential policy concern.

THE PATERNALISTIC CASE

It can be argued that mix should be promoted even in the absence of popular demand. Mixed neighbourhoods might have effects of which their potential beneficiaries are unaware. Acting with foresight, the policy maker might serve the greater good.

The paternalistic case for social mix has been made in particular and in general terms (Form 1951; Gans 1961; Keller 1966; Saldov 1981; Sarkissian 1975). Commonly today, it is suggested that mix is desirable as a means of ensuring that the poor receive the same government services as the more affluent and that in educational terms there is equality of opportunity. This argument has not been evaluated in Canada, but the US evidence suggests that there are other, and more cost-effective, ways of ensuring equality of service provision than by enforcing social mix. Indeed, forced mix may create and exacerbate social tensions so that everyone, including the poor, loses out.

A broad argument has been made that mix, by promoting mutual awareness and a shared use of facilities, can benefit the whole community. Again, most evaluations use evidence for the US (or Britain). This argument has substance when applied to ethnic mix, but the same is not true of socio-economic mix. Except among children (Andrews 1986), proximity does not have much effect, good or bad, on social interaction among groups that are dissimilar. Vischer Skaburskis, Planners (1979b, 54), for example, concluded that patterns of neighbouring and friendship formation in False Creek were much the same as those that might be expected in any new development. Evidently, mixed projects can "work." With careful planning, they will not be stigmatized, while their residents are likely to be content with their social environment (CMHC 1983a, 246; CMHC 1984a; Diaz-Delfino 1984). But, even at their best, they do not provide any clear social benefit.

Social Mix as a Policy Goal

At the scales of the block and neighbourhood, the rationale for social mix is weak. Most people do not want to live in a more mixed environment than their current neighbourhood of residence. Moreover, the paternalistic case for a policy of social mix to promote social equality is at best ambiguous. Public housing showed that forced ghettoes are unpopular, but so is forced mix. A general policy of social mix might do more harm than good.

There are two specific contexts, however, in which an active policy of social mix makes sense. First, mixed projects are widely thought to be better than segregated low-income housing. The degree of segregation found in public housing is unacceptable to most communities and a source of dissatisfaction to project residents themselves (Ontario Housing Corporation 1983). Mixed projects work better and indeed, although evidence for the smaller projects is thin, the larger projects seem to work best. Of necessity, small projects must attempt to achieve mix at a fine scale, where social tension is most likely to arise. In contrast, the larger ones can accommodate fine-scaled segregation within a coarse-grained mix. Socially and politically, mixed neighbourhood-scale projects are a workable means of delivering subsidized housing to needy households.

A policy of social mix might also be justified at the municipal scale. Given that local governments depend upon the property tax, the absence of mix can lead to municipal inequities in service provision. Moreover, segregation at this scale can impose commuting costs on those least able to afford it, while increasing the aggregate costs of providing roads and public transit. These have already become policy concerns in some Canadian cities and, since the pressures for inner-city gentrification show no signs of slackening, they are likely to become more pressing over the next decade (Ley 1985). Here, an active policy of mix is justified.

Social Variations in Housing Tenure

Because of social variations in housing tenure, a policy of social mix will require an analogous policy with respect to housing tenure. Since 1931 home ownership levels in Canada have fluctuated from 56% to 66%, with no long-term upward trend (see Table 1 to Chapter 3 and Harris 1986a). However, the rate in urban areas has generally increased, albeit unevenly, from 46% in 1931 to 56% in 1981. In this context, some classes have fared better than others (Table 18.2). These social differences in tenure position are due primarily to the affordability of homes to people in different classes. Income is important in enabling the household to save for a downpayment and in qualifying it for a mortgage. In general, household income is not only a convenient indicator of socio-economic status but also a determinant of tenure. Age, too, is significant in that, other things being equal, young adults are unlikely to have been able to save for a mortgage downpayment. When income and age are controlled, most of the class differences in ownership rates disappear (Harris 1986b).

To understand this, it is necessary to examine more closely the question of tenure "preference." The great majority of households, in every class, would prefer to own a home if they could afford to do so. [9] Only those who move often prefer to rent, while even those who wish to avoid the responsibility of home maintenance now have the option of condominium ownership. In April 1986, 14 of the 44 planners surveyed thought that there was "some" local concern about the frustration of home ownership aspirations, but the remainder could identify "little or no" concern of this kind.

Among the poor, much more is at stake than the frustration of ownership aspirations. Gentrification has bid up rents in the inner city to the point where many households are spending over 50% of their income on rent (Social Planning Council of Metropolitan Toronto 1983). Even worse, an increasing number of people who have been unable to find affordable accommodation are now homeless. Some of the more modest condominiums that have been built downtown may have helped a few moderate and middle-income households to stay in the central city. But to keep low-income households there, and to provide them with adequate housing, the urgent need is not for further assistance to home ownership but for the conservation and production of low-cost rental housing. It is significant that, although none of the planners surveyed considered that the frustration of ownership aspirations was a "serious" concern in their municipality, twelve thought that the unavailability of low-income rental housing was a serious issue. Four of the twelve represented municipalities in the Metropolitan region of Toronto, and all but two were from Ontario.

The importance of tenure to a policy of social mix has been appreciated to some extent at the local level, most notably through the design of several municipally-sponsored projects. In Ottawa, for example, the LeBreton Flats project built in the 1970s was designed to "accommodate and integrate a heterogeneous mix of home owners and renters of all income groups" among its 425 units (CMHC 1983a, 5). In False Creek and St. Lawrence, a combination of ownership, market rental and social housing units – one-half of the latter being continuing non-profit cooperatives – have ensured a mix (Hulchanski 1984). Similar considerations guided the ill-fated 6,000 to 7,000 unit Ataratiri project in Toronto. The City's stated intention had been to create a mixed-income neighbourhood through the provision of a range of tenure types ("Council Approves" 1988). Although some of these individual projects constitute distinct neighbourhoods, they are not numerous enough to have an impact upon the social composition of entire municipal jurisdictions. At this scale, municipalities have made less effort – because they have fewer powers – to influence social mix through housing tenure.

Conclusion

Although there is no case for making the achievement of social mix in residential areas a general policy goal in Canada, mix is important in specific contexts. It has played, and should continue to play, a role in social housing policy. Mixed projects are superior to segregated low-income housing. The most effective mixed projects are likely to be large, allowing mix at the neighbourhood scale to be achieved without forcing people to live next to people who are different from themselves. If exclusive low-income projects are to be built they should be small enough to have little visible or social impact upon the immediate neighbourhood. The optimum size will probably depend upon the character of the area in question, being smaller among single family homes than among high-rises. There is no conclusive evidence that points to any specific maximum size, but in many areas a project of about twenty units seems to be a workable scale.

Mix is also important at the municipal scale. Here, its immediate advantages are economic in character, but they have implications for social equity and community development. Hitherto, local jurisdictions in Canada have not been characterized by a marked degree of segregation. In some cases, however, especially among the larger metropolitan areas, this is changing. Although the process of gentrification has not turned inner cities into ghettoes for the rich, current trends indicate that mix will increasingly be a policy concern at this scale.

To achieve social mix, it is necessary to develop and/or maintain a range of rented and owner-occupied dwellings. Tenure mix is one of the easiest ways of establishing social mix within specific projects. It is also effective in inhibiting any tendency for projects to become more homogeneous over time. At the municipal level, the preservation of current levels of social mix will depend upon the maintenance of tenure choice. This will not be a policy issue in most municipalities over the next decade, but in larger centres it will become even more important. Here, the tenure and housing affordability position of low-income households is the crucial and growing problem. Given the advantages to be derived from maintaining social mix at the municipal scale, there is a strong case for promoting inner city tenure mix.

In pursuing such a course, governments will be handicapped by market forces and by entrenched tax subsidies (non-taxation of capital gains and imputed rents) to home owners. The demand for inner city housing is strong and shows no sign of slackening. By increasing the level of home ownership, implicit home owner subsidies have made it more difficult to secure community acceptance of social mix. Reducing tax subsidies would assist the goal of promoting municipal mix, but such an action would have many ramifications and is most unlikely. Other, more direct means of promoting mix at this scale must be used.

The choice of policy instruments to promote tenure and social mix at the municipal scale depends upon economic, as well as social and political, criteria. This is not the place for a thorough evaluation. From what is already known, however, it is at least possible to comment on the social acceptability of the policy options at the neighbourhood level. Judging from past experience, the regulation of condominium conversions and apartment renovations will meet with little resistance because such policies tend to preserve the status quo. Unfortunately, passive policies of this kind have already proved to be inadequate. More active policies, however, will face more resistance. Attempts to encourage the more dense occupation of existing residential areas should have some effect in areas where high prices are already forcing first-time buyers to think about subdividing their homes. Such attempts will meet with opposition, however, especially where they entail the influx of households with incomes that are below the current neighbourhood average. Small housing projects, even those that are mixed, are likely to fare no better unless they replace an undesirable land use or are placed in an area that already has a mixture of uses (Vancouver 1986, 14).

In social terms, rent supplements and/or large projects like St. Lawrence and False Creek are likely to be more acceptable. Rent supplements allow subsidized housing units to be mixed in with the existing stock and seem to arouse little opposition even when subsidized units are recognized as such. Unfortunately, they depend upon the participation of private landlords who, for the most part, have shown little enthusiasm. Mixed, neighbourhood-scale projects would arouse opposition in existing residential neighbourhoods, but on vacant or abandoned sites this is not an issue. Such sites have been created by the loss of manufacturing employment from several inner city areas, the Massey lands in Toronto being a case in point. They offer a rare opportunity to build low-income housing where it is needed and without arousing strong community opposition. The problem here – exemplified by the cancellation of the City of Toronto's Ataratiri project – is the cost of environmental clean-up. If the price can be made acceptable, perhaps by ensuring that middle and moderate-income households pay full market costs, such projects offer hope that social mix within the Canadian inner city might be maintained, to the benefit of everyone concerned.

Notes

1 CMHC's evaluation of these programs treats mix as an implicit "additional objective." See CMHC (1983a, 160-81) and Hulchanski and Patterson (1984)

2 In practice, however, because some subsidies went to those who did not need them, mixed projects were costly. But this has less to do with the fact of mix than with program design and implementation.

3 A copy of the questionnaire is available on request. I mailed it to the directors of 69 local planning departments on 27 March 1986. Departments were selected by a stratified sampling procedure that was intended to secure representation from city and suburban jurisdictions in settlements of all sizes and in all regions of the country. No particular bias was apparent among those who responded. In most cases, planning directors had delegated the task of completing the questionnaire to staff who were responsible for housing or community planning.

4 Indeed, if homemakers constitute a class of their own, all households containing one wage-earner and a full-time homemaker are socially mixed.

5 Income and home ownership statistics for social classes have been calculated from reported "class of worker" and occupational data (Harris 1986b, note 5). These "headship" data are open to the objection, discussed above, that the social position of the spouse is ignored. Personal data on the relationship between class and income (not reported) reveals a similar situation as that for households.

6 No firmer statement is possible because of the lack of historical data.

7 The index values for 1981 are not precisely comparable with those for 1971 since they pertain to family income, rather than household income. Moreover, because they are estimated for seven income categories, rather than six, one would expect the resultant index values to be somewhat higher. In that context, the apparent decline in segregation among the most affluent may in fact be more considerable than a direct comparison of indexes would suggest.

8 Five occupational groups were defined. For managers (managerial, administrative, and related), the city indexes averaged 0.19 and ranged from 0.17 in St. John's to 0.24 in Hamilton. Among the professional and technical (teaching, medicine, and technological), clerical, sales (sales and service), and blue collar (primary processing and machining), the city indexes averaged correspondingly 0.15, 0.07, 0.06, 0.15, and 0.13. When these broad categories were disaggregated into fourteen occupational groupings, index values ranged much higher, in cases exceeding 0.50. These data are not reported in detail but are available on request to the author. The indices of socio-economic segregation are for male-headed households only.

9 In a survey carried out in Toronto in the early 1970s, Michelson (1977, 137) found that 81% of all middle-income tenants wished to live in an owner-occupied dwelling. The proportion among owners was even higher.

=

CHAPTER NINETEEN

Housing and Community Development Policies

Jeffrey Patterson

COMMUNITY DEVELOPMENT is a process and an end state (Compton 1971). As a process, it concerns people gaining control over their lives – often termed "empowerment" (Maslow 1954; Single Displaced Persons Project 1983). It also encompasses support for and facilitation of "citizen participation." As an end, the term describes various services relevant to the quality of life and includes government programs aimed at improving the quality of life and ameliorating deprivation.

Why has community development often become an integral part of government housing policies and programs? Why is it of concern to housing planners? As an end, the provision of community services relevant to the quality of life, certainly the planning for them, is a natural extension of housing programs. Community development is central to government housing programs because the adequate maintenance of housing depends on the efforts of whole communities, and community efforts often complement, supplement, and increase the impact of government housing programs, reducing the public funds needed to achieve housing objectives (United Nations 1987, 6).

This chapter examines trends in availability, quality, and adequacy of community services and facilities, focusing on the roles of governments. It also examines changes in the control which Canadians exercise over housing quality and choice and their lives in the residential setting. These discussions lead to an assessment of progress in community development and the role of housing policies and programs therein.

Meeting community development objectives and the interaction between housing and community development involves all levels of government and their respective programs and policies. All levels provide services. The provinces provide many of the services and facilities (either directly, by funding, or by regulation), regulate land use, and set the framework within which local governments plan land use and facilities. Municipalities are often responsible for planning and for provision of services and facilities used in everyday life, for example, streets, sewers, water, public transport, education and recreation facilities and services, and social services. At the federal level, several government departments, as well as CMHC, are instrumental in community development. As well, the federal government influences service provision through its spending programs. With each level of government working independently toward its own community development objectives, as well as housing programs and policies in many cases, description of the Canadian community development process is necessarily complex. In this chapter, only a few program and policy areas are highlighted.

Origins of the Modern Urban Reform Movement

Canadian housing and community development policies following World War II were conditioned by the housing situation, by economic and social events prior to 1945, and by Canada's cultural and institutional traditions. Early post-war housing policies responded foremost to the shortage of housing. In 1951, 1 in 9 Canadian families would today be considered homeless in that they did not maintain their own dwellings (*CHS* 1985, Table 111). Over 1 in 5 households were crowded into too little space (defined by *CHS* as having more than one person per room). The perceived need to house and create immediate employment opportunities for returning soldiers as well as to create alternative employment opportunities for those who had been employed in munitions production also guided early post-war housing policies (Rose 1980, 5-7).

Post-war housing policies were developed to a lesser extent in response to slums and the need to improve the adequacy and suitability of housing for the working class. [1] By the time of the Great Depression of the 1930s, urban conditions in large parts of Montreal were scarcely better than those described by Herbert Ames (1897, 18-29) at the end of the 19th century. The 1930s were perhaps harder for the working class of Montreal than in any other city in the nation (Copp 1974, 140). The need for slum clearance was seen as increasingly urgent, especially in Montreal and Toronto.

While slum conditions of the working class were neither as prevalent nor as severe in Toronto as in Montreal, they nevertheless generated a will and resolve to ameliorate them in the post-war period (Lemon 1985, 81-112). The Bruce Report (Ontario 1934) recommended to Toronto's Board of Control that unfit slums be replaced with low-cost housing and that federal and provincial governments subsidize Toronto's, and Canada's, first urban renewal initiative. In the following decade, conservatives and radicals debated how to ameliorate Toronto's slums. Conservatives called for renovation programs. In 1936, Toronto City Council passed Canada's first Standard of Housing By-law. By 1940, 16,400 loans totalling $5.6 million had been made in Toronto under the federal Home Improvement Loans Guarantee Act – from 20% to 30% of the commitments nationwide (Lemon 1985, 68). Conservatives also argued that slum clearance and the provision of subsidized housing for low-income households would result in a degree of class segregation that would be "even worse than the old feudal system" (Central Council of Ratepayers' Associations, quoted in Lemon 1985, 67).

While they had no way of realizing their ambitions, reformers continued to view the renovation solution as inadequate. For them, the renovation solution was aimed more at creating employment than at improving housing conditions.[2] Reformers also sought a planned urban environment that would overcome the disorganization of slum life (League for Social Reconstruction 1935). It never occurred to the reformers that slum residents would oppose such action. It was intended that those dislocated be rehoused in superior housing at affordable rents (Carver and Hopwood 1948). The Toronto Reconstruction Council was established in 1943, and in 1946 it published a report arguing for the need for subsidized housing, calculating that need at 50,000 units (Carver and Adamson 1946). In December 1945, the reformers finally won their battle when Toronto City Council approved development of the Regent Park Rental Housing Project.

COMMUNITY DEVELOPMENT AND CANADA'S SUBURBS

A principal accomplishment of post-war housing policy, beyond the production of subsidized housing, the production of rental housing, and even the redevelopment of slums, was the creation of Canada's suburban communities and housing developments (see Chapter 12). Of equal importance to the increase in numbers of new households accommodated in the suburbs was the onset for the next fifteen years of the "baby boom" and the fact that much of the new population was accommodated in family households (*CHS* 1985, Table 111).[3]

Aided by government policy at all three levels and encouraged by lending institutions, the development industry soon became dominated by corporate developers (Sewell 1976). New suburban developments were primarily settled by people under age 45 years and in their early years of child-rearing (Clark 1966, 82-141). Settlers did not bring strong urban attachments with them. Primary loyalties were to family life and associations formed through work and child-rearing. Suburbanites sought fulfilment through family home ownership and the drive for open space (Thorns 1972, 111-25). There were costs: for example, in the loneliness experienced by women and the burden of mortgage discipline and daily commuting experienced by men (see Clark 1966; Thorns 1972).

What prompted families to move to the suburban setting? What has it meant for community development in Canada in the past and present? What are its implications for the future? Going back to the 19th century, city life for the majority was the experience of deprivation, instability, and the threat of unemployment (Mumford 1938, 143-68). This disorder provided the basis for the development of the ideal neighbourhood.

Presented in 1924 by Clarence Perry, the neighbourhood unit concept stressed its satisfaction of social objectives (Colcord 1939, 83). His principles of neighbourhood design included:

- location of the elementary school at the centre of the unit along with a library branch, a motion picture theatre, and a church;

- concentration of commercial functions into shopping centres located at the periphery;
- family backyards;
- arterial highways as boundaries for the district;
- apartments located at the periphery; and
- homogeneous income groupings within the unit.

In Perry's view, the corporate developer was ideal, as it was more difficult to realize the ideal neighbourhood unit with the piecemeal development that had characterized urban development at the time he formulated his principles. Developers with large land holdings were seen as potential sources of progress in the promotion of social objectives in neighbourhood planning.

The neighbourhood unit concept was one response to the experience of urban turbulence according to Emery and Trist (1973, 57-67). Simplification, reduction of complexity, and withdrawal were understandable responses to turbulence. These patterns are evident in the alternative post-war residential environments that emerged first in the United States and United Kingdom. These environments characterize to one degree or another the suburban developments of Montreal, Toronto, and Vancouver. Don Mills, Ontario, became the ideal type in Canada. As well, application of the same principles spread to smaller communities.

PUBLIC POLICY AND SUBURBIA

The experience of the fifteen years preceding 1945 did not prepare governments for the scale of growth that was to ensue. However, governments did accommodate the new growth. The programs and policies put into effect under the NHA to stimulate housing construction – joint mortgages followed by mortgage insurance, rental construction programs, and assistance to private rental entrepreneurs – accommodated the new scale of development. The provinces had already begun to establish a planning framework under which the new development took place. Provinces and municipalities also combined efforts to construct trunk sewers, water mains, and sewage treatment plants to which new subdivisions could be linked.

The pace and scale of growth strained provinces and municipalities alike, but new policies allowed them to cope. Increasingly, developers were required to install services and build local streets prior to the acceptance of subdivisions for sale of lots. Large corporate developers were able to integrate these new functions into their operations. In provinces where the large corporate developer was not typical, particularly in Quebec and the Atlantic provinces, servicing often did not reach the level that was achieved elsewhere.

Lacking access to revenues from commercial and industrial assessments, small suburban municipalities found it most difficult to finance new services.

One outcome was the impetus to form regional governments (Colton 1980, 52-73; Lemon 1985, 108-11; Rose 1972). The federated Municipality of Metropolitan Toronto became the first such regional government in 1953. Other cities followed in the next two decades including Montreal, Winnipeg, and Vancouver, although they used different models and allocations of functions to the regional level. Ontario and Quebec established regional governments in large, fast-growing cities. Alberta solved the problem by encouraging and allowing existing cities to annex new suburban territories.

Other outcomes included the emergence of fiscal planning and of lot levies in new suburban developments. Fiscal planning requires that new housing forms and developments generate sufficient property tax revenues – still the main source of income for municipalities across Canada. Lot levies, which in 1988 approached $10,000 per new dwelling in some municipalities in British Columbia and Ontario, have permitted municipalities to develop community facilities that had often not been present initially in earlier suburban developments. While one result of these phenomena is higher quality suburban developments, another is the exclusion of increasing numbers of less-affluent Canadians from newer suburbs.

As well, the federal government, including CMHC, assisted provinces and municipalities with grants and loans to create the water and sewage treatment plants and trunk services which were needed by the new subdivisions. These contributions seldom had a social or community development objective. One exception, the Municipal Incentive Grant program, paid municipalities $1,000 for each medium density, modest-sized housing unit granted a building permit between 1975 and 1978.[4] The objective of the program was to encourage municipalities to allow housing for modest-income residents that might not otherwise be built.

CMHC and provincial and municipal governments also undertook public land assembly following 1945, although the federal government ended its participation in 1978 (Spurr 1976, 275).[5] These land assemblies were undertaken for several reasons (Spurr 1976, 247-57); one of them, facilitating the construction of social housing, is integral to community development objectives. Large numbers of social housing units were built on land assembled by public authorities in Toronto and other Ontario municipalities, along with Winnipeg and Edmonton; and smaller numbers were built in many other municipalities.

Only Saskatoon, and to a lesser extent Edmonton, have concertedly attempted to influence the course of suburban development using public land assemblies. Ontario's Home Ownership Made Easy (HOME) program mixed housing types at the community scale and limited construction to modest housing affordable for low to moderate-income purchasers. In the late 1970s CMHC established new guidelines for disposal of lands that it had helped assemble. These guidelines require that profits resulting from land development schemes be retained in the community in the form of services. Community development objectives figured more prominently in developing publicly-assembled land than private suburban developments; for example, a concern for demographic and social mix and some social housing has usually been a part of the former (Comprehensive Land Use Management Program).

Public land assembly and development programs can influence the course of urban development. In only a few localities, however, has this ever been done (Mount Pearl, Newfoundland, for instance). With rapid increases in housing prices in the early and mid 1970s, there was much public discussion of the potential contribution of public land assembly. In 1972 the CMHC Task Force on Low-income Housing recommended a new joint land assembly program.[6] In 1978 a study of the supply and price of residential land was commissioned jointly by CMHC and the provinces (Greenspan 1978). Following the Task Force Report, however, discussion of large public land assembly projects ended abruptly when the federal government decided not to contribute to such efforts. A few provinces, notably Ontario, Manitoba, and Saskatchewan, and their municipalities, have continued activity in land assembly projects, but at a reduced scale of activity.

URBAN RENEWAL AND PUBLIC HOUSING IN THE INNER CITY

Canada's new suburbs were not the only areas where development responded to the turbulence that had characterized cities earlier in the 20th century. The same patterns of adaptation, simplification, segmentation, and withdrawal were applied to the inner city with the development of public housing projects and the undertaking of urban redevelopment. Public housing development and urban renewal placed the responsibility for achieving community development objectives on government and tested government's ability to achieve these objectives.

A milestone in this history was the Regent Park North project in Toronto. When requested by the City of Toronto to amend the NHA to allow the federal government to contribute to ongoing rental housing subsidies, CMHC offered only to underwrite a part of the cost of acquisition and clearance of the Regent Park site. This "urban renewal" activity allowed the City to develop rental housing on the site. Development of the site began in 1946. Design principles incorporated into the project plan included limited penetration of the automobile into the neighbourhood, removal of all commercial establishments, uncluttered green space around the housing units, and uniformity of houses; similar principles also dominated suburban development design. Rents were subsidized on a sliding scale. Thus was born Canada's first public housing project in which rents were publicly subsidized and geared-to-income.

In 1949 the federal government amended the NHA to permit the development of joint housing projects with the provinces and to share the cost of providing rental subsidies on a continuing basis, thus facilitating the development of housing projects on the Regent Park model in large cities across Canada. Montreal's first public housing project, Jeanne Mance, was developed in 1956 with over 900 units. Strathcona was developed in Vancouver, and Lord Selkirk and Mulgrave Park were developed in Winnipeg and Halifax, respectively. Regent Park South, Moss Park, and Alexandra Park, all developed on the sites of former slums, were added to the Toronto public housing stock by the mid 1960s.

The 1949 NHA amendments and the development of public housing projects in cities across Canada was a limited victory for the urban reform movement. The federal government also announced in 1949 that it was terminating its direct rental housing development program and that federal rent controls were being dismantled. Where possible, it sold units to occupants, many of them of moderate income with an ambition to own their own homes.

There were also constraints on public housing development itself. At first, CMHC insisted its mandate was to provide only housing, and there was little allowance for space in the projects for social and community services. More important, the development of public housing tended to be tied to the clearance of an equivalent number of existing housing units. In response to the issue of whether the federal government would agree to develop a public housing project on previously vacant suburban land, the Minister responsible for housing wrote CMHC's president in 1956:

> ... we would be justified in using public funds for housing only where private enterprise fails to meet the need.... That is why the Government feels that the clearest case to be made for public housing is where it is related to slum clearance. [7]

Housing and Community Development in the 1960s and 1970s

In the early 1960s governments escalated their attempts to alter the pattern of Canadian urban development. Urban redevelopment projects proliferated, assisted by grants and contributions from the federal government for planning and undertaking projects. Suburban patterns were altered as well. Apartments, both publicly and privately developed, were built in large numbers in the postwar suburbs. The social and demographic nature of suburbs was altered dramatically.

During the 1950s many Canadian cities acquired one or more large public housing projects, usually on slum clearance sites. Urban renewal became a mature federal-provincial program in the 1960s. By the time that a moratorium on new project approvals was declared in 1969, 161 urban renewal studies had been carried out in practically every large city as well as in many smaller communities (CMHC, City Urban Assistance Research Group 1972, 2a: 24). Some eighty-four urban renewal projects had been approved. While a few big projects in large cities attracted most of the public attention, the majority of projects took place in medium-sized and small communities. Table 19.1 summarizes urban renewal disbursements by the federal government, approximately one-half of total expenditures by the three levels of government. Total spending for projects authorized to 1969 was over $226 million. Projects in cities of over 100,000 population accounted for 77% of total spending; spending in Halifax,

Table 19.1
Disbursements under the Urban renewal program and NIP
by Canadian urban centre ($000)

City	Urban renewal 1948-1973	NIP 1973-1978
Halifax-Dartmouth, N.S.	12,345	3,711
Saint John, N.B.	19,113	1,498
St. John's, Nfld.	4,304	4,099
Other places in Atlantic provinces	2,002	17,292
Montreal, Que.	30,312	5,559
Other places in Quebec	40,601	38,335
Hamilton, Ont.	22,976	733
London, Ont.	3,587	2,469
Ottawa, Ont.	20,452	2,413
Sudbury, Ont.	9,930	1,229
Thunder Bay, Ont.	4,322	2,291
Toronto, Ont.	18,441	6,562
Windsor, Ont.	2,610	2,186
Other places in Ontario	10,149	41,014
Winnipeg, Man.	7,189	8,792
Regina, Sask.	178	1,464
Calgary, Alta.	6,948	4,223
Other places in Prairie Provinces	175	22,527
Vancouver	7,292	6,325
Victoria	1,697	1,500
Other places in British Columbia	2,531	12,097
Urban areas over 100,000 inhabitants	174,797	59,945
Other places in Canada	51,292	126,447

SOURCE Unpublished data made available by CMHC.

Saint John, Montreal, Ottawa, Toronto, and Hamilton alone accounted for about one-half of this activity.

By the late 1960s urban renewal had become controversial. Sizeable areas near the centres of Halifax, Montreal and Toronto were cleared for redevelopment. While a portion of cleared land, almost all of it in the case of Toronto, was used to develop new public housing, many families had been forced to relocate elsewhere, and the numerous single persons who resided in rooms found other rooms wherever they could, as they were not eligible to be housed in public housing. Thousands of tenants and home owners were threatened with eviction or expropriation, often at financial loss. Poor people and their advocates came to view urban renewal's objective as the removal of poor people. Home owners often received compensation at less than full market value for their homes under antiquated provincial expropriation laws. While expropriation laws and policies for compensation upon removal were as much at issue, blame was attached to the renewal program itself.

Urban renewal dispersed existing communities, detracting from community development objectives. The benefits of renewal went to the minority of dislocated people who had the opportunity to move back into the public housing and who chose to do so. For those who continued to rent housing in the private sector, urban renewal often led to higher rents in addition to disruption of life. There was a gradual realization of the need to minimize dislocation and rent increases, and to work with tenants and home owners *in situ* to achieve community development objectives. The need to provide community amenities and services to residents also grew with time.

In the 1960s federal programs and policies also changed in regard to social housing programs. NHA amendments in 1964 permitted loans to non-profit housing companies, and a new public housing section – allowing CMHC to make loans to municipal and provincial housing companies – provided an alternative to the cumbersome federal-provincial partnership arrangement. The Government of Ontario responded to the latter change by establishing the Ontario Housing Corporation. Other provinces immediately followed suit, choosing to build public housing projects with loans from the federal government rather than using the partnership approach. Municipal governments were never able to take advantage of the new public housing financing arrangements on a large scale, probably because of the requirement to make a 10% contribution. Almost 23,000 public housing units were built in the subsequent five years, and 162,000 were built before the federal government terminated the generous financial arrangements in 1978 as part of an overall reduction in its housing commitments.[8]

Like the concurrent urban renewal activity, increased public housing activity generated public controversy. The provinces generated large numbers of units rapidly by building on a large scale and on suburban sites. "Project" living came to be unpopular in the community in general as well as among residents themselves. In its report on the relocation of tenants in Toronto's Alexandra Park area, the Social Planning Council reported that many tenants refused to be rehoused in the public housing project even though it was financially advantageous (Social Planning Council of Metropolitan Toronto 1970, S17). Public housing projects were also unpopular among neighbouring suburban home owners.

Dissatisfaction with the urban renewal and public housing programs was felt by federal politicians. In the case of urban renewal, members of Parliament and ministers were lobbied by opponents to end funding for approved plans. The federal government had launched its "war on poverty" in 1967, but its urban renewal program was harming the interests of the poor (*Trefann Court News* 23 November, 1967). Ottawa would have to review and reformulate its housing and urban development programs. In the cities themselves, urban renewal was generating opposition to entrenched municipal councils in Montreal, Toronto, Winnipeg, Edmonton, and Vancouver on a scale not seen since the 1930s.

Five Years of Policy Review

The years 1968 to 1973 were marked by a more or less continuous review of Canada's housing and urban development policies, and during this period the Ministry of State for Urban Affairs was established. In April 1968 Paul Hellyer, Minister of Transport and Deputy Prime Minister, was also given responsibility for federal housing policy and CMHC. In September, he launched his Task Force on Housing and Urban Development and gave both subjects a higher profile than they had ever had. Hellyer and his task force visited all of the contested urban renewal sites. Briefs supportive of new program initiatives in urban development were received from numerous quarters. Hellyer submitted his report in January 1969, but he resigned from Cabinet when it became evident that quick action, maybe any action at all, would not be forthcoming on most of his proposals.[9]

Hellyer's successor, Robert Andras, began immediately to attempt to meet expectations that the visibility of Hellyer and his report had created. Economist N.H. Lithwick of Carleton University was retained to undertake a review of federal urban policy. In early 1971 CMHC established external task force reports on low-income housing and on urban assistance. Reporting and implementing any recommendations for new programs and legislation would take time. To maintain the momentum for reform, a $200 million innovative housing program was launched under which as much experimentation as was legislatively possible was undertaken. Assisted home ownership and rental projects, and cooperative housing projects, later to be sanctioned legislatively, were initiated. Loans for housing for Métis, non-status Indians, and status Indians residing in western cities were also initiated. While the federal government was still not formally able to make grants for housing rehabilitation to individuals, ways were found to support these efforts as well.[10]

NEW PROGRAMS AND POLICIES

The 1973 amendments enabled CMHC to make loans to cooperative housing societies. Loans to them and to non-profit companies could be for as much as 100% of appraised value, thus eliminating the need for investor equity and removing obstacles to use by municipalities and small groups. The new programs recognized that the residents of low-income housing projects, especially large ones, bore a stigma that was not healthy for the residents or the community as a whole, and they provided that future low-income housing projects should be built as mixed-income projects; 25% was considered an ideal proportion of low-income tenants.

Finally, the amendments enabled grants to home owners and landlords for housing rehabilitation (RRAP), as well as a new program, NIP, to replace the urban renewal program. Provisions for the new program emphasized housing rehabilitation and conservation and assistance for provision of basic services and community facilities. From the perspective of the federal government, NIP, by limiting assistance for slum clearance, also had the advantage of reducing costs (Crenna 1971). Because of continued uncertainty regarding the best direction for urban development policy, NIP and RRAP were to expire after five years.[11]

The overall thrust attempted to build on what were growing to be recognized as good community development principles. Residents could best be supported in their *in situ* locations. Measures to facilitate rehabilitation were required for dwellings needing major repairs. Communities, many of them under-serviced or with services operating out of facilities in need of improvement, needed funds to upgrade neighbourhood facilities, compensating to some extent for the deprivation that might attend being of low income (Joint Task Force on Neighbourhood Support Services 1983, ES1-12). Many neighbourhoods in need of improvement were near old "smoke-stack" factories, and the environment needed to be cleaned up as well.

Housing developed and managed by cooperative societies and non-profit companies embodied these principles as well. Residents would be potentially involved in design, and they would either manage, or be more involved in management than in the case of housing provided by public housing authorities or provincial housing companies.

OTHER SUPPORTS AND SERVICES

Other changes were required, and indeed they were occurring, in the programs and policies of other federal departments and agencies and of provincial and municipal governments. In 1971 the federal government initiated some short-term job creation programs as a response to rising unemployment, notably the Local Initiatives Program (LIP) and Opportunities for Youth (OFY). Under these, several local initiatives were programs of neighbourhood rehabilitation. While no complete inventory was ever published, numerous projects initiated by neighbourhood groups in renewal areas, or in areas to be approved under NIP, were approved. Other services, including assistance for citizen participation in planning for their neighbourhoods, storefront community services, as well as more traditional services were supported by the LIP and OFY programs. The Company of Young Canadians (CYC), a federal Crown agency founded in 1968, also assigned workers to urban renewal and neighbourhood improvement project sites.

Aside from contributing to the costs of neighbourhood improvement, the

provinces assisted by altering planning legislation and policies to promote and require citizen participation in municipal planning decisions, including planning in urban renewal and neighbourhood improvement areas, and in amending landlord-tenant law to provide tenants greater security of tenure, moving from a system based on one of land tenure law to one of contracts.

Amendments to landlord-tenant law were also important from a community development perspective. Ontario led the way for the common law provinces when it amended Part IV of the Landlord-Tenant Act in 1970 as proposed by its Law Reform Commission. The remaining eight common-law provinces followed suit in the next five years, and two, British Columbia and Manitoba, largely removed landlord-tenant matters from the cumbersome court system by creating an office of provincial rentalsman. Although it has a legal tradition different from the common-law provinces, Quebec also modernized its landlord-tenant legislation.

Tenants in Quebec and Newfoundland continued to be protected by rent review systems that had remained intact since the Second World War. The eight remaining provinces adopted a rent review system in 1975 as part of the federal government's anti-inflation program. All but Alberta and British Columbia continued to have rent review at the end of 1987.

Similarly progressive changes were made in provincial planning legislation to require that citizens have ample opportunity to influence the substance of community plans. The holding of at least one public meeting is usually the requirement of provincial legislation or regulations. Most municipalities have gone further, however, especially in the case of urban renewal or neighbourhood improvement areas. "Site" planning offices away from city hall have been common in larger cities. Ongoing committees of local residents have also been common.

The new neighbourhood improvement thrust emanating from the extensive NHA amendments, taken together with the initiatives of other federal departments and agencies and provincial and municipal governments, set Canada on a course toward realizing the community development objectives that had been held since the Great Depression. Their single most distinguishing feature was an attempt to have future housing and urban development simulate what occurred historically. Neighbourhoods were to contain a mixture of income groups; projects with a narrow income range of occupants were to be small.

Toward a National Urban Policy

The ultimate objective of many policy makers was a national urban policy that would enable the federal government to intervene sensitively and in cooperation with private parties and provincial governments across the nation. A national urban policy was initiated in 1965 by the Privy Council Office and its Special Planning Secretariat, the latter being responsible for Canada's "war on poverty." One of the Secretariat's activities had been the co-ordination of a federal inter-departmental committee representing eight departments;

the objective of this committee was to co-ordinate and expand federal community development resources through what was hoped would become a community development agency. [12]

The Ministry of State for Urban Affairs (MSUA) was established in July 1971. Its terms of reference included urban policy research and the formulation of federal urban policy (Sunga and Due 1975). While MSUA was granted wide powers to formulate and evaluate federal urban policies and programs, to undertake research and to co-ordinate policies, its overall impact up to the time it was disbanded in 1978 must be judged as only minor. As one observer noted,

> The effectiveness of MSUA's research evaluative and policy development functions and the Ministry's proposals for the co-ordination of federal inter-agency and intergovernmental initiatives are clearly constrained by the extent of its participation in the decision-making processes of government (Sunga and Due 1975, 9).

The Ministry never possessed any decision-making powers, and their absence clearly limited its effectiveness. While the Ministry initiated tripartite coordinating committees in several large cities, it never attempted to replicate the community development efforts of the PCO's Special Planning Secretariat or of the CYC. [13]

Appraisal of New Federal Housing Policies and Programs

The new housing and urban development policies and programs were formally initiated in 1973. They were the last major new commitment in the forty year period under review. CMHC published an evaluation of the cooperative and non-profit housing programs in November 1983 (CMHC 1983a). A comprehensive evaluation of all social housing programs was commenced but never published (CMHC 1983b). These program evaluations are limited to the non-profit and cooperative, public housing and rent supplement programs.

The evaluation by CMHC of its new non-profit and cooperative housing programs concluded that the mixing of income groups had occurred. Surveys of tenants indicated that they were satisfied, and tenants in the non-profit and cooperative projects indicated a greater degree of satisfaction than tenants housed in either public housing or rent supplement units (CMHC 1983a, 130). Project meetings occurred in over two-thirds of non-profit and cooperative projects but only in one-third of public housing projects (CMHC 1983a, 182). Tenant suggestions were implemented in about 70% of the non-profit and cooperative projects, but in fewer than 50% of the public housing projects. [14] Preliminary and fragmentary data from a variety of sources indicate that there is room for significant improvement in the management of public and private rental projects, especially from the point of view of supplemented tenants in the latter (Edmonton Social Planning Council 1973, 4-10; Ontario Legislative Assembly 1981, iii-xix).

While experience with income mix in the non-profit and cooperative

housing programs was generally positive, continued support for the income mix objective has been tempered by its cost. As low-income tenants often comprise only about one-fourth of the total, and the mortgage interest was subsidized down to 2% for all units, the subsidy appeared to benefit middle-income Canadians (Canada 1985). Builders launched a campaign to have the federal government replace its housing supply programs with a shelter allowance targeted to those households most in need (see, for instance, Clayton Research Associates Limited 1984b). Following consultation with all provinces and territories, CMHC negotiated new agreements with provinces and territories, whereby they contributed a minimum of 25% of costs for programs they wished to deliver. CMHC confirmed that federal subsidies were to be targeted to households in core need.[15]

NEIGHBOURHOOD IMPROVEMENT AND RESIDENTIAL REHABILITATION

While NIP was terminated in 1978, it did achieve some of its objectives.[16] Like MSUA, the program fell victim to a debate concerning the nature of the federal mandate. From 1973 to 1978 some 493 neighbourhoods were designated, and $500 million was spent by the three levels of government.

The program was used in large cities throughout Canada, but one of the main differences from the urban renewal program was its widespread use in smaller communities. As is shown in Table 19.1, spending outside the nineteen cities with a population of 100,000 or over in 1981 was more than two-thirds of the program total, although only one-third of Canadians live outside large cities. The program was central to the revitalization of older neighbourhoods in smaller cities and towns.

No comprehensive evaluation of NIP was ever undertaken, either by the government or third parties, although evaluations of some individual projects were published in early 1986:

> Evaluation generally found that NIP resulted in positive physical changes in designated neighbourhoods through the addition of amenities; rehabilitation or reconstruction of housing; infrastructure improvements; down-zoning; and other related measures. Assessments were mixed, however, on the extent of achievements in areas such as resident participation; by-law enforcement; community planning; integration and targeting of other government and non-government resources; long-term municipal and resident commitment to older areas; and the capacity of municipalities and provinces to bear the costs of neighbourhood improvement without continued federal assistance (Lyon 1986, 3).

NIP overcame most of the criticisms of the urban renewal program but did not achieve its ambitious community development objectives. Total average spending of $1 million per designated neighbourhood simply may have not been sufficient to accomplish these objectives.

RRAP remains a program of CMHC. Almost 314,000 units, 71% owner-occupied, received assistance from 1973 to 1985 (*CHS* 1985, Table 74).[17] About 39%

of the owner occupant and 89% of the renter units receiving assistance were located in urban areas.[18]

While RRAP has, like NIP, been successful in overcoming many of the failures of previous programs and policies, it is not possible to evaluate it conclusively. The number of units assisted seems impressive, but less so when distributed across 479 approved neighbourhood improvement areas. A 1979 evaluation published by CMHC expressed concern with respect to the partial nature of rehabilitation caused by the limited dollar amount of RRAP loans and grants (Social Policy Research Associates 1979).

The principal criticism of urban renewal was dislocation, especially of low-income tenants. While the evidence on tenant dislocation from buildings receiving RRAP assistance is incomplete, it appears that the program has overcome this shortcoming in most but not all of the cities in which it has been used. A 1979 survey of landlords found that 80% of tenants remained in their units following rehabilitation but that from 6% to 24% of tenants were permanently dislocated as a result of RRAP (Social Policy Research Associates 1979, 102). About one-half of tenants had to pay a higher rent after rehabilitation, but only 5% to 15% received "socially undesirable" rent hikes (Social Policy Research Associates 1979, 103).[19]

Average rent increases in Quebec were 40%, reportedly the greatest in the 1979 survey. A 1984 survey of central Montreal dwellings rehabilitated at a cost of $5,000 or more is disquieting. That survey discovered that 90% of tenants had moved after two years (LARSI-UQAM 1985). Over one-third of the previous tenants were paying 30% or more of their income in rent. The rehabilitated dwellings, 62% of them subsequently converted to condominium or other co-ownership schemes, attracted tenants of higher income. Over three-fourths of the new tenants were single. The proportion with university degrees increased from 31% to 45% (LARSI-UQAM 1985, 130).

Most of Canada's large cities, however, seem to be characterized by massive dislocation of tenants and home owners alike in their central areas. In Toronto, some 9,000 rental housing units were lost as a result of conversions in the early 1980s (Silzer and Ward 1986). In the 1971-6 period, almost 5,000 tenant households in owner-occupied buildings were dislocated (City of Toronto 1980). House prices in central areas increased considerably faster than the urban central area average (Social Planning Council of Metropolitan Toronto 1987). Private upgrading of central city housing stock, not public action, emerged as the primary threat to low-priced housing stock and tenants in many Canadian cities in the 1980s.

RURAL AND NATIVE HOUSING:
NEW DEVELOPMENT AND REHABILITATION

Almost 1 in 4 Canadians lives in a rural area. Some of these areas are typified by an idyllic rural environment. Many of these rural communities, however, as well as many of Canada's small towns, are located in areas by-passed by contemporary economic development, and their housing conditions remain the worst in Canada. Historically, the housing policies and programs of federal and provincial governments took little notice of these areas. Increased attention was directed to them in the 1960s, coinciding with concern about the continued existence of poverty and extreme regional disparities in Canada.

Of special concern are the housing and living conditions of Canada's Aboriginal peoples, numbering about half a million people in 1981: 368,000 Indians; 25,000 Inuit and 98,000 Métis. "Status" Indians, most of whom live on reserves, accounted for 293,000, while the remaining 75,000 are "non-status" Indians (Statistics Canada 1984a, 20). The 1941 Census recorded only 118,000 Indians. The Aboriginal population grew rapidly over the forty-year period under review, and fertility rates remained about double the Canadian average. Of concern is a shortage of housing on reserves and the fact that the shortage could become even more acute, further weakening Aboriginal kinship networks and motivating youth to migrate to urban areas in even-larger numbers (Siggner 1979). An interest by the two senior levels of government in "rural and Native" housing generally coincided with heightened interest in urban housing and renewal.

A larger view of rural community development accompanied the new interest (see Baker 1971), and housing and residential environment were integral to this. The quality of housing in these communities suffered in part because commercial financial institutions were unwilling to commit mortgage and home improvement loans in communities whose economic future was in doubt (Herchak 1973, 11). The need for increases and improvement in the housing stock became urgent on Indian reserves. The continued viability of Indian communities, especially with respect to kinship ties, was at stake.

The federal Department of Indian Affairs and Northern Development increased funding for new housing and housing repair, although the adequacy of the response on Indian reserves remained in doubt at the end of 1987, as did the form of the response viewed from the perspective of community development. There are numerous complaints that houses constructed on Indian reserves, most of which are relatively far north, simply replicate southern technology and layout unsuited to northern climates and Indian culture. It is also reported too few Indians have been granted training opportunities in the construction trade.[20] One result is housing that is expensive compared to what might be put in place with greater involvement by Indians themselves.

Off-reserve housing in rural areas and housing rehabilitation have been the subject of a CMHC initiative known as "Rural and Native Housing," developed subsequent to the 1973 NHA amendments. The main components include home ownership and the residential rehabilitation program. The Urban Native Program offers housing assistance to Aboriginal peoples living in urban areas.

While some of the same problems in relation to technology, building form, and the need for making training in construction trades an integral part of public programs may be documented, innovative arrangements with local groups have been negotiated in a variety of locations. One instance is Mocrebec, a non-profit company established to provide off-reserve housing in Moose Bay, Ontario.[21] In 1986 CMHC introduced a demonstration program of a self-help approach to constructing new units.

As indicated above, a large portion of the RRAP has been implemented in rural areas – about 48% of the total number of dwelling units receiving rehabilitation assistance to the end of 1985 (*CHS* 1985, Tables 74 and 75). About 90% of units assisted were owner occupied. A variety of arrangements have been made with the provinces and community and non-profit companies to implement the program. The program is reportedly one of the more innovative in involving local groups.

Future Policy Directions and Further Research Needed

Provincial and municipal governments have been full partners in providing housing for Canadians, with the provinces providing human services and a planning and legal framework, while municipal governments have been responsible for planning and often for delivering services.

The federal government became involved in "social" housing for low-income Canadians and in the renewal of inner-city urban areas only gradually and as a result of initiatives by municipalities, civic boosters, and public housing advocates. These programs achieved a significant scale in the late 1960s. The reaction against urban renewal occurred because the program failed to address human development problems adequately or the need for participation. It caused massive dislocation, usually not accompanied by sufficient relocation assistance.

New programs and policies, many formally adopted in 1973, overcame objections to previous policies. Accomplishment of longer-term goals and objectives, however, remains in doubt. That the neighbourhood improvement program was cancelled in 1978 and that it never reached a large scale limited the program's ability to realize all the objectives established for it. Provincial social program spending reductions of the same kind that caused the federal government to rescind NIP have added to these shortcomings. In most instances, however, the program did not result in the same errors that led to resistance to the urban renewal program. Combined with RRAP, the housing stock does seem to have improved without significant dislocation of either owners or tenants. Provincial and municipal governments have maintained the neighbourhood improvement program initiative to varying degrees across the country, especially in concert with continued use of RRAP funds, but on a smaller scale than in the 1973-8 period.

The exception to this generally positive conclusion is related to the tendency toward "gentrification" of some inner city areas in Canada and the loss of low-income housing stock associated with this phenomenon. Nevertheless, the presence of gentrification, the loss of low-cost housing, and the fact that all three levels of government have been unable to stem it is cause for concern. It indicates an inability to respond to current housing issues. It also confirms the necessity for constant monitoring of housing needs and trends and for altering programs and policies in response to current situations.

One manifestation of the loss of low-income housing stock across the country is an increase in homelessness in Canada in the 1980s.[22] When focus is shifted to such issues, the achievement of Canada's housing and community development objectives has remained elusive. A number of questions remain with respect to existing housing and community development policies and programs.

Adequate improvement in public housing management continues to elude policy makers. There is currently a focus in some provinces on increased tenant participation in public housing management. An issue identified in some studies concerns tenants whose rents are supplemented in housing projects developed by private builders. Tenant integration requires more than simply leasing an adequate dwelling. Although several provinces were devoting more attention to housing management issues by the end of the 1980s, more research needs to be conducted into programs required.

The form and pattern of development of Canadian cities have changed considerably since NIP was initiated. Market pressures that reduce the amount of housing available for low-income inner-city residents are as great a concern today as the need for housing rehabilitation in cities and neighbourhoods not experiencing such market pressures. The adequacy of existing rehabilitation programs and ways to improve them need to be explored. Ways and means of maintaining the existing stock in inner-city neighbourhoods, especially in Montreal, Toronto, and Vancouver, urgently need exploration.

Notes

1 Adequacy refers to the need for repairs and for bath and sanitation facilities. Suitability refers to the need to reduce overcrowding.

2 E.J. Urwick, Ontario Housing and Planning Association, 1 and 5 June 1939, cited in Lemon (1985, 68).

3 In 1951, 89% of Canada's households were families. Between 1951 and 1961, 81% of all new households were family households. By contrast, the proportion of new family households between 1971 and 1981 decreased to 58%.

4 Approximately 160,000 units received this assistance, mostly small apartments and row housing and mostly for rental tenure (Correspondence to author from CMHC).

5 In excess of 18,000 hectares in approximately 160 projects were assembled over the life of the program.

6 It was recommended that the federal government and interested provincial and municipal governments make five-year block funding commitments for public land assembly for up to 100% of the cost of assemblies, the federal government to be repaid when the lands were developed. See Dennis and Fish (1972, 346).

7 The Honourable R. H. Winters to Stewart Bates, 8 June 1956. The project was Lawrence Heights in North York, Ontario, a suburb of Toronto.

8 Total of units built under NHA Sections 40 and 42 from 1964 through 1979.

9 Submission to Cabinet, 24 February 1970, Appendix C.

10 CMHC, George Devine, "Data Profiles for Seven Rehabilitation Projects" (undated).

11 The first use of "sunset" legislation in the NHA.

12 The eight departments included Citizenship and Immigration-Indian Affairs Branch, National Health and Welfare, Forestry-ARDA, Northern Affairs and Natural Resources, Labour, Industry-Area Development Agency, Atlantic Development Board and the Company of Young Canadians; CMHC was not a part of the original committee.

13 The latter was dissolved by the federal government in 1975.

14 Questions related to participation in decision-making and formal social interaction were omitted from the survey of rent-supplemented tenants.

15 Interview with Sean Goetz-Gadon, Executive Assistant to the Ontario Minister of Housing. Core housing need varies by urban area and approximates Statistics Canada's low-income cut-off lines.

16 A Community Services Contribution Program began in 1978 as a partial replacement for programs which were terminated: neighbourhood improvement, land assembly, and municipal infrastructure. Contributions under this transition program terminated in 1981.

17 RRAP/Rental was terminated as a program in 1989.

18 An additional 18,042 units in existing projects acquired by non-profit companies and cooperative housing societies received RRAP grants.

19 Socially undesirable is defined by the authors as in excess of 20%.

20 Interviews with chief, Moose Band, and executive director of Frontiers Foundation, August 1987.

21 Interview with Mocrebec president, Randy Kapashesit, August 1987.

22 For instance, the number of emergency hostel spaces in Metropolitan Toronto increased from 1,375 in 1982 to 2,328 in 1987. See Memorandum, Commissioner of Community Services to Community Services and Housing Committee, Council of Metropolitan Toronto, 25 September 1987; also see Canada Department of External Affairs, *Canada Position Paper, IYSH, CMHC 4135–2187*, 5, submitted to Canadian delegate to the 10th Commemorative Conference, United Nations Centre for Human Settlements, Nairobi, Kenya, February 1987.

CHAPTER TWENTY

====

The Supply of Housing in Resource Towns in Canada

John H. Bradbury

RESOURCE TOWNS occupy a special niche in any profile of Canadian settlements. Often isolated by virtue of their attachment to particular resource extraction sites, they have unique social and economic problems as well as special housing conditions and needs (Himelfarb 1976). The local housing stock has to adjust to meet the needs of residents across what is often a limited industrial life span of the community. Specific supply, maintenance, and tenure mechanisms are required if the stock is to adjust to the pronounced fluctuations that typify the economic base of such communities. Special problems arise when growth spurts lead to temporary housing that becomes permanent; during downswings, housing equity becomes problematic as the population departs to search for employment and accommodation elsewhere.

The two major trends in the post-war period were the move toward the privatization of housing and the normalization of local government relations in single industry towns across the country. The post-war period was marked by a general expansion of resource towns in new frontier areas: a trend that was halted in the recession of the early 1980s (Bradbury 1984a). While some of the particular and unique conditions in resource towns have been studied and the debate on policy priorities for resource towns has been longstanding, only in recent years has national housing policy begun to focus on the systemic problems (Bradbury and Wolfe 1983; Canada, Task Force on Mining Communities 1982; Canada 1985; Shaw 1970; Wojciechowski 1984). [1]

What is a Resource Town?
Resource towns are located throughout Canada wherever there is need for accommodation attached to a resource extraction activity. For the most part, they are isolated settlements whose *raison d'être* is the resource company with which they are associated. In many cases, the latter are local branches of multinational companies with operations in several different localities (Canada 1979; Lucas 1971).

However, not all resource towns are in remote areas; there are "mature" resource regions with well-established transport networks and socio-economic linkages. The mining and forestry region of northern Ontario, the mining and processing centres in the Kootenay region of British Columbia, and the asbestos towns of southern Quebec exemplify such regional complexes of resource settlements. There are differences among these regions in terms of transportation systems, the potential for commuting, and structural additions to the local economic base.

Resource towns generally are thought to include all primary sector communities (agriculture, forestry, fishing, and mining); however, this chapter is concerned largely with single-enterprise resource-extraction communities dominated by a large firm in which the housing is constructed by a resource company, a subsidiary, or a contracting firm and sold or rented to the workforce. In some instances, a private housing stock may be built by individuals, and a separate stock may be built by the state and the service sector to house employees.

Resource towns of this form are unique in the Canadian urban milieu because of their relative isolation and their location adjacent to a mineral deposit or in the vicinity of a forestry reserve. Their development, survival, and growth differs from other single-industry towns located within the heartland of the country and closer to larger urban centres. Their dependence upon one industry makes them especially vulnerable to technological change and market restructuring, in addition to the possible impact of a decline in the richness of the resources available. This increases the vulnerability and ultimately the planned life of the settlement.

The impacts of dependency, remoteness, and vulnerability on the conditions of housing and housing tenure vary with the size of the resource town. A population of 10,000 persons forms a threshold above which some forms of diversification and economies of scale will influence the viability and longevity of the settlement. The smaller towns are typically more dependent, single-industry-enterprise operations in isolated areas, commonly referred to as "company towns," "single-sector communities," or "single-industry towns" (Stelter and Artibise 1977). For present purposes, a typical resource town is defined as follows:

- employment dependent mainly on one industry;
- small in size: average size about 3,500 persons;
- household incomes higher than average;
- housing costs higher than average;
- mix of tenant and ownership relationships;
- dwelling values lower than average;
- turnover and mobility higher than average;
- abnormal age structures and sex ratios;
- linked directly with global market structures;
- often isolated with poor transportation linkages; and
- strongly influenced by corporate policies.

Dwellings in resource-based communities are generally newer than elsewhere in Canada[2] because of the relative newness of the resource settlements.[3] In 1981 this was particularly true in Alberta, British Columbia, and the Northwest Territories. Only in the Territories was the housing stock for the total population newer than in its resource towns. This exemplifies the impact of the development of the mining and lumber towns in these regions in the post-war period, reflecting the construction of new townsites that superseded older pre-war communities in towns like Tumbler Ridge in British Columbia or Fort McMurray in Alberta. The pre-war resource towns of northern Ontario, Quebec and the Maritimes, in contrast, continue to be characterized by older housing.

In the Quebec-Labrador mining region and in British Columbia and Alberta, there are wide variations in the incidence of home ownership. Perhaps more than any other settlement type, comparisons among resource towns illustrate the significance of corporate control over housing ownership. Following company policy, some settlements have almost 100% rentals, while others have home ownership rates more like the national average. In townsites such as Gagnon (in northern Quebec), for instance, renting was commonplace (68%); further north in the same region, Schefferville, Labrador City, and Fermont had higher incidences of ownership.[4]

Some resource industries are characterized by seasonal peaks of work and therefore need temporary workers for whom the housing stock must be sufficiently flexible to accommodate them. Furthermore, the rise and fall of population in the long run parallels the boom and bust cycle of the industrial base. In the late 1970s and early 1980s, this was especially noticeable in "vulnerable" mining towns such as the iron-ore towns scattered across Quebec and Ontario, in nickel-belt towns in Ontario and Manitoba, and in coal and copper towns in western Canada.

Migration patterns too have an impact on the housing market and socio-economic life of Canadian resource towns. In remote areas, there is little or no local labour-shed on which to draw when industries expand; they must rely on in-migration. In a downswing, these workers and their families tend to return to their original home sites or move on to new places of employment, thus placing added pressure on the housing stock and job market in another part of the country.

On the whole, resource-based communities exhibit higher than average incomes when compared to the rest of Canada. However, such data must be tempered by the fact that housing, food, and transportation are typically expensive.[5] Furthermore, average household incomes in resource towns are misleading since the families of those who become unemployed tend to leave and locate elsewhere.

Historical Evolution of Resource Towns
The earliest generation of mining camps and settlements in Canada were largely make-shift. Such settlements were commonplace in the 19th century and in

several new development projects in the 20th century (Dietze 1968; Schoenauer 1982, 1). Companies provided a minimal level of accommodation and services. Some of these sites were subsequently abandoned; others survived. Nowadays, however, such settlements are frowned upon or banned by provincial governments or local by-laws (McCann 1978).

A new generation of new towns were laid out in the first few decades of this century by town planners who wanted to create quality housing and physical environments in isolated frontier areas. These are typified by the work of Thomas Adams at Temiscaming. That Adams' experience was largely British and metropolitan in origin did not deter the first efforts of the "city beautiful movement" in the wilderness in Canada (Armstrong 1968). The experience was to influence future planning and housing in other resource frontier zones as companies expanded into new mining and forestry areas.

After 1945 new designs that built upon the experiences of Adams and others and were further modified by emerging post-war suburban plans (the prototype being Radburn, New Jersey) were directly transposed from southern and non-Canadian experiences into the resource frontiers. Similarly, housing designs were directly taken from the various copybooks and CMHC designs that were prevalent at the time (Walker 1953; Robinson 1962). Modelled on typical suburban designs in more temperate climatic zones, these dispersed settlements often had housing distributed along boulevards and curvilinear streets. Housing included single-detached dwellings, some multi-family housing, and single-men's quarters. Such designs fitted the perceived social and economic requirements at the time, but were expensive to build and maintain (Schoenauer 1982).

These designs were subsequently superseded by planned communities in which the traditional dispersed set of service buildings and municipal structures was replaced by a town centre made up of a more compact set of public and commercial buildings. Transportation networks between houses and the "downtown core" were by walking paths – of little use in the snow and inclement weather – and by street patterns focused on the central core. This core then became the social and economic centre of town, located in one or more large climatically controlled mall-like structures and surrounded by a parking lot that often became a desert (Schoenauer 1982, 2).

Contemporary plans include several subtle modifications in housing and community design, which take into account the location of the settlements in isolated areas and which have attempted to alleviate some of the problems of climate and "impermanence." Planners have created modular and mobile housing that could be moved from one site to another.[6] The construction was such that there was little or no internal or external evidence that they could be taken apart and moved. Hence, they differed from what is commonly thought of as mobile housing, that is, trailer homes in mobile home courts or parks (Blanc-Schneegans 1982; Paquette 1984). Furthermore, the planners in such communities endeavoured to make the best of the locational and site aspects of the settlements. In the case of Fermont, Quebec, for instance, the planner chose a southerly facing site and created a windbreak effect with a long multi-purpose, five-storey building that provided both housing and commercial services.

Stages of Development

Lucas (1971) observed that resource towns pass through four stages of development: youth (construction), adolescence (recruitment of citizens), transition, and maturity. Later developments, particularly the experiences of several coal and iron-mining towns in the recession of the early 1980s, led some observers to add two more possible, though not inevitable, stages to Lucas' model: the "winding-down" phase and "closure" (Bradbury and St-Martin 1983). Each of the six phases has specific housing conditions and requirements depending on the particular circumstances of the community, the company, and the status of the resource on the world market.

In the first two stages initial construction and recruitment, turnover is high within the new and typically young, male workforce. Workers are highly mobile and transient; some may stay on, but most leave once construction is finished. Their housing needs are temporary, but the stock created for them may be used by the first residents of the new town. Up to one-quarter of the workforce of these new towns during these stages is male seasonal workers in their twenties and, in some townsites, bunkhouses were created to serve this particular clientele.

In the transition stage the settlement is presumed to change from a company dependent system into an independent community. Although it is difficult to generalize here, this may take from five to ten years depending upon the nature and the stability of the industrial base. There are, for instance, cases where townsites have been "artificially" transformed into the third phase by government legislation such as the Instant Towns Act in British Columbia. In such cases, the townsite is provided with local government, commercial services, and a mix of housing types in an attempt to move the settlement immediately into a mature and stable phase.

In this third stage the resource company may divorce itself from running the town and from acting as a landlord. Home ownership is promoted as a stabilizing mechanism in what have traditionally been regarded as unstable communities with high labour turnover. Home ownership, therefore, is an important element in the attempt to move toward maturity (although it is not seen as a vital dependent variable in Lucas' model). This phase also features an increase in the levels of open government and the avoidance by the company of overt interference or participation in local municipal affairs. Even so, the company still remains involved especially in the affairs of housing, where distribution and allocation mechanisms, including mortgage provision, lot leases, and buy-back agreements, are often provided by the resource companies in a manner not unlike that in the older company towns.

In the fourth stage, with an aging population and a "forced" emigration of young adults, the settlements experience reduced turnover and a more normal demographic profile. In this phase, housing characteristics change again. As young families complete their childbearing, their increased family sizes put pressure on the existing housing stock until young adults leave to go to school or a new job. The problem of adjusting housing consumption to needs is constrained by the limited choices of housing types and the small absolute numbers of units that characterize resource towns. Similarly, the presence of retired persons may create friction in such communities, especially when the availability of housing is predicated upon a family member actually being employed by the resource company. Many resource towns have few retirees for this reason. Not being gainfully occupied, they consume housing which the company may wish to allocate or sell to another home owner.

Those resource towns that experience the fifth and sixth stages of "winding down" and "closure," face major problems around housing. In permanent closures, real estate loses value dramatically unless re-purchased at some pre-established rate of compensation, either by the company or the state. The effects are most severe and disruptive when they occur in a mature resource community. Recurrent temporary mine closures or recurrent production fluctuations can have an effect similar to the announcement of a permanent closure. The impact of fluctuations can be great enough to increase personal instability during a winding-down phase and dissolve community attachment, causing further out-migration and the loss of housing and equity (Bradbury and St-Martin 1983).

In periods of winding-down in the resource industry, vacancy rates vary with the form of housing. Cheaper and less "permanent" housing is often vacated first. The layoff strategies in downswings often start with the job categories occupied by unskilled persons and shorter-term residents, largely because layoffs and firings occur through seniority (Hess 1984). Inasmuch as job categories match housing types, stepped housing movements will occur: mobile homes and apartments occupied by unskilled and semi-skilled workers become vacant first. Basement apartments are vacated next, followed by permanent multi-family housing, and lastly, single family dwellings. House sales in such circumstances also tend to follow a cycle paralleling or even preceding incidents and events which have in the past precipitated downswings; these include price fluctuations in the resource base, union-company wage and salary negotiations, and periods of seasonal or cyclical plant closure.

Privatization, Equity and the Housing Market in Resource Towns

Relationships among the resource company, the local community, local government, and the provincial and federal agencies responsible for administration are an integral part of housing-related problems at the political level. The various levels of responsibility in resource towns, especially where the infrastructure and the townsite have been created by the resource company, become a matter of who will pay for what. Since 1945 the trend has been from traditional company towns, where the company was the supplier of employment and housing, to more open communities with local government and with housing made available for sale to residents.

Herein lies one of the major areas of change in resource towns across Canada: the creation of local housing markets which do not have a "normal" structure of buyers and sellers. The trend which has become most marked in resource town housing is the privatization of the stock and the move towards creating more "permanent" communities. House values are susceptible to sudden devaluation if the town is closed or if the industry enters a short-term recession.

In some towns, resource companies have created partial housing markets to cope with the fluctuations and with the apparent demand for home ownership. In some circumstances, for instance, companies have created partial markets in which house sales are permitted and others where tenure and sales are strictly regulated (Bradbury 1984b). Furthermore, in some towns, companies have sought to allocate housing types to different employees – whether management, technical, or skilled workers. Thus, even within the overall objective of increasing the levels of home ownership under a privatization program and the generation of a new class of owners, the distribution of the stock is not free from company influence (Walker 1953, 103).

Home ownership and equity became more problematic in the economic downswing which occurred in the late 1970s and early 1980s, as the post-war period of optimism and frontier expansion came to an end. Housing equity is an important component of retirement income or of the funds required to purchase a new home elsewhere. The loss of equity therefore represents a serious issue (Pinfield and Etherington 1982; Bates 1983).

In part to deal with this question and to retain control over the housing stock, "buyback clauses" and sales agreements between the company and first or subsequent owners were set in place in many such towns. In one sense the housing market and the trend towards privatization were controlled by the existence of buyback agreements, but only to the extent that resource companies or their housing subsidiaries maintained a hold on the housing stock.

There are various kinds of buyback clauses. Most systems set a selling price which roughly matches the equity value, together with improvements, minus some form of depreciation. Some operate exclusively for the first owner with modifications to the liability and status of subsequent buyers (Pinfield and Etherington 1982). The first owner may be guaranteed a sale to the resource company within the first five to ten years of occupancy. Under these circumstances an adjusted price – depending upon the company's perception of house values and the growth of a home owner's equity – will be paid upon resale to the company.

However, once house sales are no longer guaranteed by a buyback clause, a second pricing system takes over characterized by a local market structure which varies with phases of the local or regional business cycle. In such settlements, prices vary with the level of demand for the local commodity and the status of the boom and bust structure of the regional economy. House prices in peak periods may be as much as 20% above low cycles. However, in downswings and periods of winding down or closure, prices will show considerably greater variations and ranges.

In nationwide downswings affecting resource towns, questions arise as to if and how home owners are to be compensated for lost equity. In such cases the state, the companies, and the communities have been called upon to exercise judgment over the costs and responsibilities of the loss of equity to the workforce and to the associated business community in those settlements. The evaluation of such costs and the equitable distribution of compensation, when compensation is considered, has been a difficult political problem. That it should be resolved primarily in the political field represents a tacit recognition of the place of private enterprise in the resource frontier settlements in Canada and an awareness, at least on the part of the resource companies, that risks must be shared by all parties.

The opposing view suggests that the companies have passed along many of the risks of cost sharing of housing, particularly to the citizens of the townsites, and have thus divorced themselves from responsibility in the equity issue. So, the state has been called upon to provide compensation as well as to provide judicial mechanisms to cover community costs in periods of rapid downswing and consequent loss of equity. A parallel argument suggests that the resource companies should assume the risks of housing costs and housing losses as part of the overall costs of production. This would involve the development of rental-only settlements and a degree of financial support from government agencies such as CMHC. At present, CMHC policy limits mortgage insurance in resource communities to home owners.

Investment risk, then, is a topic which has concerned the creators and the occupants of Canadian resource towns. As administrator for the Mortgage Insurance Fund, CMHC remains concerned about claims losses in resource-based communities. As of 1986 the pressing need was for a comprehensive policy for these areas; the issue being the appropriate measure of risk for public investment or guarantees in locations of uncertain viability. [7]

Post-War Resource Town Development and Policy Orientation

In the 1950s and 1960s several Canadian provinces enacted "New Towns" acts specifically for the more isolated towns that were sprouting on the resource frontier. These acts specified the principles of town planning, town layout and the levels of responsibility of companies, local governments, and provincial arenas of power through the medium of by-laws and "Letters Patent." Furthermore, the Acts specified mechanisms for the transfer of political power to local municipal councils, although in several cases the councillors were drawn from company management in the early years of the townsites (Bradbury 1978).

In the first few years of these townsites, the costs of creating the infrastructure were high, building in a heavy tax burden for future residents (Bradbury 1978). Municipal elections brought together an interesting mix of local issues and company concerns, including the issue of political control over the development and costs of the settlements and the distribution of housing. There were wide variations in local policies and in the distribution of power but for the most part the transfer of political control was a slow process because of the omnipresence of company interests and differential corporate financing of community development. By the 1980s municipal elections brought together a wider range of interests including local business groups as well as union and company representatives. This process proved to be an intriguing mixing vessel for the different groups in many "company towns," but it did not bring significant change. In general, municipal government decisions remained tied closely to resource company financing and policy. The exceptions were those settlements which had diversified their economic base so that local tax support was not derived substantially from a single company.

As well as changes in the policy toward running townsites, there were numerous modifications over time to the way houses were owned, rented, and distributed. In older company towns, housing together with the local company store, movie theatre and union hall was an important part of the physical artifacts and symbolism of company dominance in the resource townsite. However, after 1950 the law and order of the corporate physical environment was replaced by towns designed by engineers and architects. Housing was created as a commodity – rather than purely as an item of accommodation to be distributed by the resource company alone – although there were several subtle modifications made by the companies to facilitate housing distribution. Furthermore, the physical structures and symbols of corporate presence were deliberately torn down and burned in several townsites: for example, Port Alice, Vancouver Island, British Columbia and Natal and Michel in southeast British Columbia. In their place, a neat suburban system was created with law and order defined by by-laws rather than through the frontier rules of "company towns" (Walker 1953).

Table 20.1 illustrates the importance of direct government involvement in housing and land development. It lists the forty-three resource-based communities in all provinces except Quebec, where the federal government participated with a provincial government to produce serviced lots to allow for the creation or expansion of the town's housing sector.

The post-war period also saw planning introduced as a means of shaping the physical and social environments of resource towns. Planners hoped to create stability and a sense of permanence by using macro and micro-scale designs imported from southern suburban experiences (Robinson 1962; Roberts and Paget 1985). It was clear by the early 1950s that companies and provincial governments were concerned about the corporate image of unkempt company towns, and the design of new towns was expected to reshape the physical and social

Table 20.1
NHA Land assembly activity in resource towns,
by location and date of initiation*

Province	Section 40, NHA†		Section 42, NHA†	
Nfld	Baie Verte	1972	Arnolds Cove	1974
	Burin	1967	Bonavista	1976
	Carbonear	1967	Daniels Harbour	1975
	Fortune	1967	Wabush	1974
	Grand Bank	1967		
	Harbour Breton	1967		
	Marystown	1966		
	Trepassey	1968		
N.S.			Port Hawkesbury	1974
N.B.	Nackawic	1966		
Ont	Atikokan	1950	Elliot Lake	1976
	Espanola	1968	Hearst	1969
	Longlac	1967	Hornepayne	1976
	Timmins	1967	Nakina	1976
			Wawa	1975
Man			The Pas	1975
			Thompson	1975
Sask	Uranium City	1961	Hudson Bay	1974
Alta			High Level	1973
			Lac LaBiche	1975
			Slave Lake	1973
			Smoky Lake	1972
			Spirit River	1971
B.C.	Cumberland	1974	Fraser Lake	1976
	Duncan	1957		
	Kimberley	1953		
	Ladysmith	1969		
	Mackenzie	1971		
	Masset	1969		
	Powell River	1974		
	Prince George	1957		
	Sparwood	1969		
	Trail	1951		

SOURCE CMHC Research Division.
* "Resource towns" taken from a list of 426 communities defined by DREE (1979), refined to 279 centres by CMHC. Includes single sector communities, single industry towns, single company towns, and excludes prairie service centres, centres based on federal employment, communities north of 60th parallel and Indian reserves.
† Section 40, National Housing Act authorized federal/provincial partnerships (75% federal) to acquire, plan, service, develop and market land for residential and ancillary uses. Section 42 authorized federal loans for the same purposes to be made to provinces, municipalities and their agencies.

fabric of settlements. With the cooperation of the resource companies, the planning profession endeavoured to modify the physical structure of townsites to improve social relations and to effect a stronger sense and awareness of permanence (Parker 1963).

In the 1970s several resource companies opted for an alternative strategy in northern settlements: namely, the use of commuter settlements. The use of existing housing in nearby settlements enabled some companies to avoid building new townsites. Even so, the overall problems of housing values and housing needs were not overcome, for in the long run most commuter settlements were still subject to economic upswings and downswings. In situations where resource operations were stable, such as at Rabbit Lake in Saskatchewan, community stability was matched by a satisfactory housing situation. Commuter settlements may be the appropriate format for the immediate future; but they must have a firm economic base for the housing market to be stable, and they are only possible when resource development occurs within commuting distance of an existing settlement.

Government participation in resource town planning and in the financing of housing has increased since 1945. Table 20.2 shows a set of examples depicting various aspects of the public role in financing, from provincial regulations and housing programs to the NHA and CMHC-insured lending and mortgage programs. From 1954 to 1957 CMHC made direct 80% loans to resource companies for employee housing. After 1957 the resource companies were required to obtain private capital for employee housing. In January 1963 a one-industry-town policy was introduced. This covered housing for company employees only, in towns dominated by one industry. The guarantees made the companies responsible for either the total loss on a home for a period of ten years, or for a maximum loss of $10 000 per unit for a period of twenty years or the life of the mortgage.

In April 1978 the one-industry-town policy was superseded by the resource industry guarantee policy. Under this, towns were designated special resource towns when they were growing at 20% or more per year. In July 1979 this policy was amended to include resource towns with zero or a declining population, with the designation lasting two years. This policy required the resource company to provide a guarantee that they would buy back the property upon default of the mortgage. The Resource Town Lending Policy was suspended in 1983 because of the high rates of defaults in some resource towns. As of 1986 new mortgage insurance approvals in resource towns were made by CMHC on a case-by-case basis.

As of 1986 provincial governments were becoming increasingly prominent in community development as the trend away from company control evolved. In British Columbia, for instance, regional plans are required before resource development can proceed. Measures by the province are aimed at "capturing" extraction operations within the local municipal boundaries so that local

Table 20.2

Government participation in selected resource town housing, financing and controls

Name	Province	Townsite date	Government participation in housing finance
Lynn Lake	Manitoba	1951-53	Federal participation, but the resource company required to guarantee mortgages.
Leaf Rapids	Manitoba	1970-71	Manitoba government directly involved in town development. Federal government provided housing funding through insured loan program of CMHC.
Fermont	Québec	1971-73	No federal government programs were used in housing in Fermont.
Lanigan	Sask	1968-70	Earlier village established circa 1930. Federal government assistance through NHA mortgages and cooperative housing. Costs are defrayed in mortgage payments by resident owners.
MacKenzie	BC	1966	B.C. Housing Authority constructed townhouses for rent. CMHC financing coupled with company guarantees of loans. CMHC Land Assembly program used (Table 20.1). "Instant Towns" legislation used by Provincial government. NHA mortgage program used for private house construction – paid for by residents.
Manitouwadge	Ontario	1954	Federal participation through an Ontario government housing project; the company owned housing and wrote off investment at a "30% rate on a declining balance."
Tumbler Ridge	BC	1981	NHA mortgages and provincial assistance.

SOURCE Fletcher and Robinson (1977) and Rabnett Associates (1981).

property taxes from the company generate the majority of local revenue. This can prove to be hazardous, however, as the fiscal fortunes of the community are directly dependent upon the profitability of the resource company. There are examples of unsupportable municipal debt in declining resource communities that eventually revert to senior levels of government. In this regard, provincial governments will bear the responsibility and consequences of judging the long-term risk associated with infrastructure development in resource towns.

Conclusion

The post-war period has seen remarkable changes in the form and context of housing in Canadian resource towns. The old regime of the "company towns" has disappeared, and in their place, new planned towns on the resource frontier have been built. Each of these settlements has been graced with new houses and with new housing policies. Each townsite has grown under a different corporate and government regime which has coloured the type of housing and the distribution mechanisms.

The major trends in Canada as elsewhere have been the privatization of housing and the normalization of local government relations in most single industry resource-based settlements (Neil and Brealey 1982; Neil, Brealey and Jones 1982). Privatization has occurred unevenly, however, and several communities still have some characteristics of "company towns" and high levels of rental tenure. The change to private ownership of housing has occurred within the overall milieu of a dominant or monopoly firm which has tended to retain some levels of control over the production and distribution of housing, because housing constitutes an important part of productive capital investments. Both the industry and workers are caught in a dilemma. In upswings, private ownership of housing is apparently satisfactory. However, in downswings, housing prices tumble and home owners lose their equity and savings. The alternatives are to spread the risk over government and companies, to adopt commuter strategies, or to revert to rentals in housing owned by government or the resource company.

It is also important to distinguish between the housing distribution system and the housing markets within resource towns and those in "normal" townsites. In resource towns, because of the dependence on a single industry base and because of the upswings and downswings in some raw materials industries, there is an unusual market structure for housing. An artificial market structure operates in some areas; and in others, companies have deployed a system of buy-back clauses to overcome problems in the distribution of housing and to alleviate difficulties in the retention of equity in a downswing. The resource town housing system is thus unusual in the Canadian context because of the mix of different housing policies and tenure types. The resource companies continue to dominate in the housing arena, despite political measures that promised to change these relationships by "normalizing" local government relations.

Notes

1 For an opposing view on the evolution of resource town policy in Canada, see Robson (1985) and Saarinen (1986).

2 The general pattern of housing values also reflects this overall newness, but the data must be tempered by the knowledge that in several settlements the values are greatly influenced by structural devaluations brought on by the decline or demise of the economic base and of the community itself.

3 Furthermore, resource towns, at least in the post-war period, tend to be constructed in a short period of time in blocks and rows of houses of similar design, a factor which influences the overall trend of newness.

4 In the case of Schefferville, the townsite experienced a critical market downswing, and mortgages were less popular because sales of houses were marginal. Both prices and equity were substantially lowered by closure of the townsite.

5 There are some slight variations in air transport costs between remote settlements and central urban areas in several provinces.

6 In this latter phase, the planning strategies applied in Fort McMurray, Tumbler Ridge, and Hemlo were deliberately designed to incorporate considerable mobility. However, even with this innovative modification, the settlements may still succumb to an economic downswing and a loss of equity in housing.

7 A new federal policy with specific provisions for special-risk communities was introduced in 1987.

═══

CHAPTER TWENTY-ONE

Lessons Learned from Canada's Post-war Housing Experience

John R. Miron

IN PART, this book is a historical overview. Beginning at 1945 or earlier, each chapter details post-war changes in housing outcomes, in the shifts in demand, supply, and allocation that gave rise to these outcomes, and the causes of these shifts. Also explored are the many public policies that shaped or directed these changed outcomes. From its own perspective, each chapter attempts to answer the question "what can be learned about housing progress from the post-war experience?"

In drawing lessons from past experience, the authors of this volume identify mechanisms and policies that did or did not work. They also identify the underlying conditions that gave rise to this consequence. These lessons are useful in thinking about current or future housing problems. They suggest where and under what circumstances a particular mechanism or policy might or might not work again. However, just because a policy failed (or succeeded) in the past does not necessarily mean that it will do so again in the future; the processes through which the policy acted may change, or the underlying conditions that gave rise to the policy consequence may not remain the same. Understanding how and why housing progress occurred helps us to assess whether post-war experience can be applied to present or future problems.

Lessons Learned About Producing Housing

In this book, the authors comment only briefly on lessons that can be learned regarding the housing industry, since this topic is covered in a companion study on the industry undertaken by CMHC (Clayton Research Associates Limited 1988). What follows is restricted to ideas discussed in Section II (the supply side of housing). Here, I assume that the housing industry consists of the collection of builders producing new or renovated housing stock for sale and landlords in the business of providing rental housing services from the stock. Initially, I exclude home owners (that is, self-providers of housing services and stock) from this discussion, preferring to treat them separately in the context of do-it-yourselfing.

PERFORMANCE

One lesson to be learned from this monograph is that the residential construction industry in Canada is big, robust, and healthy. For many Canadians, the private sector part of this industry has been able to deliver efficiently the housing they want. While it is true that the housing needs of other Canadians were not as well met, and that the public sector also contributed in important ways, we should not understate the importance of having had a competitive and efficient market for housing production.

This same industry developed the post-war suburban landscapes that now surround our cities and in which a majority of Canadians presently live. We may now despair of the aesthetic, visual, and social homogeneity of those suburban neighbourhoods, their squandering of energy, the difficulty of providing social services to them, and other problems. Nonetheless, they were remarkable success stories. [1] As Canada's population doubled, these neighbourhoods provided safe, clean, comfortable, and healthy accommodation together with good traffic control, ample parkland, and conveniently clustered community and commercial facilities. [2]

At the same time, we have learned that the challenges facing this industry are changing. However successful at past challenges, the industry may now be entering a period of transition that will require different skills and techniques. The large amount of new housing produced in the last four decades, together with an expected demographically driven slump in net new demand, suggests that the amount of new construction will subside. Renovation of the now-aging existing stock, and especially high-rise rental units, is becoming more important. It is not yet clear how pronounced this transition might be, nor how easily the industry will cope.

ORGANIZATION

We have also learned that the housing industry is a paradox in Canadian industrial organization. On the one hand, it is large: residential construction is about 20% of total Gross Fixed Capital Formation. On the other hand, unlike many other large industries, business is not concentrated in the hands of a few large firms (see Chapter 8). [3] Although the post-war period has witnessed the emergence of some major firms within the industry (and, in some cases, their demise), the industry has generally been characterized by a large number of small suppliers.

This paradox is all the more surprising given all that has been learned about the entrepreneurial risks involved, particularly for builders. Almost universally, home building involves risks. One advantage of a large firm is that it can spread the risks of a particular venture over a wider base of operations. Why did the riskiness of enterprise not lead to more large firms in the industry? And among those that did emerge, particularly in the 1970s, why did several eventually move from the residential sector into other real estate development?

Here, one has to be careful to specify the advantages and disadvantages of bigness within the industry. The advantages of bigness primarily take the form of declining unit costs of maintenance and servicing; for example, larger buildings can usually be maintained at a lower cost per suite than can small buildings, and having more suites or buildings enables the landlord to negotiate better with utilities and service and repair trades. At the same time, these advantages are largely local; having buildings in two cities that are far apart is not as advantageous as having them in the same area. At the same time, there are also portfolio-holding risks: for example, holding all one's rental stock in Calgary during the oil bust of the early 1980s. Another important disadvantage of bigness in property management is the problem of cost control and supervision. As a firm becomes bigger, it becomes more difficult to ensure that employees act efficiently.[4]

There are advantages to large-scale construction in a given local market. However, given variations among local markets in financing, in zoning and building code requirements, in land subdivision, in labour practices, and in building technology and practice, it is not surprising that firms largely concentrate in one local market. Again, as well, cost control and supervision are problematic for firms that operate in several local markets at the same time. Then, too, there is the problem of portfolio-holding; firms that operate primarily in just one local market must live with the ebbs and flows of that market.

In part, the period since 1945 has been characterized by a shrinking in the disadvantages of bigness – a lesson quickly learned by some of the emerging major developers. The increasing standardization of building codes, improved diffusion of "best practice" building technologies, an increasing use of prefabricated parts and building systems, the development of specialized subcontractors and consultants, and improved management techniques and information systems helped make big firms more competitive and manageable. In addition, there was a demand for the integrated neighbourhoods and large-scale suburban planning that big developers were able to provide.

Also, we have learned that the changing economics of land assembly were important to the emergence and survival of large firms. Up until the mid 1970s residential land prices increased modestly faster than either inflation generally or the costs of holding land (that is, mortgage costs) specifically. This gave a comparative advantage to companies that assembled land for large developments and moved these projects downstream and into the market. In the late 1970s the situation changed as the boom in land prices subsided and holding costs increased sharply. The financial leverage that earlier propelled the growth of large firms became the undoing of some of them. The return of some local markets in the late 1980s to the rising land prices and robust housing demand that had characterized the 1960s and early 1970s demonstrates the cyclical nature of land development that was an important lesson learned by large firms.

TECHNOLOGICAL INNOVATION AND RISK TAKING

Chapter 8 argues that Canada was well-served by its housing construction technology. The efficiency of site-built wood-frame platform construction that characterized low-rise construction has improved steadily since 1945, partly as a result of building research funded by CMHC and other public agencies.[5] We have learned that this gradual improvement proved more successful than either the modular building systems or manufactured homes once thought to be the ways of the future. Although the amount of manufactured components used in site-built housing has risen, technological change has been gradual. The story is similar for developers of high-rise accommodation. There were technological breakthroughs, but radical shifts such as manufacturing entire houses generally failed to take hold. Furthermore, greater standardization led to the development of specialized subcontracting that reduced the need for developers to keep abreast of all technological changes and in part undermined the *raison d'être* of large firms.

At the same time it is of interest to note the role of CMHC and NRC in developing and promoting innovations in housing construction. The lack of adequate research and development expenditure within the residential construction industry has been widely noted. Did CMHC and NRC inadvertently help ensure that large firms did not develop in the industry by taking away one important reason for their existence? Could the lesson to be learned here be that a government policy introduced to help small firms perpetuated inefficient organization of the industry?

There is also a lesson to be learned about risk exposure and containment. In speculative construction, builders risk taking a long time to find a buyer. In mortgage financing, landlords take risks in borrowing "long" (that is, through mortgages) in order to lend "short," that is, in rental leases. A similar risk is inherent in land banking where developers realize leverage when land prices rise faster than financial holding costs, but where losses also can be multiplied. From relatively cautious practices in the late 1940s, a surging demand for housing led to an increased risk-taking by the industry in the 1960s and 1970s. In the early 1980s risk containment became a keyword under the sobering influence of recession. With the growth in size, larger development firms spread risks by diversifying into other forms of real estate development or into other regions of the country.

Risk containment has also been a concern to the private mortgage insurance market in Canada. Insurers have several options available to help manage risks: giving coverage to a range of properties, insuring in a variety of geographical markets, varying the terms and fee for insurance among categories of risk units, diversifying insurance fund investments, seeking reinsurance, and using sophisticated hedging strategies.[6] There has been a growing use of these various strategies, as insurers have learned of their benefits and as the increasing sophistication of financial markets has so permitted.

PUBLIC AND SOCIAL HOUSING

Metropolitan areas began the post-war period with what might best be thought of as a direct approach to the problem of inadequate housing. The approach was to use large-scale slum clearance to remove "urban blight" and to replace it with subsidized, large-scale, socially-segregated, publicly-owned and/or managed, high-rise housing for low-income households. Not widely anticipated were the problems that such redevelopment created: the anger, frustration, and loss of senses of community and control that arise from eviction, the inadequate compensation of landlords and sitting tenants for costs arising from eviction, the temporary or permanent displacement of existing residents, and the higher rents often faced by dislocated tenants who were ineligible or unwilling to move back into the newly-constructed public housing.

We have learned much about the value of being more sensitive in the design and delivery of low-rent housing. We have experimented successfully with small-scale developments, low and medium-rise building forms, private for-profit and third-sector ownership and/or management, socially mixed housing, and other means of delivering housing subsidies, including shelter allowances.

That there is a niche for non-profit and cooperative housing is an important lesson learned. Third sector housing (non-profits and non-equity cooperatives) can have some comparative advantages. It can be nominally less expensive to produce: in part a result of sweat equity and in part because it can be carefully designed to meet the needs of its clientele. In general, it may provide tenants with a better security of tenure than otherwise is found in the private rental sector. Finally, such housing can provide for more local community interest and involvement in financing, design, construction, and operation.

RENTAL SECTOR

This chapter would be remiss if it did not comment on the current state of the private rental sector. This is a real conundrum in housing policy. On the one hand, there were areas in Canada in the mid 1980s (notably Quebec) where this sector was healthy: exhibiting at least modest rates of vacancy, rent increase, and new construction. In other areas of the country, however, new construction was negligible; the existing stock (much of it high-rise) was thought to be slowly deteriorating; and we had either near-zero vacancy rates with rents straining upwards or high vacancy rates with moribund rents. As several chapters in this book attest, just what has caused these conditions is not clear; nor is it clear that a policy prescription is required. We have learned that the health and operation of the private rental market are the result of complex interactions (see also Jones 1983, 52-9). The critical factors may be demand-based (for example, sluggish growth of average tenant income), supply-based (for example, riskiness of new rental construction), or policy based (for example, restrictive zoning, building codes, or rent regulation). It is easy to attribute an unhealthy market to just one of these factors, but the empirical evidence is mixed. As a result, it is not clear just what can or should be done to rectify the condition. This is an area that requires more research.

In this regard, we have also begun to learn of the potential of the condominium sector in the supply of rental units. Although originally intended to be a vehicle for owner-occupancy in multi-unit buildings, condominiums also provide small landlords with a relatively liquid investment, well-defined monthly costs, and the ability to realize capital gains by reselling the dwelling to a home owner at some future date. This has been occurring both informally (that is, where a building has a mix of renters and home owners) and formally (that is, where condominium ownership is used to syndicate a rental building). [7]

DO IT YOURSELF

Finally, although we lack good data, we have learned that the do-it-yourselfers have likely been instrumental in achieving housing progress across Canada: whether in rural areas, small towns, or large cities. Although it is difficult to define exactly what constitutes a renovation expense, Canadians may have been spending more money each year on renovation by the mid 1980s than on new construction; and this is in spite of the relative newness of much of the housing stock. Some of this work was done in compliance with local building and zoning regulations, but presumably some was also in violation. Do-it-yourself is an affordable, if not always best-practice or safe, alternative to commercial renovation and, sometimes, even new construction. In attempting to suppress certain undesirable activities, heavily-regulated urban areas with their official plans, building codes, zoning by-laws, and strict enforcement, also suppress the informal sector. However, it is not obvious what governments can or should do to encourage this sector – other than to improve the level of information made available to do-it-yourselfers.

Lessons Learned About Consuming Housing

The first part of this discussion looks at lessons learned with respect to home ownership as a financial asset. The second part considers lessons learned when housing is viewed more broadly as a consumer good.

HOUSING AS A CAPITAL GOOD

Looking at many of post-war Canada's existing home owners, the principal lesson to be learned may well have to do with the redistribution of wealth possible from home ownership and the leverage entailed in mortgage financing. Although house prices tended to increase everywhere, the increases were most marked in the major cities. In part, the prices of existing homes rose because it became more expensive to purchase the materials, fittings, and labour required to produce new housing. However, also important was the rising price of metropolitan land: jointly a result of population growth (net immigration, natural increase, and increasing longevity) and increasing real incomes, with both driven by a surging metropolitan economy.

For many urbanites, home ownership became their single best route to wealth accumulation, and it may have come to colour their attitudes, and those of their children, toward lifetime savings and investment strategies. Housing was not simply acquired to be consumed over one's lifetime, with little regard for resale (or, at least, resale at a gain). Instead, a view may have arisen that housing was an investment that, especially in metropolitan areas, was increasingly liquid and would pay future gains that were as important, in an economic sense, as the benefits arising from occupancy.

Expansion and liberalization of the mortgage market allowed for a broad participation in metropolitan housing markets. This had two important effects. One was to enhance the demand for owner occupancy, and thereby the gains to be made from investing in it. The other was to spread the gains across a wide cross section of the market: by income, by age, and by family type. However, these capital gains were just a redistribution of wealth, not a net addition to it. Existing owners benefitted at the expense of new owners, and owners of more-expensive homes at the expense of other owners who upgraded. We simply do not know just who in Canadian society presently have been the net beneficiaries and who the net losers from all this redistribution. Also, because post-war population growth was just one factor, albeit important, that fuelled the capital gains boom, the boom may not necessarily end with the anticipated slowing growth or decline in Canada's population over the next half-century.

Also important here was a lesson learned regarding the changing significance of "filtering" in the housing markets of many large cities. The early post-war suburban housing boom led to the outward movement of more affluent households. The older, inner-city stock – much of it depreciated but still of good quality – became occupied by the less-affluent households that had remained (or immigrated). It can be argued that filtering spread the benefits of net additions to the suburban housing stock broadly among income groups. The process of inner-city revitalization – gentrification – that began in the mid 1960s in some cities arrested the filtering process and may well have reduced overall the benefits arising to less-affluent households.

HOUSING AS A CONSUMER GOOD

Running through this monograph is a lesson learned about the ongoing dichotomization of Canadian households into housing haves and have-nots. On the one hand, housing has become more a consumer good, and less a necessity, for many Canadians. On the other hand, a growing group of Canadians have been poorly or unaffordably housed. Among the haves, typical housing consumption is currently at a level that in some respects exceeds any plausible minimum standard of decent accommodation. The haves, if anyone, represent the success stories of the post-war Canadian housing market and housing policy. Never before have so many Canadians been accommodated in comfortable, warm, healthy, and safe housing, nor had such access to community infrastructure and social services. At the same time, a persistent and growing number of Canadian households (and would-be households) have not been well served; they either simply cannot find housing at all, or can ill afford what they do find. As argued in Chapter 4, the dichotomization may in part be a perverse and unanticipated impact of the underlying focus on home ownership in some federal and provincial policies affecting housing.

Underlying parts of Chapters 8 and 11 is a view that housing is increasingly seen, among the haves at least, as an increasingly elaborate consumer good, like a VCR or an automobile with ever-more options. If housing is just another consumer good like automobiles, why do we need government housing agencies, ministries, or departments? Historically, we needed them in part because households (and governments) have perceived housing differently, in some sense, from other goods. Have the attitudes of consumers changed? Is housing now somehow less important or less cherished than it used to be, and if so, is there still a role for governments to play in the production, supply, or allocation of housing? It is impossible to answer such questions directly as empirical data on changes in attitudes toward housing are scarce. Nonetheless, these are interesting questions, in part because underlying conditions that might determine such attitudes have changed. There may be an important lesson here in terms of the need for government and public policy to respond to changing attitudes.

Housing takes on a special meaning to its inhabitants for two contrasting reasons. One is that housing, or the "home", symbolizes the history of the family and the process of living therein. In other words, consumers treat housing differently from other goods because it comes to represent their hopes and dreams, their successes and failures, and the major events of family life. The second reason that housing may have been treated differently is because of its uniqueness and relatively illiquidity. Owner-occupied housing was typically expensive and not easily disposed of. To the extent that we "cherish" that which is too costly to throw away or sell at less than fair value, housing may have had a special meaning to us. We also may cherish a dwelling because of something that is unique about it: for example, an architectural detail, the layout of rooms, nearby amenities, or the special location of the dwelling within the community.

Housing as Symbol of Family Life

An important change over the post-war period has been the growing populations of large cities and metropolitan regions throughout Canada. Urbanization made feasible new forms of living arrangements and housing along with ways of providing community and social services. It nurtured the formation of non-family households whose sense of the importance and value of "home" may differ from the traditional family household. In addition, for some of these households, and some family households as well, activities that used to be an integral part of home life (for example, food preparation, or elementary medical and recovery care) can now take place outside the dwelling or be brought in.

The family household also changed in ways that may have affected the meaning and perception of "home." One change has been the rise in paid workforce participation among mothers; another has been a greater part-time workforce participation among teenagers and students.[8] Together, these may have reduced the amount of time that families spend together in their dwelling. Another important factor has been the declining importance of childrearing in the family life cycle: because of increasing longevity generally, declining fertility and increasing childlessness after 1960 as well as the narrowing age range of mothers at child birth. With an improved health care system, and a resulting greater use of hospitals, fewer Canadians were born, or died, at home. In this sense too, the home may have become a less important symbol of family life. To the extent that attachment to "home" is a function of the family activities and memories that occur there, such changes affect the attachment of families to their housing.

Also important has been the decline of shared accommodation. The relatives and lodgers who used to live with families in larger dwellings added something to the quality of home life. While the experience may not always have been positive and families may well have been happy to come to live alone, sharing may have enriched the experience of domestic life by broadening the resident set of personalities and perspectives. Its decline may thus have adversely affected the richness and quality of home life in ways that reduced the special meaning of home.

Although difficult to document, it is also widely believed that post-war Canadians became geographically more mobile. If so, this may have meant that individuals have come to spend less of their life in any one dwelling, foregoing the opportunity to develop a longstanding attachment to one home. Also important in this increased mobility was a closer matching of dwelling size to family size that meant less room for sharers. Interestingly, the attachment to "home" may have begun to re-express itself in relation to the family camp, cottage, farm, vacation, seasonal, or weekend home. Households might move from one principal dwelling to the next (sometimes at great distances), but they often kept the camp/cottage/farm for "the family."[9]

Changing Liquidity and Uniqueness of Housing

The divergence between house and home may, in part, also have been abetted by the increasing use of prefabricated components. As argued in Chapter 11, dwellings can be perceived as boxes into which, since 1945, have been stuffed an increasingly rich array of appliances, furnishings, and fittings. In so doing, the box itself can become relatively unimportant. Given sufficient funds and enough space, it may be possible to take a box, stuff it appropriately, and make it look from the inside essentially like any other box. If so, then part of the "uniqueness" of a dwelling may have been lost.

It can be argued that we cherish things that, having outlived their usefulness, are not easily re-sold. If so, post-war improvements in the efficiency of the housing market may have reduced the special significance of the home. One can point to the more-efficient housing markets that arose with increasing urbanization, and better-organized systems for advertising and selling owner-occupied property. Also important was the felling of restrictions on mortgage finance and the gradual integration of residential and other consumer financing. Finally, the waves of house price inflation that swept through various parts of Canada generally in the post-war period and particularly in the 1970s also helped improve liquidity in the housing market.

In part, an increasing sophistication with respect to financing options may reflect an emerging view of housing as just another consumer good. Since 1945, as lenders came to offer more products, households have grown more willing to accept increased risks in return for obtaining better housing or financing. When home is sacred, households might well be expected to be risk-averse. When it is not, they are willing to take more chances. As the range and variety of forms of consumer credit increased, so too did the variations in mortgage funding. Whether by choice or necessity, households came to use a broader range of risk options.

Lessons Learned About Government's Role

At all levels of government in post-war Canada, there were new attempts to redesign the way in which housing was produced, demanded, or allocated. In a sense, we have the most to learn about the roles and impacts of governments, because they were relative newcomers to the housing market.

ADEQUACY AND AFFORDABILITY

Arguably, the most important lesson learned here concerns intricacies in defining housing adequacy. Governments at all levels have wrestled with this. The fundamental questions remain unanswered. How does one define a set of minimum standards for housing? How and why should these standards differ depending on the characteristics of the households? For which potential households are these standards to apply? Who should set the standard? That definitions of adequacy do vary is not surprising given the manifold ways in which housing contributes to our happiness, health, and well-being, our sense of place and community, our access to amenities and services, our neighbours, our status, and our aspirations for the future. Also important here is one's perception of society's goals, and of how society should proceed to address those goals. If anything, the preceding chapters emphasize that we have learned of the need for governments to be explicit about what they are trying to achieve with their policies.

Perhaps nowhere is this need better evidenced than in relation to the promotion of home ownership. Since 1945, governments at all levels have promoted home ownership as good for Canada.[10] In Chapter 3, it is argued that home ownership may help promote social goals such as efficiency, redistribution of wealth or income, quality of life and personal security, and security of tenure.

What is not clear is the magnitude of these benefits or the opportunity costs in terms of other goals. Do the benefits outweigh the negative impacts? This is a question for which, even today, there are only simplistic answers or crude calculations.

Related to this is a lesson learned about the difficulty of defining what is affordable. Some households have to cope with situations that, according to an arbitrary rule of thumb, require spending an unreasonable or undesirable amount on housing. Even leaving aside the subjective question of how one defines the limit to affordability, it is difficult to measure the funds that a given household has available to spend on housing, to identify the housing alternatives open to the individuals comprising that household, or – and this applies especially in the case of home owners in an inflationary environment – to measure the real cost of the housing they consume.

It would be inappropriate to end this section without considering the questions of whether there has been an over-investment in Canadian housing since 1945 and what has been the associated role of governments. An American study concluded that from 1929 to 1983 the rate of return to housing capital in that country had been about half that of non-housing capital and that an efficient housing stock would be only about 75% of its 1983 size (Mills 1987, 601). The study concluded that the favoured tax status of owner-occupied housing in particular and a tax system that biases capital financing in favour of debt over equity in general could account for much of the over-investment in the United States. It is not clear whether over-investment is present to the same extent in Canada, but if so, this suggests the importance of seriously reviewing the system of income taxation.

SOCIAL MIX AND COMMUNITY DEVELOPMENT

Another lesson learned has to do with social mix. Some early experiments in post-war housing (both private and public) were criticized for being too homogeneous or segregated. On the public side, for example, large-scale urban renewal and public housing schemes were found to be inferior to solutions that favoured neighbourhood renovation (for example, NIP/RRAP), carefully designed infilling, and medium-density housing. Throughout Canada, planners have attempted to encourage a mix of incomes and age groups at the neighbourhood or community level. In some cases these efforts at integration have been divisive or expensive. Encouraging social mix is thought to create several benefits to Canadians: for example, improved social justice, equality of opportunity, compassion, diversity, and sense of community. However, there is surprisingly little evidence of the extent to which current social mix policies actually do result in such benefits. At the same time, these policies can and do clash with the concerns of local residents over such things as loss of property values or personal safety. We have learned the importance of finding answers to the following questions. What are the social costs of not promoting social mix policies? How large are these? What alternative means are available to promote social mix, and how effective are these? Governments need to think explicitly about what they hope to achieve, and how and why it will create net benefits.

Since the early 1980s there has been a renewed policy emphasis on "targeting" which has been taken to mean that subsidies should be restricted to the needy. While laudable in terms of program efficiency – after all, no one wants to subsidize people who do not need subsidies – social mix may be a casualty. If one wants to encourage mix and non-needy households resist intrusion into their neighbourhood, there are only two policy options. One is to force the integration by fiat (possibly at great political cost). The other is to provide incentives for the non-needy to accept the integration. One effective way of doing this – subsidizing the housing of the non-needy – is closed off by restrictive targeting. Social mix may be important. It is also elusive. We have learned that it is important to think about how and whether to promote it.

Another lesson learned concerns the need to reduce discrimination against women and mother-led families. Since 1945 institutional lenders have become more amenable to the inclusion of the incomes of working wives in calculating mortgage eligibility among husband-wife families. This helped two-earner families achieve home ownership at a younger age and made it possible for a broader group of households to achieve home ownership at some point in their lives. And as argued in Chapter 3, home ownership is an important tool for governments seeking least-cost income maintenance schemes for the elderly. Also important have been the first steps of governments, through public housing and anti-discriminatory regulation in the private rental market, to ensure that lone-parent families (typically female-headed and poor) have access to adequate and affordable housing. [11]

STRATEGIES AND TOOLS

Another important lesson learned is that the appropriate policy solution is partly determined by conditions specific to the locale. Because Canada is large and geographically diverse, global solutions need to be flexible enough to work in a variety of local situations. At the broadest scale, comprehensive national strategies – whether indigenous or imported from abroad – have to be designed to meet local needs. At a more localized level, it means that within provinces or regions policies and programs need to be geared to specific local housing conditions. There have been some attempts to redress this problem – for example, by targeting policies to specific local areas or problem situations. However, it is the source of a continuing dilemma. One possibility is to tailor a policy (either by individual design or the offering of options) to each locale, but this can detract from administrative efficiency or regional equity. [12]

There is a need to consider how a policy might best be implemented given Canada's multiple levels of government. A federal role has been especially effective in the areas of (1) reducing impediments to the efficient operation of the housing market: for example, in improving the liquidity, availability, and supply of mortgage financing, and in defining national standards for building materials and construction techniques; (2) pursuing housing policies related to macroeconomic goals such as full employment; (3) assisting, coordinating, and training provincial and local agencies in developing and implementing local housing policies.

Governments have also learned of the benefits of involving those affected by their housing policies in the development and delivery of solutions. In some cases, this has meant involving the households that will live there and their neighbours in planning and design. In other cases such as rural/remote housing, it has meant not blindly imposing mainstream expertise or experience based on solutions that have worked in urban areas.

Governments at various levels are continuing to learn about the advantages and disadvantages of the various policy tools, that is, tax expenditures, direct expenditures, regulation, crown corporations, and loan guarantees. In the case of tax expenditures and policy, the sensitivity of rental dwelling construction to the tax treatment of rental losses and capital gains (illustrated by MURBs) has been especially noteworthy. It can be argued that governments have increasingly turned to regulation as fiscal constraints have limited other courses of action and as governments found it difficult to control the dollar amounts involved in direct spending programs.[13] At the same time, poor regulation can breed new direct or indirect costs. This has led to contradictory moves: on the one hand reducing regulatory constraints (for example, federal policy to eliminate regulatory requirements that were restricting liquidity in the mortgage market); on the other hand increasing regulatory activity (for example, land subdivision or rent regulation).

At the federal level post-war governments have from time to time used the residential construction industry to achieve macroeconomic objectives such as full employment, economic growth, and price stability. Sometimes, the "tools" employed were direct (for example, construction or mortgage subsidies) and sometimes indirect (for example, fixing maximum NHA mortgage rates above or below market. Whatever the merits of the macroeconomic objectives, we have learned that such policy tools can have undesirable "boom or bust" implications for the housing industry; in other words, stability is important to the development of an efficient construction industry.

LAND-USE PLANNING AND REGULATION

Another lesson learned deals with the uses and limitations of land-use planning. In Canada, modern planning techniques were almost non-existent before 1945. Building code and land subdivision regulations were found in only a few locales. Land-use planning and development controls became widespread only in the 1950s and 1960s. They were implemented to serve several objectives: for example, consumer protection, social mix, improved efficiency, reduction of harmful impacts on the environment, minimization of externalities, and preservation of farmland. While regulation has undoubtedly solved some problems, it has also created others. Early post-war suburbs tended to be socially homogeneous, have little physical infrastructure and few services, and be constructed at low densities. Later suburbs tended to be designed for a mix of incomes, with a higher level of infrastructure, and at high densities. There have been complaints about the uniformity (that is, lack of diversity) of suburban development, high densities, and enforced mixing. A growing uneasiness about post-war land-use planning is another lesson learned; the complexity of urban life and human aspirations makes it difficult, if not impossible, to regulate into existence a satisfying suburban environment.

We also have learned of the contradictory objectives of zoning (see Stach 1987). In the early post-war years, there was a sense among home owners that zoning restrictions were written in stone. If one bought a lot in this new development, one could be assured that all surrounding lots would be restricted forever to the same usage. However, this fixity later partly dissolved. Planners used uneconomic zoning as a bargaining tool to get developers to provide other concessions. Developers viewed current zoning restrictions as an initial negotiating position, and bewildered residents saw the changes as attacks upon their neighbourhood and their own property values. As the complexity of the regulation spread (in land subdivision, zoning, development, and demolition control), negotiation and resolution increasingly bewildered developers too.

Out of that experience came another lesson. Beginning with the slum clearance and urban expressway projects of the 1950s and 1960s, some consumers in metropolitan areas began to band together to protect their neighbourhoods. With the spread of socially mixed housing and increasing traffic densities, the phenomenon soon also engulfed the suburbs. Residents' associations became a significant new political force that created both benefits and costs for society. One lesson learned was the importance of the neighbourhood in determining the quality of life possible in a particular dwelling and the feasibility of political action to preserve it. Another lesson learned was the difficult problem faced by governments that have to balance the interests of existing residents and their would-be new neighbours.

A lesson still in the making has to do with the cost of regulation. Over the post-war period, the regulation of housing has increased dramatically. Some, perhaps all, of this regulation has been beneficial. However, in the last decade or so, there has come to be a growing awareness that regulation is not costless. In the case of housing regulation, the available evidence is far from complete. However, there are continuing claims that the costs of regulation (nominal costs plus time delay) may be high and that, while there may well be benefits to society (for example, improved health, safety, or efficiency), every regulation should be reviewed to ensure that its benefit exceeds its cost. For consumers, one implication is that regulation that is inefficient in the above sense pushes up the cost of housing and/or reduces housing affordability.

COSTS OF SOLVING HOUSING PROBLEMS

Another lesson learned is that the cost of "solving" housing problems can be high. Virtually every housing program developed in post-war Canada has, directly or indirectly, a significant cap on total expenditure. For example, most public rental housing has been restricted to low-income families with children and/or senior citizens. Until recently, non-elderly singles and couples typically could not apply, regardless of financial condition. As another example, entry to programs such as shelter allowances is typically limited by availability (for example, they are open only to the elderly) rather than need. Universality is almost unheard of. Almost all housing programs have been targeted at specific groups or situations. In part, targeting is used to separate "problem" households from others. In part, though, targeting is also an arbitrary measure designed to limit the government's financial exposure. This raises questions of horizontal inequity because two similar "problem" households are treated differently when one is arbitrarily prevented from participating in a housing program.

In part, we have learned that the high cost of solving housing problems is related to patterns of subsidies among and within households. Suppose the separate accommodation of a low-income individual (such as a student or senior citizen) is subsidized by a family living elsewhere. Suppose that this individual then moves into a publicly-subsidized unit. The subsidy paid by a government typically supplants the amount formerly paid by the family. A similar substitution occurs when an elderly parent moves out of a child's home (where a low or zero nominal rent reflects an implicit subsidy) to a subsidized senior citizen apartment. In part, the high cost of housing programs to governments is a result of such subsidy substitution.

There is a related lesson about deinstitutionalization that we have been slowly learning in the past few years. With the relative reduction in Canada's institutionalized population come new housing policy responsibilities. Persons with disabilities, for example, need parallel support services if they are to make their way as normal members of a community. In general, the services they would have received in an institutional setting are not available in the community. Some of these services are best provided centrally, requiring that clients be located nearby. Other services are best provided in-home. In other cases, clients are best settled in special housing or group homes. The integration of these people into the community is one important aspect of social mix.

We have also learned something about the adaptability of the existing stock. Since 1945, there has been a switch in emphasis in major urban areas from demolition to renovation.[14] In our cities are many examples of older buildings and neighbourhoods preserved or restored, of ambitious attempts at low-rise intensification, and of conversions and deconversions of structures that reflect the needs of new inhabitants. At the same time, renovation can sometimes be simply too costly an alternative. Also, location is ever important; having a fine old structure may not be enough if it is in the wrong place.

Most governments and consumers naturally want housing produced as inexpensively as possible. So too do housing producers who see a loss of consumers if their housing becomes less affordable. However, we have learned that it is easy to be led astray by a short-run view of what constitutes inexpensive housing.

Housing is a capital stock with a long life; for housing to be efficiently provided, we have learned to take a correspondingly long-run view of its costs of construction, maintenance, and renovation. Public rental housing, for example, is now generally built according to standards for durability that meet or exceed those in the private sector. In building codes, there has been experimentation with cheaper alternatives for all residential construction; some substitutions proved workable, but others (such as switching to aluminium from copper wiring) proved unfeasible because of fire or safety hazards or limited durability. Provinces and municipalities have also explored land-use and zoning changes that increase densities and reduce land costs. The problem here is in distinguishing between what is cheap and what is efficient. It is sometimes argued that well-built housing, while initially more expensive, costs less to maintain over the long run, is more adaptable to other uses in the future as the need arises, or has a longer useful life. However, this is a dubious generalization. In some cases, it is expensive to renovate older structures. At the same time, renovation and demolition-plus-new-construction impose different social costs (and benefits) in the surrounding neighbourhood and in the community as a whole. It is difficult and unwise to generalize; the desirability of building quality and adaptability into the stock must be assessed in terms of the potential costs and benefits individually in each case.

HOUSING AND INCOME MAINTENANCE

Another lesson that is still being learned concerns the impact of post-war income maintenance schemes. In part, the housing affordability problem arises because some households do not have a sufficient income to afford the basic requirements of life. Since 1945 per capita incomes have risen sharply, especially in comparison to shelter costs. In that respect, housing affordability problems should have declined. However, averages mask important changes for particular households. As important were changes in income maintenance schemes that assure a household a steady disposable income over its lifetime: for example, unemployment insurance, family allowances, workman's compensation, old age security, the guaranteed income supplement, Canada/Quebec pension plan, medicare, and a variety of private pension and long-term disability insurance plans. In addition, the entry of married women into the labour force helped stabilize family incomes. The result was that relatively more households could expect to find housing affordable throughout their life span. For this group, housing policies aimed at improved affordability became less urgent.

The curious twist is that post-war prosperity created a new class of poor. By providing subsidized housing, medicare, and a variety of other subsidized goods, governments encouraged the formation of separate households among groups (such as the elderly and lone parent families) who had not previously been prone to living alone. However just the cause, these new households typically had low incomes relative to shelter costs and augmented the affordability problem. Governments are now only beginning to learn the curious lesson that affordability might be inversely related to subsidization.

Finally, a lesson still in the making concerns the fundamental goals of Canadian society and the desirability of housing subsidies *versus* cash grants or other tools as a means for achieving them. Analysts and policy makers, for example, continue to wrestle with the question of whether there is a unique role for housing policy or whether housing affordability problems are simply a manifestation of inadequate income (see, for example, Bourne 1986).

ROLES OF PUBLIC AND PRIVATE SECTORS

This insight points to a broader lesson. In the early post-war period, governments saw their role as fixing up market shortcomings and helping those who were left behind by the market. In effect, even though from 1945 into the 1960s from one-third to one-half of all new construction was assisted by NHA financing, government policy essentially worked at the margin (that is, leftovers) of the market and did not directly compete with it. In the ensuing decades, a more activist role emerged in which governments increasingly began to interact, if not compete, more extensively with the private market. Senior citizen and low-income family rental housing, for example, in some sense competes with an available private stock. At first, the competition was weak. What governments built was far better than what was in the private market. Later, as the worst part of the private stock was culled out or withdrawn and the extent of subsidization declined, the differences between the two stocks became blurred. Increasingly, governments may find themselves competing with the private sector for the same group of clients. One lesson learned is that governments will increasingly have to decide just whether, why, where, and how they will choose to compete with the private sector.

Finally, a lesson has been learned about uncertainty and its implications for housing policy. The tools of housing policy consist of carrots and sticks. We may perceive a problem – even a solution – but be unable to correct it. In part, this could be because of a misperception of the problem. In part, the carrots and sticks may not be sufficient for the job. In part also, it may also be because of an interaction between the private market and public policy. We set out, for example, to house existing elderly households better, only to find out that the number of elderly households increases all the more rapidly as a consequence. If there is one overall lesson, it is that policy makers should understand what they hope to achieve, the feasibility and means of achieving it, and the market conditions within which the policy will operate.

Notes

1 Vischer (1987) further discusses respects in which post-war suburban design principles have withstood the test of time.

2 In saying this, I do not mean to gloss over the real problems that modern suburban and residential design have posed for many women. See Hayden (1984).

3 Gluskin (1976, 134) estimates that the ten largest publicly-owned builders in Canada produced only 7.4% of the new dwellings sold in Canada in 1976. Gluskin also reports that Cadillac-Fairview Corporation, at the time the largest corporate landlord and focused on the Toronto market, possessed under 6% of Toronto's total stock of rental apartments (1976, 115).

4 Of course, the same might be said of large firms in any industry. The landlord industry is particularly sensitive because the variety of differences among buildings make it difficult for senior managers to develop good predictions of expenses appropriate to any one particular building.

5 For a description of the historical evolution of house building technology and design up to about 1920, see Doucet and Weaver (1985).

6 These are described in more detail in Boyle (1986).

7 In the mid 1980s informal and formal rentals appear to account for almost one-half of all residential condominium units. In regard to the formal sector, Skaburskis and Associates (1985, Table 2) found that of 61 condominium projects in a random sample drawn from across Canada, 14 had to be replaced because they were syndications. According to the 1984 FAMEX, roughly another one-quarter of the condominium stock is occupied by renters in the informal sector (that is, where the tenant knows that the building is a condominium).

8 The labour force participation rate among persons aged 15-19 years rose from 42% in 1970 to 57% in 1988. See Statistics Canada (1989, 242).

9 There is a paucity of good data on owned second homes in Canada. While such dwellings are presumably a small proportion of the total stock of private dwellings, undoubtedly an additional substantial proportion of families can lay claim to renting such accommodation or to sharing accommodation with owners who are relatives or friends.

10 Many analysts also put a focus on home ownership in assessing the extent of housing progress. See, for example, Myers (1982).

11 One should also mention the introduction of a variety of short and medium-term housing for battered wives in this regard.

12 It is of interest to note, for example, the devolution of ownership of public rental housing over time: from WHL, through the provincial housing corporations, to municipal agencies. However, what might be best for the administration of public rental housing may not be best for other facets of housing policy.

13 This phenomenon has not been limited to Canada. Popper (1988) describes the increasing extent of centralized regulation in various parts of the US, particularly since the late 1960s.

14 In part, the change in emphasis may simply reflect the fact that much of the old urban stock that was not repairable had been demolished by the end of the 1960s. The stock remaining thereafter was typically of better quality.

═══

CHAPTER TWENTY-TWO

Current and Future Challenges and Issues

John Hitchcock

THIS MONOGRAPH suggests two dimensions to the current housing policy environment relative to post-war experience: type of problem and degree of complexity. These dimensions permit us to contrast the situation facing policy makers at mid-century and that facing them near the end of the century. In the early post-war years, housing problems seemed physically large but conceptually simple; the backlog of housing problems from depression and war might be likened to the Augean stables, and the task of housing policy was one of the labours of Hercules; the task was enormous but clear, and it was expected that, when the task was done, it was done. The progress in housing quality over the post-war period has been unmistakable. While much can be attributed to general prosperity, a significant contribution was made by improvements in mortgage financing as well as by policies related to specific housing issues such as social housing and rehabilitation. The problems we now confront are more complex: resembling the labours of Sisyphus more than Hercules.

This monograph argues that housing is a necessary, but not sufficient, element of a livable, successful community and that we must look at the conjunction of housing and neighbourhood conditions and services, considering the health, home support, social, and daily convenience needs of different households. As an example, the housing needs of ex-psychiatric patients cannot be divorced from their non-housing requirements. The relative lack of official recognition of, and response to, the needs of this particular group during the past decade has not been to our credit, though to take a charitable view, it suggests the difficulty of linking housing and other needs within our existing political and organizational environment. Policy must consider a matrix of housing and service needs, and we have to learn how to live with the organizational complexity inherent in that.[1]

Some people may be tempted to think that an eagerness to view things broadly simply reflects success in solving earlier tough problems, that perhaps we are dealing now with the less important ones – as if Hercules were casting about for a make-work project. However, housing policy has never fully acknowledged the links between housing and economic development, land-use control, social services, and income security, in part, precisely because the links are complex. What we now confront is not so much "new" complexity as the revelation of a tough problem that has been here all along.

However, there are also genuinely new complexities. Compared with the 1940s, for example, the range of household types that exist in significant numbers is now larger. There is also increasing recognition of the different kinds of "special needs" that exist in the general population.

Today, we are dealing with a very different urban system. Post-war urban growth was initially fuelled by overall population growth and rural-to-urban migration. As these declined in importance, the growth prospects of an urban region have become more dependent on the regional economic and income differentials that affect interurban migration, adding new volatility to changes in economic and demographic characteristics across the urban system. Variations in economic conditions are now also intimately linked to the global economy, heightening the probability that Canada's urban regions will be abruptly affected by events abroad.

This volatility has two implications. First, local demand can change rapidly relative to the local supply of housing, straining the capacity of regions and local government to respond. Second, there no longer appears to be one over-riding national housing situation. Each region, and perhaps each urban subsystem within each region has, or may soon have, a different housing problem. This sensitivity to change is amplified by changes in household structure, since the formation of non-family households is more sensitive to economic conditions than is the formation of family households.

Forecasting urban change – never easy – becomes more difficult in this kind of environment. Also important is the strain on our policy-making machinery imposed by this regional diversity. In its early years, CMHC provided policy leadership and the bulk of the financial resources directed to housing. It dealt with a limited set of concerns, and in a climate of a commonly-recognized national housing problem. As the chapters in this book attest, [2] it requires mental gymnastics today to consider simultaneously the issues of rural housing in Nova Scotia (Rowe 1986), the problems of lost home-owner equity in declining resource towns, the unique characteristics of Montreal's rental market (pointed out by Choko 1986), and revitalization in selected metropolitan areas. Regional differences have always been with us, but the expanding range of differences and rapid changes in trends add to the difficulty of understanding and responding.

Finally, Chapter 7 underlines the increasing regulation of housing, reflecting an increasing awareness of the complex ways in which housing has an impact on other areas of concern, such as safety and environmental protection. The increased regulation has, in turn, added to the complexity of the task of framing and implementing housing policy.

Permeating this volume is a vision of the role of housing policy which suggests that housing should not be thought of simply as a problem to be solved;

instead, it is an ever-present range of concerns that will always occupy us, with particular concerns varying in salience over time. It is sobering to consider two areas of concern current in the late 1980s: management of the existing stock of housing and affordability. As the rate of overall population growth and rural-to-urban migration have declined, the existing stock has become more important. We are less able to accommodate change by means of new construction at the periphery, as was done in the first decades of the post-war period. It is no longer sufficient to determine that we need x new housing units of type y. We now have to consider carefully where those units are to be located and how they will be created.

We also have had to reorient our thinking from large-scale new construction on "greenfield" sites to a range of alternative ways of supplying additional housing and to recognize that past policies to deal with the existing stock of housing on a neighbourhood by neighbourhood basis have had a mixed record. The urban renewal programs and their successors left much to be done (or undone), and our current puzzlement over revitalization promises no quick fix in this respect either. Because structures seem so solid, we are lulled into thinking of the existing stock as something fixed. If we think in terms of occupancy, however, it is easier to picture our existing stock for what it is: an ever restless sea (and perhaps the baby boom as a rogue wave). Even if nobody moves, households age and their needs change.

Affordability has never ceased to be important, but its importance has increased as the salience of inadequate physical condition has declined. In spite of laudable efforts to produce social housing in the post-war decades, affordability problems appear to be as compelling and difficult as ever (in part because of unanticipated changes in the existing stock). These issues will be on the agenda for a long time – a task worthy of Sisyphus. The first challenge that the authors of this volume set for us, then, is to adjust our assumptions and expectations to reflect what "housing" means in the late 20th century.

The main linkages and complementarities among the chapters of this volume are reflected in five areas: tenure, management of the existing stock, adequacy, affordability, and support services. Within each area, there are at least five generic challenges.

- the intellectual challenge, that is, to understand the phenomena under investigation;

- the need to relate that understanding to the quality of people's lives, that is, to see the relationship of complex processes to "flesh and blood";[3]

- the challenge continually to re-evaluate conditions, drawing on a never-complete intellectual understanding;

- the continuing challenge to improve the structures and processes for implementing policy; and

- the challenge to commit resources to solve housing problems, knowing that there are competing demands for those resources.

Tenure

We have reason for satisfaction with our success in providing opportunities for home ownership, a form of tenure whose benefits include security of tenure, control, and savings. On the other hand, there appear to be questions about progress with respect to private rental tenure. The percentage of households owning now declines consistently from the highest to the lowest income quintile, whereas two decades ago there was less imbalance across income groups. For many low-income Canadians the choice is now private rental or (for a small percentage) social housing. Because many problems appear to be localized within the private rental sector, it is convenient to make it the focus of attention. However, the source of problems does not necessarily lie within this sector.

THE PRIVATE RENTAL SECTOR

The change in demand implied by low incomes among renters can be seen as a portent of decline in the private rental sector. This concern is echoed by other commentators in North America: "Notwithstanding an increased array of governmental policies to promote the rental sector, its future is clouded in most industrialized countries" (Howenstein 1981, 108). In keeping with our general challenge to see housing as a continuous voyage rather than a single problem, however, we should be cautious in accepting the view that the rental sector is a "Humpty Dumpty" that has fallen off the wall – never to be put back together again:

> By ... 1982, it was widely accepted that private renting was in a seemingly inevitable decline and that it would no longer provide the "normal" lifetime tenure for a large proportion of the population.... But any explanation ... which relates the decline of private renting and the growth of the other tenures to an analysis of the changing social relations of housing must also acknowledge that some of the conditions which brought about the growth of social rented and owner occupied housing are now undergoing major changes. There is nothing inevitable about the rise of these tenures or their continued significance – despite the popularity of naturalistic explanations of their growth (Harloe 1985, 310, 314).

That the rental sector is currently in decline seems plausible, but our emerging awareness of complexity should induce caution. Table 4.3 indicates marked changes in the income profile of renters from 1967 to 1981. The proportion of all renter households that were in the lowest two income quintiles increased from 44% to 57%. On the other hand, rents have not risen as rapidly as prices in the owner-occupied sector. With a relative decline in price, is it surprising to find an increase in renter households with low income? Is this the same thing as saying that the rental housing sector has become a "residual" one?

Table 22.1 shows a difference in the 1967-81 change between family and one-person households. The proportion of single family renter households in the lowest two quintiles increased by almost 15 percentage points, while the comparable figure for one-person households was under 8 points. In 1981 the proportion of one-person renter households in the lowest two quintiles was less than that of family households. Renters include those in social housing, and until recently, one-person households other than seniors were largely ineligible for social housing. Table 22.2 indicates the growth in the social housing stock since 1951, particularly in the 1960s and 1970s. Some of the changing income profile and the changing mix of that profile as between family and one-person households resulted from an increased demand for housing among low-income households made possible by an enlarged inventory of social housing.

Another row from Table 22.2 indicates that the total number of assisted units, including both social housing and a variety of forms of assistance in the private rental market, has increased in the past two decades. The implication of such data is not that the rental sector has been in decline, but that it has not declined because it has been "propped up" by subsidies in both the private and public sector. Is the increased support for rental housing an indication of failure in the autonomous operation of a market, or is it a success in terms of a public response to need?

Table 3.1 provides an added perspective on the rental sector. Renter households increased as a share of all households from 1961 to 1971 and declined slightly from 1971 to 1986. As the total number of households increased over this period, the number of renter households also increased; this does not suggest that the importance of the rental sector has declined.

A different view of decline emerges if we consider the proportion of the population served by rental housing. To some extent, the decline in tenant incomes is a reflection of the marked decline in renter household size – implying, on average, fewer earners to contribute to the total household income.

POSSIBLE ALTERNATIVE FORMS OF TENURE

While acknowledging the ambiguities in these trends, there is enough evidence to challenge us to re-examine the assumption that private rental should bear the main burden of housing those below the median income, with social housing filling in gaps. This involves questions of political values as well as programs and mechanisms. With respect to the latter, we should take up the challenge given by David Donnison in the first Canadian Conference on Housing:

> [Y]ou should beware of too great a unification of ownership and too great a standardization of policies in the public sector. The government, just like the developers and lenders in the open market, tends to give priority to some needs and neglect others. It may be the newcomers to the city, mobile households, the single or the childless, but we may be sure that minorities of some kind will be neglected

Table 22.1
Renter households by income quintile: Canada, 1967-81 (%)

	1967	1973	1977	1981	Change 1967-1981
(a) Family renter households					
Quintiles 1 and 2	43	50	54	57	+15
Quintiles 3 and 4	43	40	36	34	-9
Quintile 5	15	10	10	9	-6
Total	100	100	100	100	0
(b) One-person renter households					
Quintiles 1 and 2	31	36	36	39	+8
Quintiles 3 and 4	44	42	43	43	-1
Quintile 5	25	22	21	18	-7
Total	100	100	100	100	0

SOURCE Statistics Canada, *Housing Facilities by Income and other Characteristics, 1982,* Tables 6-8; cited in Patterson, 1985.

Table 22.2
Assisted housing and rental starts: Canada, 1951-1981

	1951-1960	1961-1970	1971-1980
Rental housing starts	271,159	721,267	826,433
Assisted rental starts (social + other rental)	47,086	116,289	313,106
% assisted	17.3	16.1	37.9
Social housing rental starts	8,838	73,621	152,902
as % of rental starts	3.3	10.2	18.5

SOURCE Patterson, 1985, Appendix A).

by democratic bodies accountable to majorities. I hope you will ensure, therefore, that housing which can be allocated on social grounds is built and managed by different bodies whose policies and priorities, though carefully coordinated, are not all the same (Wheeler 1969, 238).

To capture our sense of the challenge, Donnison's phrase, "allocated on social grounds," should be understood to mean housing that is intended to serve a social purpose, at least in part, but where there is no *a priori* assumption about ownership or allocation mechanisms. Many policies affecting the private rental sector that have been employed in the past could fit within this general category.

To explore ways to lighten the burden now placed on the private rental sector, we need to think of tenure not as a set of discrete options (owner occupancy, condominium, cooperative, and rental) but as a range of possibilities created by various combinations of (1) land ownership/leasehold, (2) building ownership/leasehold, and (3) type of agent/agency responsible for management. In a phrase that had currency years ago, in addition to social housing we need to consider a significant portion of our housing stock as a "utility." As a rough order of magnitude, "significant portion" might range from 15% of the rental stock to 30% of the total stock, depending on definitions, assumptions, and region. This term captures the idea of a mechanism intended to assure the availability of affordable housing, while minimizing public subsidy and direct management. For this purpose, the choice of method should be pragmatic, not ideological, seeking ways in which much of the full capital and operating costs of housing might be paid by its users.[4] The cooperative program, employing an inflation-indexed mortgage, has been but one interesting experiment in this direction.[5]

Experience with the quality of management in both public and private rental housing suggests room for improvement. The non-profit program, on the other hand, has been better in this regard. While one must be cautious in identifying the magic ingredient (for example, the importance of income mix is not clear), varying the agency responsible for management does seem to make a difference. There may be other combinations of the three tenure dimensions (land, building, and management), in which private management might achieve similar results. It should be recognized that, while alternative forms of tenure present potential advantages, they make it more difficult for governments at all levels to monitor the situation and introduce changes. An essential part of the challenge is to cope with this complexity efficiently.

MAXIMIZING CHOICE WITHIN TENURE CATEGORIES

Within the standard forms of owning and renting, there is a need to maximize choice so consumers can gain the benefits of that tenure without having to accept too many undesirable trade-offs. Tenure is often tied to distinct financial benefits, building forms, and locations (for example, central city *versus* peripheral locations). While there are differences inherent in these two forms of tenure, policy makers must be ever vigilant to limit the magnitude of the differences.[6]

One explanation for the perceived decline in the private rental sector is that ownership can provide investment opportunities (including tax advantages) in addition to housing services, thereby increasing demand for owned housing relative to rental.[7] Minimizing differences might provide increased demand for private rental tenure, and thereby increase the range of rental options. Chapter 3, for example, notes the correlation between ownership and savings; older owners have a higher net worth than non-owners. If these two variables are causally linked, it raises the question of whether deduction of a portion of rental payments from income for tax purposes should be allowed. Such a deduction would not eliminate the differences in financial benefits made possible by non-taxation of the capital gains of principal residences, but it would at least offset the advantage that home owners receive from the non-taxation of imputed rent.

Chapter 3 also notes how inflation has influenced investor demand for rental housing. Reducing the differences between tenures could dampen the rate at which investment decisions shift between tenures, allowing leeway for the system to adjust, and perhaps permitting the maintenance of a wider variety of options within each tenure form.

Finally, Steele has noted a less publicized aspect of the processes that bear on tenure choice: rules (whether formal or informal), which exclude future rental income from part of an owner-occupied unit in assessing loans, may have a bearing on the feasibility of producing rental units from the existing stock. Ensuring that rental income is included would encourage production of rental accommodation, while facilitating the acquisition of owner-occupied property.

We can think of other variants that would reduce differences or increase choice. Tenant management or participation in management might be explored to reduce the advantage of ownership in terms of control over one's environment. The other side of this coin is to develop ways to assist owners in the management of their property. Some maintenance problems of senior owners and others may be a result of deliberate dissaving (the opportunity for which is an advantage of ownership), but some result from inadequate managerial skills or physical disabilities. Imaginative use of condominium tenure in the existing as well as new stock might address this aspect of management. Rent regulation and various forms of social housing have altered the degree of security of tenure available to those in rental tenure. Other options for housing as a utility, as suggested above, could extend this further. Comparable options need to be explored for owners. Housing allowances for owners may be desirable on the grounds that they could cushion an owner against temporary financial setbacks, preserving ownership and avoiding the transaction costs of moving.

There are arguments for developing agencies that could assume "an equity position" in private housing in order to permit the continuation of ownership. In England, "sheltered housing" for the elderly employs a number of legal/financial mechanisms to share capital and operating costs as well as management between the occupant and relatives or some institutional body (Sherebrin 1982). In a shared equity arrangement, an agency might purchase half of a dwelling, an owner-occupant might retain the other half. The agency could benefit from capital gains, in the same way as other owners, thus justifying its continuing investment. The agency's role need only be passive if the property is properly maintained, but it might intervene in various circumstances. This concept might be extended to non-senior families as a hybrid form of social housing. It is one way of dealing with the fact that certain housing types are normally associated with certain tenure types (for instance, it is easier to rent an apartment than a house), and thus might help establish greater symmetry between tenure types. This mixed ownership approach could facilitate use of "sweat equity" as a means of reducing nominal operating costs, since "house form" buildings lend themselves to this more easily than do apartment buildings. At least for younger family households this mixed form might well be preferred to apartment rentals.

Management of the Existing Stock

Compared to the immediate post war era, housing markets must now find, reclaim, or reconstitute a larger proportion of housing units from within the existing stock or from the reuse of developed land. Development of "raw" land typically involves large pieces of land in single ownership where externalities can be internalized, and decisions on trade-offs can be made before any interested parties (the new occupants) arrive on the scene. With the existing stock, however, there are not only more interests per unit of space but also more types of interests to be accommodated. Other things being equal, the longer since the original raw land conversion, the more complex are those interests. Those involved in working with resident groups can attest to the variety of reasons why current residents prefer things the way they are. One can be cynical about this and complain about selfishness and the NIMBY ("Not In My Back Yard") syndrome, but the fact remains that there may be non-trivial reasons for residents "in place" to oppose changes, so we should expect opposition as a legitimate response. This is not to say that existing residents should always prevail over future ones; there is a challenge in finding institutions and mechanisms for dealing with the conflicts that exist and that will continue to exist. Merely decrying resident opposition does not advance our ability to meet that challenge.

Are new institutions and programs needed? At least in areas where changes are particularly rapid, if we are to make use of existing land and/or the existing stock at a sufficient scale to meet future needs, purely voluntary market transactions may be inadequate. The painful experience of urban renewal is still fresh in our minds, however, and there is an understandable reluctance to reintroduce the political tensions in a similar program.

Much discussion has suggested the need to override local zoning or other regulations in order to provide for newcomers, but the principle of local determination is not easily put aside. Is there an inevitable collision course here between two desirable principles? The problem is further complicated because the housing market is extensively regulated at all stages of production and reuse.

At any particular time, the market reflects the balance of existing regulations, taxation, and other forms of government intervention. We cannot arbitrarily say "let the market solve the problem" because it would be difficult to define what "the market" should be. This is a dilemma and a continuing challenge.

Another problem concerning use of existing housing/land is the preservation of existing moderately priced housing. The conversion of older rental units into luxury rental or condominium units, and deconversion of house-form dwellings from multiple rental into owner-occupied units both eliminate moderately priced units and exacerbate the affordability problem. Deconversion, in particular, may be difficult to detect using the normal indicators of building permits and demolitions (City of Toronto 1986), and is difficult to control. We come to rely on converted units from the previous generation, and when they disappear it is a shock to discover that the existing stock does not provide a solid foundation. The way in which new rental units in Toronto, for example, have been offset by losses of rental stock through deconversion might be likened to walking up a down escalator.

These appear to be examples where normal functioning of the housing market creates rather than solves problems. With respect to the existing stock, current public policy initiatives seek ways to forestall such changes. They are designed, in effect, to frustrate – not facilitate – the actions of the market. This is surely a challenge. Are there ways that market forces can be harnessed to preserve housing? If not, we should face the fact and use public intervention to reorganize the market situation. There may be ways to accomplish this by using new forms of tenure. We could, for example, take housing units into public ownership and then resell to the private or non-profit sectors – using public ownership as a means for modifying or limiting rights in particular pieces of property before "relaunching" into the market place. There is a difference, in other words, between perpetual public ownership, and the use of public ownership as a transitional mechanism for providing a public context for a private market.

Arguably, the most successful interventions in the housing market have been those associated with the market for capital. Successive policy initiatives by the national government have developed mechanisms or institutions for directing money into residential development at minimum cost by, for example, reducing risk through use of insurance, and by helping bring private mortgage insurance institutions into existence. The introduction of price level adjusted mortgages and mortgage-backed securities indicates continuing innovation. The capital market, however, is not tied to location; the additional dimension (or dimensions) that land and location entails (brought into focus by the importance of the existing stock) presents a challenge to our ability to create new institutional mechanisms that will not only encourage the kind of development we want but also where we want it.

Finally, it should be remembered that the increasing importance of the existing stock of housing is mirrored in the increasing importance of maintaining/replacing infrastructure. Streets, bridges, and sewer systems have their own life span, and as with the existing stock, replacement or rehabilitation is a more complicated matter than new construction on green field sites. Various levels of government responded to the needs for new sewers, for example, in the 1970s to serve new subdivisions. They may well have to develop new programs to assist infrastructure replacement in the 1990s, [8] and begin to think about how cities can be built to make them more "recyclable" in the future.

Adequacy

The four post-war decades have witnessed strides forward both in the physical condition of buildings and in space occupancy. This is testimony both to the housing policies of the post-war years and to the general prosperity of the era. The current challenge of adequacy, has three components. The first is to maintain the ground that has been won. The relentless aging of our housing always poses the challenge of physical maintenance. The second component is maintaining a reasonable balance between standards and expectations, costs and resources. The third component of the challenge is to recognize the aspects of adequacy that lie beyond the dwelling unit itself.

Our housing stock has its own demographic transitions; the building boom of the post-war era is analogous to the baby boom that it served. A large percentage of our stock was constructed in a narrow band of time. It is this narrowness that will confront us with pressures for repair all at once. It is also important to note that this narrow band of time covers a large band of space. Most of what we now think of as Toronto, for example, (strictly speaking, the Census Metropolitan Area) is a post-war creation, even though the smaller core settlement has existed over a much longer span of time. Within this broad picture, there are particular situations of concern in the high-rise rental stock, where it is feared that needed repairs may raise rents and exacerbate a difficult situation within the rental sector generally. It is to be expected that similar problems exist for other high-rise buildings, such as those in the condominium sector, and in some future period we can expect a substantial bill for infrastructure replacement.

With respect to the second component, it is notable that the question whether our standards are excessive has been asked more insistently in the past ten years than in the previous thirty. This may reflect, in part, the sheer accumulation of regulatory "refuse" over the years. There is a continuing demand for regulation, and it is easier to make new regulations than throw old ones away.

The urge to question excessive standards with respect to the housing stock, however, also reflects the peaking of our post-war increase in prosperity as well as declines (in real terms at least) in income among selected population groups. The proliferation of bachelorette units in Toronto illustrates the dilemmas surrounding the setting and maintenance of standards. Briefly stated, the demand for low-rent (per dwelling unit, not per unit floor area) accommodation in Toronto stimulated the production of rental units that were smaller than the operative zoning by-law permitted, though most fell within the housing standards by-law. The varying standards for minimum unit size in the housing and zoning by-laws in Toronto reflect the differing basis for the standards – one derived primarily from a user/tenant-oriented public health standpoint, the other from a principle of community adequacy. This issue demonstrated both the difficulty of agreeing what the standard should be and of enforcing occupancy standards at the municipal level. This particular debate is far from decided, though the legacy of the recent recessions and other forces bearing on the incomes of single persons (as well as the availability of moderately priced housing) may finally be resolving the conflict in favour of a lower minimum standard. If this is the result, however, we still do not know whether to laugh or cry. In the post-war period we have not advanced in our ability to say where the line should be drawn between reduced standards for legitimate economic reasons and increased standards to protect individual welfare. There is a challenge for theoretical and analytic work to assist policy makers and regulators in drawing the line. [9]

Chapter 12 notes the fact that improving formal standards or regulations for housing, such as those contained in the National Building Code, has become more difficult on a Canada-wide basis, though there remains a national leadership role to encourage continuing improvement. The challenge is to acknowledge both the differing levels of prosperity, local history, and other factors accounting for varying regional differences in expectations and needs – both suggesting "micro standards" that vary over space, rather than one "macro" standard. Of course, the political complexities of setting such standards in an equitable way are large.

Finally, once you have improved housing standards to a certain point, adequacy can no longer be considered as a unidimensional indicator; it has to be recognized as a relationship between a "what" and a "whom." Chapter 12 asks, for example, "To what extent should the definition of physically adequate housing be independent of the needs of the occupying household?" The problem can be illustrated by the simultaneous aging of housing stock and of occupants in post-war suburban municipalities. There comes a point when housing becomes increasingly inadequate in terms of its internal configuration, management requirements, and locational attributes, even though the unit has been maintained in good condition and is occupied by the same household throughout its life. One could argue that the inadequacy has been created by the household rather than the house.

The problem is more complex still, since there are really at least three terms to the equation: the household, the housing unit, and the neighbourhood. Thus, a responsibility for housing adequacy, taken in its full sense, is an imposing mandate. It entails a continuing effort to understand the relation between occupants and (1) the physical housing unit, housing adequacy, (2) the availability and quality of physical facilities and services of various kinds (including transportation and human services), neighbourhood adequacy, and (3) what Chapter 12 calls environmental adequacy (particularly concerned with levels of pollution and crime).

This more complex formulation of adequacy introduces various alternatives for housing improvement without immediately indicating which alternative is best. If the relation among these elements is inadequate, what do we change? An elderly person who moves from a detached home in one area to a seniors' residence in another may gain housing adequacy and lose neighbourhood adequacy. The challenge for future housing policy is to recognize formally what every resident knows informally: housing as experienced encompasses a large number of elements both within and outside the housing unit. All of these elements have to be adequate if the housing as experienced is to be adequate. There is a challenge in translating this conception of adequacy into meaningful indicators and some limited range of feasible options. The fact that we face this new challenge implicitly acknowledges post-war progress with respect to the physical condition of the housing unit itself.

Affordability

Chapter 14 notes that there has been little change in the proportion of income spent on housing in the post-war period, and the gloomy assessment is that "Canada has fewer grounds for optimism about the solution of this housing problem in the 1980s than it had in the 1940s when it embarked on its first subsidized public housing project." It argues for the need to establish a relevant set of indicators for affordability and a longitudinal measure of improvement (or lack of it) in the housing opportunities available to individual households. Moore and Clatworthy (1978) pointed out the advantage of being able to measure not only whether a household was crowded but also for how long it was crowded, and whether the direction of change was up or down. A related notion is evident in a paper by Lilla (1984) giving a longitudinal perspective on income distribution in the United States. Lilla contends that the extent of time a particular person suffers from inadequate income (as defined by others) is a more useful measure of the severity of the poverty problem than the number with low income at any one time. Longitudinal statistics indicated that roughly half of the population that was poor could be described as temporarily poor, and a small proportion was poor for as long as ten years. Analogous measures would be useful to assess affordability problems, particularly if they could indicate the direction of change.

The search for better indicators is sometimes dismissed on the grounds that it is a delaying tactic. Those "in the front line" know there are serious affordability problems now; many, however, do not have front-line knowledge but they have a legitimate need to understand clearly what other people are experiencing. Commitment of resources ultimately requires that those who are not experiencing the problem directly have to be able to understand the statistics in "flesh and blood" terms, to adapt a phrase from Chapter 14, to reassure themselves of the legitimacy of expressed needs. In the last decade, there has been a tendency for normative standards for shelter-to-income ratios to creep upwards from 25 to 30% and more, and one suspects that policy makers facing economic constraints have become a bit cynical about standard indicators. The task, then, is to define measures that, while applying to an aggregate, are somehow reconcilable to the level of individual households. As housing analysts familiar with the data implications will know, this is no mean challenge.

RELATION TO ADEQUACY

While it is conceptually distinct, affordability cannot be measured accurately without simultaneously considering adequacy. It is meaningless to compare the situation at different points in time in terms of a given standard of affordability if changes in suitability and physical condition are not accounted for. The record with respect to the condition of housing is generally good, but there is an irresistible temptation to deal with affordability without having regard to neighbourhood and environmental adequacy. This sort of general issue has particular poignancy when decisions need to be made as to whether low-income households should receive a subsidy (or an additional subsidy) to permit them to reside in central areas close to employment and social services of various kinds. Housing in less-favoured locations is cheaper.

RELATION TO INCOME SECURITY

The linkages between housing policy and income security need to be sorted out. If sources of income available to classes of households are below generally defined standards of income adequacy (as is the case for many of those on welfare assistance, for example), the resulting affordability problems are predictable and are not uniquely associated with housing. Conspicuous among those with affordability problems now are single parents, a group less evident in the types of households with affordability problems forty years ago. Changes in household formation and dissolution, then, have an independent bearing on the situation. Relative to thirty or forty years ago, the percentage of households with affordability problems may be depressingly similar, but single parent families represent a new problem – one that has not been fully addressed. To use housing policy to deal with it may have short-range advantage in terms of quickness of response, but by now it should be evident that there is a need for imaginative development of social policy across concerns such as access to economic opportunities, income security, and family law, as well as housing.

AFFORDABLE WHERE?

Policies with respect to housing markets have to recognize that housing market processes may cause affordability problems as well as solve them. An obvious, yet too often neglected, aspect of affordability problems is that they vary from place to place: among metropolitan areas and within metropolitan areas. Some housing markets, notably those in the larger and faster growing metropolitan areas, are more expensive than others. This is hardly revolutionary, but it indicates that affordability is tied to regional differences in economic activity. Economic growth has its social costs, and frequently one of these costs is increasing affordability problems for low and moderate income groups. Put simply, people migrate to employment prospects more easily than housing units "migrate," and the consequent increases in prices are not matched by equivalent increases in income across all income groups. This kind of affordability problem is a consequence of the fact that housing is tied to land; it comes from the price premium paid for a location. Somewhat different processes, associated with central city revitalization, have reduced the availability of housing for low-income groups; again locational premiums can be said to play a role in the creation of affordability problems.

We need to ask not just whether housing is affordable, but whether housing in a specific place is affordable. We need to ask whether, when affordability problems occur in that place, alternative locations can provide the housing services needed.

In the absence of satisfactory alternatives for households no longer able to afford to live in the central city, we have a limited range of alternatives.

- Reduce the level of demand that causes the location premium. It has been proposed, for example, that governments remove incentives for development in central areas that are gentrifying and stimulate demand in areas that are less active. It would be difficult to do this in a targeted and sensitive manner, however.

- Increase the supply of housing in the affected location, though frequently it is impossible to increase the supply sufficiently to lower the price, in part because new housing is more expensive than old.

- Look for efficient ways to offset the premium for affected households by use of housing allowances; however, by increasing effective income we may increase demand. Since the problem is already too much demand (in the sense of creating affordability problems, at any rate), this may be equivalent to putting out a fire by throwing gasoline on it.

- Offset the location premium by use of the social ownership of land. As suggested in the section on tenure, this could include a wide variety of public or non-profit sponsors and could be coupled with a variety of forms of building management and ownership. Social ownership of land makes it possible to forego charging increases in the location premium that occur subsequent to purchase. [10] There seems to be no way to use this approach, however, without some form of bureaucratic selection of those who are to benefit, and this always seems to have its costs.

None of the alternatives is ideal, but we are not going to be able to address this particular affordability challenge unless the problem of the location premium is frankly acknowledged – something that national and provincial policy makers seem reluctant to do. The problem frequently ends up in the lap of municipalities, and they typically cope by means of land-use controls – the only real tools at their disposal.

Formal and Informal Support Services

Demand/need for housing embraces both the housing unit and various housing services that a particular location provides. There are trade-offs between improving housing by relocating a household where services are provided and providing services previously unavailable at the original location. While this has always been the case, the evolution of post-war housing and communities has made this more important for three reasons. First, significant improvement in housing condition over the past forty years have allowed us to give greater attention to questions of neighbourhood and environmental adequacy. Second, the proportion of single parent, elderly, and non-family households in the 1980s is higher than in the 1940s and 1950s, and these kinds of households tend to be more dependent on their surrounding environment than those comprised of nuclear or extended families. [11] Third, the suburban form of communities, with lower densities and greater dependence on the car has meant that many services are further removed from households and less accessible by public transport than has been true of the pre-war city. Making services available in the suburban context, therefore, requires deliberate effort.

Support services, then, may be provided at the scale of the building in multiunit buildings, the neighbourhood, or the municipality, and they may be of a formal or informal nature. Chapter 13, for example, points out the way that "plexes" in Montreal facilitate informal interaction by permitting living arrangements that are close but not too close. The management of multi-unit buildings, whether rental, cooperative, or condominium also falls within this realm. Chapter 18 suggests that management in non-profit buildings has, on average, been better than that in private rental accommodation. This needs systematic verification but suggests that progress has been made with respect to management; in fact, the various forms of non-profit housing in Canada are an international success story. It is disturbing, therefore, that recent policy changes affecting the income mix approach to non-profit housing may not make it possible to build on this success.

The challenge of support services is to recognize that both formal and informal community service networks are essential considerations in framing housing policy. One can envision scenarios in which housing (in the sense of housing services) might be greatly improved without directly dealing with physical shelter, for example, by introducing daycare or frequent public transit service within easy walking distance of a dwelling. This is easy to say but difficult to do, given typical political and bureaucratic management structures, and given the typical organizational divisions between provision of services ("software") and production of housing ("hardware").

We need to improve our understanding of the role that community and neighbourhood play in the housing experience. A number of chapters in this volume express hunches not only about the importance of community but also about changing preferences for community setting. They suggest, for example, that central city residential settings offer more community support than suburban ones. While plausible, this is not the same thing as demonstrating that preferences have changed all that much (Hitchcock 1984). One frequently encounters references to the "greying of suburbia," for example, reflecting the phenomenon of "aging in place" in post-war suburbs and the concomitant changes in needs for services; but what are the implications of this phenomenon for future public policy about urban development? In cities where suburban expansion continues apace, is there evidence of flagging demand for the basic post-war suburban pattern? Are the housing analysts wrong, or does a lack of genuine choices prevent people from expressing their true preferences for central, city-like neighbourhoods, in the market? This question has been on the North American agenda for some time, and it is a matter of urgency to answer it.

Concluding Themes
The federal government has taken the lead in housing matters in the past and continues to have a role in funding. While it has devolved much responsibility for program delivery to the provinces, this has been done at the initiative of the federal government and has put an indelible stamp on the kind of programs that the provinces mount. Furthermore, while the provinces have had at least twenty years experience in the housing business, that experience (and expertise and enthusiasm) is not evenly spread across the provinces.[12] Whatever the formal constitutional arrangements may be, the federal government will still be held accountable for equitable treatment of all Canadians and so will always be seen as having a moral responsibility for national housing policy. Principles of equity and political credit suggest that the federal government will, over the long run, need to retain capacity for program implementation as well as funding. The activities that need to be undertaken to establish progress in the future concern housing software, aspects of location, and reuse of previously developed land, all issues that are difficult for federal policies to address comprehensively. New forms of tenure, for example, can be facilitated by federal policies, but they can only be fully implemented by agencies at the scale of the regional housing market. In this respect a provincial centre of gravity for housing policy makes sense. At the same time, municipalities know best what the problems are; much innovation in post-war housing has come from the central cities of the major metropolitan areas, even though they have few resources for implementing ideas on their own.[13] Recognizing both the inherent conflicts among the three levels of government and their need for joint action, is it possible to evolve productive relationships rather than simply repeating old routines?

At the regional market level, the challenges are many and complex. It is difficult to conceive the various alternative relationships between, say, improvement of services in area A for the elderly *versus* construction of new or reconstituted housing in area B; it is more difficult to establish the cost parameters of such alternatives; and perhaps it is most difficult to develop governmental mechanisms that can treat such alternatives – typically the work of different ministries – as part of one coherent problem. As noted earlier, accidents of time and geography focus what are essentially regional housing market problems upon particular jurisdictions at particular times. Of particular concern is the inherent linkage between policies related to the existing built perimeter – neighbourhood improvement, infill, accessory units, redevelopment, and similar concerns – and those affecting new construction on raw land. In many metropolitan areas, there is no way to address the alternatives that those linkages raise within a common frame of reference, since one jurisdiction may be completely built up, while another contains all the developable "raw" land. The importance of housing policies related to the existing stock, however, gives new urgency to the challenge of developing a comprehensive view.

The final two themes can be stated briefly. First, in most policy areas discussed above, there is a tension between the problems solved and the problems created by governmental regulation of one kind or other. There seems no easy fix. In spite of what some may say, much regulation governing housing has a valid and important purpose. Equally evident, however, is that regulation imposes costs. Second, a number of the points made earlier stress the increased need for current, reliable data, available at both aggregate and disaggregate levels. This is analytically demanding and expensive but necessary.

The challenges that face us would undoubtedly be familiar to the housing analyst of, say, 1950. Yet their shape and inter-connections are different; they are more complex, focusing on the kinds of services housing provides rather than just physical condition and on the relationship of the prices of those services to incomes; they are more concerned with the problems of managing a large stock of existing housing; and because of these complex interrelationships, they are more concerned with the inter-connections among governments and government departments, and among governments and private and third sector agencies. It is a large agenda.

Notes

1 Similarly, Chapter 3 highlights the importance of the link between housing policy and income security: notably the role of home ownership in providing a mechanism for saving and protecting savings. Chapter 6 extends the link by speculating that saving via housing might be more efficient than registered retirement savings plans.

2 This was perhaps even more forcefully shown in the seminar discussions leading to the preparation of this volume.

3 Difficult as this is, it is essential in our society that politicians and the public have an intuitive grasp of the appropriateness of policies.

4 As is discussed under affordability, this would mean particular attention to the first dimension of tenure: land ownership.

5 At an annual conference of the Canadian Association of Housing and Redevelopment Officials, attended by the writer, there were several discussions as to whether the then new cooperative program, making use of an indexed mortgage, was a market or a social program. We take the willingness to accept this ambiguity as a hopeful sign.

6 Chapter 6 gives us an example of the way this has occurred in the past, with the differential eligibility of owned and rental property for mortgage insurance.

7 The decline in interest among investors in rental property comes in part from producers moving to meet the demand for owner occupancy.

8 See National Research Council, Committee on Infrastructure Innovation for research issues.

9 See Chapter 14 for a suggested two-step approach to this problem.

10 This approach can apply to single units or to entire neighbourhoods as suggested in the proposal by Wolfe (1985) for a community land trust.

11 Single person households, for example, have no one to turn to within the household for aid or support; single parents have no adult support.

12 The cooperative housing sector, for example, fought hard to prevent devolution of cooperative programs to the provincial level because of fear of lack of provincial support (*Canadian Housing*, 1987, 18).

13 Feldman and Graham (1979) review the "plight" of the municipalities.

APPENDIX A

Glossary

Prepared by John R. Miron
with the assistance of Nancy Thompson and Leigh Howell

AHOP Assisted Home-Ownership Program (1973-8, benefits continuing 1984)
Federal program to promote home ownership among lower-income families with children by reducing mortgage financing costs. Loans and grants were provided to new home buyers for the purchase of modestly-priced housing. About 40,000 households were assisted over the 1973-75 period before the plan was modified to provide strictly GPMs. Just over 94,000 GPMs were approved from 1975 to 1979.

ARP Assisted Rental Program (1975-8, advancing in some cases to 1995)
Federal program to assist in the production of new affordable rental housing in the private sector. Annual operating subsidies and, later, interest-free loans were granted for the purpose of keeping rents affordable. Over 122,000 rental units were produced under ARP from 1975 to 1980.

Bruce Commission Report (1934)
This report assessed housing conditions in Toronto's blighted areas and recommended the need to eliminate Toronto's slums and to provide low-cost housing (meeting minimum acceptable standards) for those displaced.

BETT Program Building Energy Technology Transfer Program (1980-6)
The BETT Program was administered by the federal Department of Energy, Mines, and Resources to accelerate the development and adoption of energy efficient equipment, materials, techniques, and systems used in the construction of buildings.

CAP Canada Assistance Plan (1966-)
CAP is a federal program of cost-sharing with the provinces for social services, including welfare assistance.

CHRP Canada Home Renovation Plan (1982-3)
Federal plan to stimulate employment in the building sector and encourage the upgrading of housing. Grants covering 30% of the renovation costs (up

391

to a $3,000 maximum) were made available. Approximately 121,000 home owners received assistance under CHRP.

CMHC Canada Mortgage and Housing Corporation (1946-)

Canada's national housing agency was first established as the Central Mortgage and Housing Corporation and renamed in 1979. CMHC is the crown corporation responsible for administering federal housing legislation (such as the NHA).

CMRP Canada Mortgage Renewal Plan (1981-3)

Federal plan to cushion the impact on home owners of higher interest rates upon mortgage renewal. CMRP was replaced by MRPP.

COSP Canada Oil Substitution Program (1980-5)

Federal program to encourage conversions from oil-fired space-heating and water-heating units to energy systems which do not use fossil fuels. Grants covering half of the material and labour costs involved (up to a maximum of $800 in the case of single residential units and $5,500 for residential buildings with two or more units) were available. Almost 1 million households received grants under COSP.

CPP Canada Pension Plan (1966-)

Federal pension plan (also providing survivor, death, and disability benefits) based on contributions averaged from maximum earned income, which was created to supplement OAS and GIS in all jurisdictions aside from Quebec, where QPP is in effect.

CRSP Canada Rental Supply Program (1981-4)

Federal program to stimulate the production of new rental housing in the private sector. Sponsors of eligible projects could obtain 15-year interest free loans of up to $7,500 per rental unit constructed. Over 21,000 rental units were built under the program.

Canadian Conference on Housing (1968)

This conference held in Toronto was sponsored by the Canadian Welfare Council (now the Canadian Council on Social Development) with funding from CMHC. Proceedings were published under the title *The Right to Housing*.

CCURR Canadian Council on Urban and Regional Research (1962-76)

During its first eight years of operation, CCURR undertook a research program that was funded by a grant from the Ford Foundation. Other activities of CCURR were funded by an annual grant from CMHC. After 1970 CCURR was funded by MSUA.

Canadian Farm Loan Act (1927-59)

Federal act introduced to provide subsidized long-term loans to farmers for farm improvements. From 1935 to 1939 just over 2,000 new homes were built with this assistance.

CHIP Canadian Home Insulation Program (1977-86)

Federal program to provide one-time grants to home owners to improve the thermal efficiency of the existing housing stock. Approximately 2.5 million households received CHIP grants.

CHOSP Canadian Home-Ownership Stimulation Program (1982-3)

Federal program to stimulate the economy and create employment opportunities through the provision of $3,000 grants to first time purchasers of new homes. About 260,000 home buyers were assisted under CHOSP.

CHDC Canadian Housing Design Council (1956-87)

CHDC was established by CMHC to promote improvements in housing and community development through sponsoring research, seminars, and a housing design awards program.

CHS *Canadian Housing Statistics* (1955-)

CHS provides information on residential construction and mortgage lending activity in Canada based on data from CMHC and Statistics Canada. *CHS* replaced its predecessor, *Housing in Canada*, in 1955. From 1955 to 1960 *CHS* was published on a quarterly basis. Since 1961 it has been published annually with monthly supplements.

CIPREC Canadian Institute of Public Real Estate Companies (1970-)

Created to improve accounting principles and standards in publicly-traded companies, CIPREC has since grown to include many privately-held firms and to represent Canada's largest property developers.

CCA Capital Cost Allowance (1954-)

CCA provisions define allowable depreciation expense for income tax purposes.

CA Census Agglomeration

CAs are designated for census purposes and consist of smaller urban areas centred on an urbanized core with a population of 10,000 to 99,999 at the time of the previous census. Their areal extent is largely defined using labour market criteria and includes a central city and surrounding areas that are closely linked to it.

CMA Census Metropolitan Area

CMAs are designated for census purposes and consist of large urban areas centred on urbanized cores usually with populations of at least 100,000 persons at the time of the previous census. Their areal extent is largely defined using labour market criteria (for example, commuting patterns) and includes a central city and surrounding municipalities that are closely linked to it.

CPAC Community Planning Association of Canada (1946-)

CPAC is a national voluntary organization which promotes public participation in urban and regional planning issues.

CSCP Community Services Contribution Program (March 1979, payments extended to March 1984)
Federal program to provide assistance for municipal capital projects such as sewer and neighbourhood improvements. $400.3 million was distributed in accordance with the priorities of individual provinces and their municipalities.

CYC Company of Young Canadians (1966-75)
CYC was established as a federal crown corporation to promote community development efforts through voluntary service in conjunction with the federal government's "war on poverty."

Condominium tenure
Form of property ownership whereby dwellings, typically in multiple unit structures, are individually owned while ownership and management of common elements are shared. In 1966 British Columbia and Alberta were first to pass enabling legislation. By 1970 all provinces but one had a condominium act; Prince Edward Island held off until 1977. The Territories adopted a condominium ordinance in 1969.

Consultation Paper on Housing (1985)
This report published by the federal government was intended to mark the beginning of a fundamental review of Canadian housing policy and led to a statement on new housing directions (*A National Direction for Housing Solutions*).

CPI Consumer Price Index
This index tracks the retail price of a standard basket of consumer goods and services in major markets on a monthly basis (see also FAMEX Survey).

Cooperative Housing
Cooperative housing is a form of ownership tenure whereby multiple-unit dwellings are collectively owned and managed by their occupants (see also NHA Sections 34.18 and 56.1).

CHF Cooperative Housing Foundation of Canada (1968-)
CHF was formed by the Canadian Labour Congress, the Cooperative Union of Canada, regional cooperative councils, and the Canadian Union of Students to promote the development of non-profit cooperative housing.

Core housing need
The core housing need model is currently used to identify households unable to afford adequate, suitable housing in their community without spending more than 30% of their household income for shelter.

Curtis Report (1944)
The *Report of the Subcommittee on Housing and Community Planning of the Advisory Committee on Reconstruction* examined post war housing requirements and called for a more active role by government in the housing sector.

The development of a national housing program for low-income earners and the need for comprehensive town planning were among its recommendations.

Dennis Task Force (1971)

Programs in Search of a Policy (Dennis-Fish Report) was the product of the Dennis Task Force commissioned by CMHC to evaluate federal low-income housing programs. The report drew attention to the government's emphasis on supply-side solutions and contributed to the creation of MSUA. The report was independently published in 1972.

Direct Lending (1954-)

To ensure universal access to mortgage financing, the federal government is empowered to provide direct loans for home purchase in instances where the availability of mortgage funds is limited. CMHC is thereby authorized to act as a lender of last resort. This role was not used extensively. A brief experimentation with a broader role in the mid 1950s ended with the restricted Small Homes Loans Program introduced in 1957.

DHA   Dominion Housing Act (1935-8)

Under the DHA, a $10 million fund was established to help prospective builders and home owners secure loans. Loans were provided jointly by the federal government and authorized lenders. Approximately 4,900 dwellings were financed in this manner before DHA was replaced by NHA in 1938.

Don Mills

Don Mills located in Metropolitan Toronto was the first large-scale, corporate-developed suburb in Canada. Construction began in the 1950s and included high-rise, townhouse, and detached housing.

Double Depreciation Plan (1947-9)

Accelerated depreciation charges over a 10-year period granted under the Income Tax War Act to encourage the construction of rental housing projects. Almost 500 dwellings were completed using double depreciation only while an additional 7,600 units were created in conjunction with NHA loans and rental insurance guarantees.

Emergency Shelter Program (1944-8)

Federal program to provide temporary rental accommodation to alleviate post-war housing shortages. War-time surplus huts and other available structures housed returning veterans entering university and families living on welfare. Just over 10,000 conversions were completed by federal and municipal governments and universities.

False Creek Project

The False Creek housing project located in downtown Vancouver is an innovative inner- city redevelopment scheme built in the early 1980s, which combines subsidized and private sector housing.

FAMEX Survey Family Expenditure Survey

Survey undertaken by Statistics Canada in selected metropolitan areas on a biennial basis in part to update the CPI basket of goods.

Farm Improvement Loans Act (1944-87)

Federal act introduced in 1944 to set a maximum interest rate and provide guarantees on intermediate and short-term loans granted to farmers for farm improvements, including new residential construction and improvements.

FST Federal Sales Tax

Also known as the general manufacturers' sales tax, this tax under the Excise Tax Act applied to goods manufactured, produced, or imported into Canada with some exemptions over the years. Among exemptions of significance to housing were thermal insulation materials, energy conservation equipment, and construction equipment. Building materials became subject to the tax in June 1963 at rate of 4%. The rate for new building materials was raised to 8% in April 1964, and to 11% in January 1965. It was subsequently dropped to 5% in November 1974, then raised repeatedly during the 1980s reaching 9% in January 1990. Often, the tax rate on building materials was less than that for other manufactured goods. The FST was replaced by the Goods and Services Tax (GST) in January 1991)

GPM Graduated Payment Mortgage

GPMs are designed to ease access to home ownership during times of rapid inflation. Initial mortgage payments are low in nominal terms and increase at a predetermined rate. Under AHOP, GPMs were issued by private lenders and insured under the NHA.

GRS Graduated Rent Scale (1944-)

GRS introduced by the federal government in conjunction with its public housing program. Public housing rents initially ranged from 16.7% to 25% of tenants' incomes. Provincial housing authorities began to introduce their own GRS schemes by the 1960s in which rents were typically set at 25% to 30% of gross family income.

Greenspan Report (1978)

The *Report of the Federal/Provincial Task Force on the Supply and Price of Serviced Residential Land* was commissioned in 1977 jointly by the federal government and eight provinces to make findings about the rapid increase in land and housing prices that had occurred from 1972 to 1975.

GDS Gross debt service ratio

Calculation made by mortgage lenders by dividing the monthly sum of principal, interest, and property taxes into the applicant's gross monthly income.

GIS Guaranteed Income Supplement (1966-)

GIS is a federal income subsidy for needy elderly (see also OAS and CPP).

Hellyer Task Force (1968)

The Federal Task Force on Housing and Urban Development (chaired by Paul Hellyer) constituted a rethinking of housing and urban policies. One immediate outcome of Hellyer's task force was the moratorium imposed on large-scale public housing projects. Its findings (published in 1969) and subsequent reports (see Dennis Task Force and Lithwick Report) contributed to the creation of MSUA and the changes in federal housing programs brought in under 1973 NHA amendments (for example, RRAP and NHA Sections 15.1 and 34.18).

HOME (see provincial housing programs)

Home Conversion Plan (1943-6)

Federal plan to alleviate postwar housing shortages in major urban centres. Large houses leased by the government were divided into multiple self-contained units. Over 2,000 conversions were completed under this program.

Home Extension Plan (1942-4, 1946-8)

This federal plan guarantees home extension loans made by lending institutions according to provisions similar to those under the Home Improvement Loans Guarantee Act. Under this plan, 125 loans were approved for 149 conversions.

Home Improvement Loans Guarantee Act (1937-40)

Under this federal act to provide a loan insurance plan for improvements or extensions to dwellings, about 126,000 loans (totalling $50 million) were approved including approximately 4,000 conversions that increased the rental housing stock.

HIL Program Home Improvement Loans Program (1954-86)

Federal program to guarantee availability of private home improvement loans. Initially, eligible recipients could secure loan up to $6,250 ($2,500 per unit and $1,250 for each additional unit) at a fixed interest rate to be repaid (for loans over $1,250) over 5 years. Increases in the maximum loan amount (eventually to $10,000) and amortization period (to 25 years) as well as the use of market interest rates were introduced at various times up to 1979. Over 450,000 dwellings were improved under this program from 1955 to 1981. Use of this program declined starting in the late 1960s and then more dramatically beginning in 1976 after RRAP was introduced. No new commitments were given after October 1986.

HIP Home Insulation Program (1976-81)

Federal program to improve the thermal efficiency of the existing housing stock in Prince Edward Island and Nova Scotia.

Home Purchase Programs

These programs have been available in most provinces and territories to assist low to moderate income earners in achieving home ownership. These programs assume various forms (for example, the provision of grants and interest rate subsidies) and include more unconventional approaches such as Nova Scotia's Self Help Housing Program, that combines the labour skills of prospective home owners (sweat equity) with the technical assistance of professionals, and the Alberta Shell Housing Program, that offers loans and subsidies based on the value of the shell house (that is, a house finished on the outside but not inside).

Home Warranty Programs

Ontario is the only jurisdiction in Canada with a mandatory warranty program. Its New Home Warranty Program introduced in the late 1970s gives new home buyers limited protection against defective materials and construction, including the return of deposits up to $20,000 if a builder enters bankruptcy prior to home completion. In March 1987 limits were placed on the extension of closing dates, and home buyers were given the option to cancel their contract after this time.

HIFE Household Income, Facilities, and Equipment Sample

HIFE is a public use microdata sample prepared by Statistics Canada on a biennial basis. It consists of a stratified sample of data collected from the same households in four separate surveys (Household Facilities and Equipment, Labour Force, Consumer Finances, and Rent surveys). After 1987 the sample was developed annually.

HMIS Housing Market Information System

HMIS is based on the Starts and Completions Survey (SCSS), the Market Absorption Survey, and the Rental Market Survey. These surveys conducted by CMHC monitor new residential construction, the absorption of newly-constructed housing units and vacancy, as well as rent levels in the rental market.

Innovative Housing Program (1970)

Federal demonstration program to promote social, cost, and technical innovations in housing, especially those directed to low-income households. Various subsidies were provided from a special $200 million fund, and many of the innovations were later sanctioned under 1973 NHA amendments (for example, AHOP, NHA Section 34.18, RNH Program, and RRAP).

Integrated Housing Plan (1944)

Federal plan introduced to encourage new dwelling construction (especially owned homes) by offering builders a guaranteed minimum selling price which helped to secure interim financing. CMHC bought all dwellings not sold within a year of completion and allowed veterans the first opportunity to purchase them. In 1947 and 1948 as many as 491 builders took part in this plan and produced more than 5,000 units annually—over 5% of all housing starts.

ICURR Intergovernmental Committee on Urban and Regional Research (1968-)

Joint federal-provincial clearinghouse for information and research on urban and regional matters. Funded 50% by the federal government and 50% by nine provinces.

Jeanne Mance Project

This 796 unit public housing project approved in 1956 was Montreal's first urban renewal project. It is located on a 20-acre site east of the downtown core.

Joint Loans (1936-54)

Loans originally provided jointly by the federal government and authorized lenders on a 25%-75% basis under the DHA (1936). Lending terms and conditions were set by the federal government. Mortgages were issued by approved lending institutions but held jointly by the lender and the federal government. By advancing a percentage of the mortgage amount to the lender at a below-market interest rate, the federal government effectively subsidized the mortgage loan: loans were also guaranteed by the federal government. DHA was replaced by the NHA in 1938, and joint loans were continued until the 1954 amendments to NHA replaced this scheme with loan insurance.

Land Lease Program

Manitoba is the only province which has leased land to housing cooperatives on a subsidized basis. A lot lease program for home owners was available in Ontario under its Home Ownership Made Easy (HOME) Plan as well as in British Columbia, Manitoba, and Newfoundland (see Provincial Housing Corporations)

Lawrence Heights Project

Developed in the mid 1950s and located in suburban Metropolitan Toronto, this housing project was the first major public housing project built on vacant land in the urban fringe.

Le Breton Flats Project

Demonstration housing community initiated in the mid 1970s and located on a 66 hectare site near Ottawa's downtown core.

LD Program Limited Dividend Program (1944-81)

Federal program to create low-rental housing by providing conditional loans which stipulated limits on investment return and controls on rent. Loans were made to companies and individuals for the construction of new housing or to purchase existing housing. During 1946-64, 330 loans were approved, providing 28,037 dwellings. In response to the program's declining attractiveness by the mid 1960s, loan conditions were enhanced (that is, loan levels increased to 95% of value; the 5% investment return limit was increased; and mandatory controls on rent were restricted to a 15-year period) and project eligibility broadened to include hostels, dormitories, as well as self-contained units.

Lithwick Report (1970)

Urban Canada: Problems and Prospects, represents the efforts of a research group headed by Professor N.H. Lithwick and commissioned by the federal government in 1969 to report on urban conditions in Canada. The report's findings contributed to the creation of MSUA.

Matthews Report (1979)

The *Report on Canada Mortgage and Housing Corporation* examined CMHC's role in the provision of housing and projected housing needs for the 1980s. It was the product of a task force set up by the federal government.

MUP Maximum Unit Price (1978-)

MUPs were introduced by CMHC under NHA Section 56.1 as a cost control mechanism and serve as an upper limit on the quality of social housing.

Milton Parc

Milton Parc is a renovated neighbourhood in Montreal which was saved from redevelopment through extensive lobbying by a Montreal citizens' group. Existing tenants were organized into housing cooperatives and non-profit sponsors were found for several projects. Both federal and provincial funds were used for these purposes. Construction began in October 1980.

MSUA Ministry of State for Urban Affairs (1971-9)

MSUA was created by the federal government to encourage positive urban development and to foster closer relationships with municipalities and provinces on urban matters.

Mobile Home Programs

Mobile home programs have been available in Alberta to assist low-income families who are in immediate need of housing, and in Manitoba to ensure financing arrangements are available to prospective mobile home purchasers.

MBS Program Mortgage-Backed Securities Program (1986-)

Federal program to make additional funds available to lenders for conventional financing and to encourage longer term mortgages. CMHC guarantees the timely payment of capital and interest on securities backed by a pool of NHA-insured mortgages. MBS investors have to put up a minimum of $5,000. In its first year of operation (starting December 1986), $456 million in securities were issued, almost twice what had been projected.

MICC Mortgage Insurance Company of Canada (1963-)

MICC is the only remaining private insurer of high-ratio mortgage loans in Canada. Unlike the publicly-operated MIF, MICC serves only the larger urban areas and does not provide insurance for social housing projects.

MIF Mortgage Insurance Fund (1954)

This CMHC-administered fund provided under the federal Mortgage Insurance Program (MIP) spreads the risk of default among borrowers. MIF is financed through premiums paid by recipients of insured mortgages.

Although MIF is intended to be self-sufficient, it has experienced both solvency and liquidity problems largely resulting from the impact of 1973 NHA amendments (for example, defaults under AHOP and ARP) and the Alberta real estate collapse in the 1980s.

MIP Mortgage Insurance Program (see MIF)

MRPP Mortgage Rate Protection Plan (1984-)
This federal plan provides home owners with an opportunity to purchase protection against excessive rises in interest rates upon mortgage renewal.

MLS Multiple Listing Service
Home sale registry system used by various real estate boards throughout Canada.

MURB Program Multiple Unit Residential Building Program (1974-82)
Federal program to promote investment by individuals in rental housing. Amendments to the Income Tax Act enabled those investing in MURBs to deduct from personal income rental losses incurred through capital cost allowances and soft costs. Approximately 195,000 units were approved under the MURB Program.

MIG Program Municipal Incentives Grant Program (1975-8)
Federal program to encourage the development of land for housing of moderate size, price, and density. Municipalities were entitled to receive $1,000 for each qualifying unit. Payments extended to 1982.

National Building Code (1941-)
Code developed by NRC to promote uniform building construction standards, materials, and methods throughout Canada. By the mid 1970s all provinces had adopted variants of the model national code.

NHA National Housing Act (1938-)
Marked the beginning of a greater federal role in housing. Primary intent of the initial legislation was to stimulate housing production and employment. A direct federal role in the provision of housing for low-income households was also created. Amendments to the NHA have followed periodically since then.

NHA Amendments
Amendments were in 1944 to promote new housing construction, to improve housing and living conditions; to upgrade existing housing, to encourage home ownership (especially among veterans), and to stimulate employment. Amendments were made in 1954 to promote private investment in housing and to provide assistance for slum clearance and urban renewal. Amendments were made in 1964 to encourage participation in urban renewal and low-income housing schemes, especially public housing, and introduction of a new funding mechanism for public housing initiated the realignment of social housing responsibility from the federal government to provincial governments. Amendments made in 1973 were intended to provide individuals

with "good housing at a reasonable cost," to avoid large-scale public housing projects targeted to low-income families, to broaden the income ranges of Canadians receiving housing assistance, and to initiate support for the third (non-profit) housing sector. Amendments made in 1979 were designed to encourage the production of new rental housing in the private sector and to stimulate the economy. The initiative of 1985 allowed provinces to deliver social housing programs that met federal targets, redirected social housing programs to concentrate on the most needy, removed geographic limits on urban RRAP, changed the funding of cooperative housing to index-linked mortgages, and introduced mortgage-backed securities.

NHA Part V

Under Part V of the NHA, CMHC sponsors independent housing research through its University Scholarship Program (for Graduate Studies), External Research Program (for Advanced Research), and Housing Technology Incentives Program (HTIP).

NHA Section(s) (Section numbers are updated periodically. Numbers shown are Revised Statutes of Canada 1970 or the number in effect at the time the section was enacted or used.)

15.1 and 34.18 (1973-8): The Nonprofit and Cooperative Housing Programs sections were initiated to develop modest housing for (in order of priority) low and moderate income families, especially in areas requiring new construction; senior citizens; and special needs groups. Eligible organizations were entitled to receive 90% loans with an interest rate subsidy of 8% and a 10% capital contribution. These programs are among the first wherein a federal low-income housing program does not require matching financial commitments on the part of other governments. Rent subsidies for low-income tenants were provided under NHA section 44.1(b). NHA sections 15.1 and 34.18 were replaced by section 56.1 of the NHA.

35 (1964-revised), 43 (1969-78, extended to 1983 in the Northwest Territories), and 44 (1969-): Widely used public housing construction program and a non-profit housing program for the elderly. Section 35(d)/43 entitled provinces, municipalities, and other public agencies to receive 90% loans (amortized over 50 years) for the construction of low-income housing projects where tenants pay rents according to GRS formulas. These provisions have been used to create more than 200,000 public housing units. The federal government covers half of the operating losses of these projects under section 35(e)/44.

40 (1949-78, except in Newfoundland and Prince Edward Island where program is ongoing): Federal/Provincial Public Housing Program stipulates that the capital costs and operating losses incurred by the creation of public housing projects be shared on a 75%-25% basis between the federal and respective provincial governments. CMHC assumes responsibility for approving, planning, and designing these projects. This program was

expanded in the Prairie Provinces to include housing for indigenous persons in 1965. RNH Program replaced the need for this special provision in 1974.

44.1(a) (1969-) and 44.1(b) (1975-): Rent Supplement Programs provided subsidies for low-income tenants residing in private rental accommodation (44.1(a)) and housing projects funded under NHA sections 15.1 and 34.18 (44.1(b)). Costs of these subsidies which cover the difference between rent and GRS provisions are shared equally between the federal and respective provincial governments.

56.1 (1978-86): Nonprofit and Cooperative Housing Programs replaced previous programs funded under NHA sections 15.1 and 34.18. Cooperative and non-profit housing agencies were entitled to receive maximum assistance equivalent to the difference between mortgage payments at market interest rates and at 2%. Under section 56.1 there has been a shift to private insured lending and the introduction of MUPs. From 1974 to 1984 almost 124,000 units were constructed. Funding is also available to assist in the initial development stages of these projects.

National Housing Research Committee (1987-)
National committee of governments, industry, consumer, and social organizations interested in housing which meets semi-annually to discuss and coordinate research activities.

NRC National Research Council
Federal agency responsible for developing the national building code. The Division of Building Research established in 1947, since renamed the Institute for Research in Construction (IRC), conducts research in building technologies and provides advisory services to the construction industry as well as public agencies (such as CMHC).

NIP Neighbourhood Improvement Program (1973-78)
Federal program to improve public infrastructure in designated low-income residential areas and thereby to encourage corresponding improvements in the quality of the existing housing stock. The intent of this program was to prevent the dislocation effects associated with the wholesale clearance of blighted areas. 479 neighbourhoods took part in the program and the costs involved ($500 million) were shared by all levels of government.

Neilsen Task Force
The Canadian Task Force on Program Review was created in 1984 to review federal housing programs and to put forward options related to changing the nature of these programs or improving their management. A balance of private and public sector specialists were on the Neilsen Task Force whose efforts resulted in the publication, *Housing Programs in Search of Balance* (1986).

New Towns Acts
These acts were passed by various provinces in the 1950s and 1960s primarily

to stabilize and hasten the maturity of resource/company dependent settlements. They ensured that various housing types as well as public and commercial services were incorporated in the initial stages of town development (for example, Instant Towns Act in British Columbia).

Non-profit Housing

Housing owned and operated on a non-profit basis by public or private corporations (see also NHA Sections 15.1 and 56.1).

OAS Old Age Security (1952-)

OAS is a non-contributory, indexed federal income transfer program for all persons 65 years of age or over. The program replaced the Old Age Pension Act of 1927. OAS benefit is paid in addition to CPP. In addition, since 1966 the elderly may be eligible for CPP and/or for the income-tested GIS.

OECD Organization for Economic Cooperation and Development

Association of western industrialized countries headquartered in Paris.

Provincial Housing Corporations

Nova Scotia was first to establish a housing corporation: the Nova Scotia Housing Commission (renamed the Nova Scotia Department of Housing in 1983). Most provincial housing corporations were formed to take advantage of the public housing program initiated by the federal government in 1964 (see NHA Section 35). One example of other activities by provincial housing corporations is the Home Ownership Made Easy (HOME) Plan (1967-77) initiated by Ontario's provincial government. The Ontario Housing Corporation (OHC) acquired land parcels in selected municipalities and offered lots to families who had the responsibility of constructing modest homes on them. No downpayments were required for these lots which could be leased for up to 50 years at book value or bought after 5 years' residence for the original low-end of market price. A lottery system was introduced to distribute these lots in 1973. The plan was revised in 1973 and 1975 to prevent speculation and incorporate market value in the calculation of subsidies. About 25,000 households received assistance under the HOME Plan before it was phased out and replaced by AHOP.

Provincial housing corporations and their year of establishment are as follows:

Nova Scotia Department of Housing (1932)
Ontario Housing Corporation (1964)
Alberta Housing Corporation (1967)
British Columbia Housing Management Commission (1967)
Manitoba Housing and Renewal Corporation (1967)
New Brunswick Housing Corporation (1967)
Newfoundland and Labrador Housing Corporation (1967)
Société d'habitation du Québec (1967)
Prince Edward Island Housing Corporation (1969)
Yukon Housing Corporation (1972)

Northwest Territories Housing Corporation (1972)

Saskatchewan Housing Corporation (1973)

Public Housing

In its most restrictive sense, public housing refers to housing developed under NHA Housing Sections 35, 40, 43 and 44. The total subsidy bill for these projects in 1985-86 was estimated to be just under $400 million. The term is often broadened to include all housing administered by public housing agencies.

QPP Quebec Pension Plan (see CPP)

R-2000 (see SEEH Program)

RGI (see GRS)

Radburn Plan

Design principles put forward by Clarence Stein and Henry Wright were incorporated in this model residential community plan implemented in Radburn, New Jersey from 1919 to 1930. This became the model used for many post-war suburban residential developments in Canada.

Regent Park North

A 1,400 unit low-rise public housing project located on 42 acres to the east of Toronto's downtown core was the first slum clearance/public housing scheme initiated in Canada. Construction took place from 1948 to 1958.

RHOSP Registered Home Ownership Savings Plan (1974-1985)

Federal plan to promote home ownership among resident taxpayers not owning residential property. Eligible individuals were entitled to claim tax deductions of up to $1,000 annually (to a limit of $10,000) for funds invested in registered plans provided they were eventually used for the purchase of an owned home. Adjustments made to the plan enabled purchasers of new homes acquired between 19 April 1983 and 1 March 1985 to claim a tax deduction of $10,000 minus prior contributions. Contributors were also allowed to make tax free withdrawals in 1983 towards the purchase of qualifying new furniture. When the plan folded in 1985, all contributions were to be withdrawn and the interest accumulated was in most cases tax-exempt.

RRSP Registered Retirement Savings Plan (1957-)

RRSPs were initiated at the federal level to encourage individuals to save for their retirement. Presently, annual contributions to registered plans (up to $7,500 or 20% of income, whichever is less) may be deducted from personal income and plans must be collapsed by the age of 71 years. The taxation of these contributions is deferred until retirement years when income is expected to be lower.

Rent Control

In its simplest version, a legislated freeze on residential rents with neither exceptions nor a complex formula for permissible rent increases. Canada's Wartime Prices and Trade Board imposed rent freezes in fifteen cities in

September 1940. A year later, rents in the rest of the country were frozen. This was "simple" rent control —an absolute freeze with neither exceptions nor a complex formula for permissible rent increases. Beginning in 1947 a period of rent decontrol began in Canada, and the federal government fully ended rent controls in 1951. Only the province of Quebec maintained rent control beyond 1951.

Rent Regulation

A legislated third-party review of rent increases, usually with guidelines setting out acceptable increases. Several provinces had either adopted rent regulation or were about to adopt them by mid 1975. The federal government imposed wage and price controls in October 1975 and asked the provinces to impose matching rent regulation. By April 1976 all provinces had rent regulation. Since 1976 rent regulation has been removed in British Columbia, Alberta, and New Brunswick.

Rent Supplements

Rent supplements are subsidies to assist low-income tenants in paying their rent. (see Supplements NHA Sections 44.1(a) and 44.1(b)). In addition, provincial rent supplement programs have been provided in British Columbia and Ontario.

Rental Insurance Plan (1948-50)

Federal plan to provide long-term low-interest loans to builders of low rental housing and guaranteed landlords 2% net return on their investment. About 19,000 units were constructed under this program.

Rental unit conversion programs

Programs introduced in Alberta, British Columbia, Nova Scotia, and Ontario to offer subsidies to those who convert structures to provide additional rental accommodation.

Residential General Building Contracting Industry

Defined by Statistics Canada to be the set of business establishments in Canada that derive more than 50% of their revenues from residential construction. In 1984 there were 13,885 such establishments: most in the business of constructing new single-family dwellings. Most of these were small: 86% had revenues under $500,000 in 1984, and only 472 had revenues of $2 million or more.

RRAP Residential Rehabilitation Assistance Program (1973-)

Federal program to encourage upgrading of substandard dwellings (especially those occupied by low to moderate income earners). Eligible home owners and landlords are entitled to receive subsidized loans for admissible renovation costs. Subsequent changes to RRAP have broadened eligibility in terms of geographic extent (for example, rural areas qualified in 1974) and the targeted populations involved (including Indian Band Councils and disabled persons whose dwellings require modifications to improve accessibility). Of the approximately 314,000 units rehabilitated under RRAP from 1973 to 1984, 71% were owner-occupied. Recently; RRAP has been made universal; however, tighter limits are now placed on target populations.

RNH Programs Rural and Native Housing Program (1974-)
Federal programs to assist individuals residing in rural areas and small towns (populations not exceeding 2,500 persons) with housing and renovation costs. Loans are available to finance home construction and subsidies which cover the difference between amortization costs plus property taxes (plus heating costs as of 1986), and 25% of income are shared on a 75%-25% basis between the federal and participating provincial governments. Renovation loans (in part forgivable depending on income) are available to upgrade housing to minimum standards and ensure habitability for at least 15 years. A one-time grant is available for emergency repairs to meet health and safety requirements. The renovation and emergency repair components of this program are financed entirely by the federal government in cases where provinces are not involved in delivery. The Urban Native Housing Program helps low-income Aboriginal households obtain appropriate housing in communities of over 2,500 persons.

St. Lawrence Project
Housing project initiated in the late 1970s near Toronto's downtown core. It is an innovative high-density, low-rise neighbourhood built on formerly industrial land with mixed ownership (for example, public, cooperative, and condominium) and income composition. It now houses approximately 4,600 persons, and roughly half live in non-profit units.

Shell Housing (see Home Purchase Programs)

Shelter Allowances
Programs introduced by various provinces to provide universal access to those who require assistance in paying their rent, the subsidy is determined on the basis of household income and rent paid. Shelter allowances are wholly funded by the provinces. Subsidies have been primarily directed to seniors living in private rental accommodation (for example, British Columbia (SAFER), Manitoba (SAFER), New Brunswick (RATE), and Quebec (Logirente)). In addition, Manitoba extends assistance to families (SAFFR). British Columbia and New Brunswick provide shelter allowances for the disabled.

SFD Single Family Dwelling

Social Housing
Broad term encompassing housing developed under various programs at all government levels that typically include public, cooperative and non-profit housing programs as well as rent supplement programs. Currently, this is the largest single category of federal direct expenditure on housing.

Spruce Court

Housing project, built by the Toronto Housing Company in 1914, is the first Canadian example of limited dividend housing with a mortgage guaranteed by the government. The buildings were arranged in a courtyard plan for light and air in a scheme modeled on the work of Parker and Unwin, the leading English housing reform architects who had planned and designed the buildings in Letchworth, the first Garden City.

Spurr Report

Commissioned by CMHC in the early 1970s and published in 1976 under the title, *Land and Urban Development: A Preliminary Study,* the report examines the urban land development industry in Canada during the 1960s and early 1970s. Among its recommendations is the need to create a systematic information base for analyzing urban land policy

Student Housing Program (1960-78)

Federal program to provide accommodation for college and university students by making loans available to provinces, municipalities, universities, and colleges. This program emerged in response to the rapid increase in post-secondary school enrolment produced by the baby boom generation.

SEEH Program Super Energy Efficient Home (R-2000) Program (1984-91)

Federal program to promote industry knowledge and skills as well as public demand for cost-effective energy-efficient housing. By September 1986, about 2,000 housing units had been built to the R-2000 standard.

SHU Survey of Housing Units (1974)

A stratified random sample of 62,800 households across 23 CMAs that provides valuable information about housing conditions. This unique survey, conducted by Statistics Canada on CMHC's behalf, has been widely used by researchers.

Sweat equity (see Home Purchase Programs)

Thom Commission Report

The second and final volume of the (Ontario) *Commission of Inquiry into Residential Tenancies* was issued in 1987. It argues that Ontario's scheme of rent regulation, in place since 1975, should be revoked, that landlords should be able to charge "fair market rents," and that tenants who are unable to afford such rents be subsidized.

$25 Million Loan Program (1918-23)

Federal program to alleviate an anticipated post-war housing shortage, it marks first modern involvement by the federal government in housing policy. $25 million in loans were distributed to provinces on a per capita basis for the construction of moderately-priced owned homes. Provinces were required to contribute one dollar for every three dollars provided federally. Just over 6,000 dwellings were built under this program.

UI Unemployment Insurance (1941-)
Federal program to supplement incomes of unemployed workers. Initially covering only workers in industry and excluding others such as teachers and civil servants, the program was revised in 1972 to cover almost all employees. Protection was also provided against sickness, temporary disability, and maternity leave, and to fishermen and retiring employees during an initial period of retirement. Program is funded by employer and employee contributions and general tax revenues.

UDI Urban Development Institute (1957-)
National organization that represents the land and property development industry.

Urban Renewal Program (1944-73)
Federal program for slum clearance purposes. Municipalities were entitled to receive a federal grant amounting to 50% of the costs involved. In 1953 this grant was extended to provincial governments, LD companies, and life insurance companies producing rental housing on the cleared land. Grants were also given if the cleared land was to be used for public purposes and alternate sites for rental housing were available. To supplement this program, low-interest loans for up to two-thirds of non-federal costs were available under NHA section 25. The Urban Renewal Program was curtailed following the recommendation of the Hellyer Task Force and eventually replaced by programs such as NIP and CSCP.

Veterans' Land Act (1942-75)
Federal program enabling veterans to purchase homes on favourable lending terms. Mortgages at 3.5% interest and amortized over 25 years were provided with a 10% down payment. Cash grants were also available. Originally intended to assist veterans pursuing occupations in agriculture or commercial fishing, it helped others secure accommodation near larger urban centres. Approximately 8,000 dwellings were constructed under this program over the 1946-9 period. Activity prior to 1946 largely involved procuring land and building materials.

Wartime housing
Dwellings built for WHL and leased to munitions workers and their families at rents of $20-$30 per month. The basic 53.5 m² house had two bedrooms, a bath, living room, and kitchen on one floor. In the larger 7.3 m² x 8.5 m² version, two additional bedrooms were located in a second storey. These dwellings, known as "Type C" houses, used platform wood frame construction. Local contractors built panelized wall sections on site; these were bolted together for speed of erection and possible salvaging. The only variety came in the four approved exterior siding materials: British Columbia cedar shingles, asbestos siding, fir plywood, or bevelled siding. Because of the temporary nature of the units, they were built without basements. At the end of the war, these units were sold off and most buyers added basements to them.

The Type C design was also used in the Veterans' Rental Housing program after the War and by the early NHA builders. The Veterans Rental Housing program produced 25,000 of these units from 1947 to 1950.

WHL Wartime Housing Limited (1941-48)

The mandate of this federal crown corporation was to construct, purchase, rent, and manage rental housing for war workers in areas experiencing housing shortages. By 1944 WHL also accommodated the families of servicemen under the Veterans' Low Rental Housing Program. Almost 46,000, mostly small detached dwellings, were constructed and later sold off beginning in the late 1940s. CMHC absorbed and dismantled WHL in 1948.

Wildwood

An early post-war development in Winnipeg, this model community consisted of 284 houses, many constructed using prefabricated methods, built on the 30.2 hectare site.

Willow Park

Incorporated in 1961 and completed in 1966, this 200-unit development in Winnipeg was one of the first large housing cooperatives to be built in Canada and was jointly sponsored by the Federated Cooperatives, Manitoba Pool Elevators, Cooperative Life Insurance Company, and Winnipeg District Labour Council.

Winter House Building Incentive Program (1963-5)

Federal program to encourage the building of single-detached to four unit structures during the winter months of 1963-4 and 1964-5. Payments of $500 per unit were issued providing that structures were completed over the four-month period from 1 December to March 31. This program was largely intended to reduce seasonal unemployment in the construction industry.

===

Key Event Chronology

Prepared by John R. Miron
with the assistance of Nancy Thompson and Leigh Howell
Entries in Italics further described on Glossary

1912 Alberta passed first modern town planning act in Canada. However, planning legislation had also been passed in New Brunswick and Nova Scotia in 1912. Ontario had also enacted a two-page Cities and Suburbs Plans Act in 1912.

1914 *Spruce Court,* Canada's first limited dividend housing project, was built.

1918 Twenty-five Million Dollar (*$25M.*) *Loan Program* was introduced.

1927 *Canadian Farm Loan Act* was introduced.

1929 Alberta's planning act was revised to require town plans, regional planning commissions, and zoning by-laws.

1932 Nova Scotia Housing Commission (Nova Scotia Department of Housing) was established (see *Provincial Housing Corporations.*

1934 *Bruce Commission Report* was published.

1935 *Dominion Housing Act (DHA)* was introduced.

1936 Pioneering "standard of housing by-law" was passed by the City of Toronto.

1937 *Home Improvement Loans Guarantee Act* was introduced.

1938 *National Housing Act (NHA)* was introduced, replacing *DHA.*

1939 Although municipalities had been empowered to pass zoning by-laws since before 1914 in Alberta, British Columbia, and Ontario, and since the 1920s in most other provinces, not until 1939 did the *National Research Council (NRC)* prepare the first model (national) zoning by-law.

1940 Rent controls in fifteen Canadian cities were imposed by the Wartime Prices and Trade Board.

Home Improvement Loans Guarantee Act was terminated.

1941 *Wartime Housing Limited (WHL)* was established.

Model *national building code* was introduced by *NRC*.

Rent controls were expanded to include all areas of Canada.

Federal *Unemployment Insurance (UI)* was introduced.

1942 *Home Extension Plan* was introduced.

1943 *Home Conversion Plan* was introduced.

1944 *Limited Dividend (LD) Program* was introduced.

Veterans' Low Rental Housing Program (*WHL*) introduced.

Emergency Shelter Program was introduced.

Urban Renewal Program was introduced.

Graduated Rent Scale (GRS) was introduced.

Curtis Report was published.

Integrated Housing Plan was introduced.

1946 *Central Mortgage and Housing Corporation (CMHC)* was established.

Community Planning Association of Canada (CPAC) was established.

Home Conversion Plan was terminated.

1947 McGill University established Canada's first planning school.

Division of Building Research *(NRC)* was established.

1948 *Rental Insurance Plan* was introduced.

Regent Park North, a City of Toronto housing project on a CMHC-financed urban renewal site, began Canada's organized public housing program.

Emergency Shelter Program, Home Extension Plan and *WHL* were terminated.

1949 Federal/Provincial Public Housing Program was introduced under *NHA Section 40.*

1950 *Rental Insurance Plan* was terminated.

1951 Rent controls were ended by the federal government. Quebec was the only province with comprehensive rent controls extending beyond 1951.

Newfoundland became the first province to complete a public housing project. A total of 140 units were constructed in St. John's.

1952 Federal *Old Age Security (OAS)* was introduced, replacing the Old Age Pension Act of 1927.

1953 British Columbia passed enabling legislation empowering Vancouver to use a development permit system. *Discretionary zoning systems* have since been implemented in other jurisdictions as well.

1954 *Mortgage Insurance Program* was introduced to replace joint lending schemes introduced under the *DHA*.

Canada's chartered banks could now originate mortgage loans, but their lending was restricted to new housing insured under the *NHA*.

Direct Lending by the federal government was introduced.

Home Improvement Loans (HIL) Program was introduced.

Capital Cost Allowance (CCA) was introduced to replace depreciation expense.

1955 *Housing in Canada (CHS)*, the first systematic statistical recording of housing production, was initiated.

1956 *Canadian Housing Design Council (CHDC)* was established.

Montreal's first public housing project *(Jeanne Mance)* was approved.

Vancouver's first public housing project (Strathcona) was developed.

1957 Small Homes Loans Program was introduced (see *Direct Lending*).

1960 *Student Housing Program* was introduced.

1961 *CMHC* began auctioning mortgages to foster a secondary mortgage market.

1962 *Canadian Council on Urban and Regional Research (CCURR)* was founded.

The Cooperative Union of Canada (sponsored by *CMHC*) was established to examine the feasibility of non-profit cooperative housing.

1963 *Winter House Building Incentive Program* was introduced.

Mortgage Insurance Company of Canada (MICC), a private insurer, was established.

Building materials were subject to *FST* at rate of 4%.

1964 Urban Renewal and Public Housing Programs were expanded under *NHA Section 35*.

Ontario Housing Corporation (OHC) was established (see *Provincial Housing Corporations*).

First year in which more apartment units (60,435) than single family homes (50,457) were built.

FST rate for building materials was raised to 8%.

1965 Federal/Provincial Public Housing Program under *NHA Section 40* was expanded in the prairie provinces to include housing for indigenous persons.

Winter House Building Incentive Program was terminated.

FST rate for building materials was raised to 11%.

1966 Canada's slum clearance program effectively brought to an end by the successful 4-year protest of Toronto's Trefann Court residents who were threatened with eviction.

First Condominium Acts were passed by the provinces of Alberta and British Columbia.

NHA mortgage insurance on existing owner-occupied housing was introduced.

Maximum ratio of mortgage loan to value of property for conventional loans of federally regulated lending institutions was raised to 75%.

Willow Park, in Winnipeg, Canada's first publicly-financed continuing housing cooperative, was built.

Canada Pension Plan (CPP) was introduced.

Guaranteed Income Supplement (GIS) was introduced.

Company of Young Canadians (CYC) was founded.

Canada Assistance Plan (CAP) was introduced.

1967 Canada's chartered banks could now originate conventional mortgage loans on new and existing properties. Provision were made for removal of interest rate ceiling on all bank loans.

Alberta Housing Corporation, British Columbia Housing Management Commission, Manitoba Housing and Renewal Corporation, New Brunswick Housing Corporation, Newfoundland and Labrador Housing Corporation and Societé d'habitation du Québec were established (see *Provincial Housing Corporations*).

1968 *Hellyer Task Force* was established.

Federal government imposed a moratorium on all new urban renewal approvals and on the development of large public housing projects.

Cooperative Housing Foundation of Canada (CHF) was established.

Intergovernmental Committee on Urban and Regional Research (ICURR) was established.

Canadian Conference on Housing was held.

1969 Public Housing Program under *NHA Sections 43 and 44* was introduced.

Rent Supplement Program under *NHA Section 44.1(a)* was introduced.

Lending institutions became authorized to originate high-ratio conventional mortgage loans providing the sum exceeding 75% of a property's value was insured. Interest rate ceilings were removed on *NHA*-insured loans and the minimum term of *NHA* insured loans was reduced to 5 years.

Prince Edward Island Housing Corporation was established (see *Provincial Housing Corporations*).

Landlord and tenant legislation was passed by Ontario and Manitoba.

1970 *Lithwick Report* was published.

$200 million *Innovative Housing Program* was introduced.

CMHC allowed costs of recreational facilities in social housing projects.

Canadian Institute of Public Real Estate Companies (CIPREC) was founded.

1971 *Dennis Task Force* was established by CMHC.

Major revisions to federal Income Tax Act. Losses created by capital cost allowances for rental property were no longer deductible from non-rental income. A taxpayer's principal residence became the only dwelling exempted from capital gains tax.

1972 Yukon Housing Corporation and Northwest Territories Housing Corporation were established (see *Provincial Housing Corporations*).

NHA approved lenders were authorized to include any or all of spouse's earned income in determining borrower eligibility for NHA insured loans for home ownership.

British Columbia established an Agricultural Land Commission that froze conversion of agricultural land to housing in the Lower Fraser Valley. At about the same time, Ontario froze urban development along the Niagara Escarpment and in the Parkway belt running through and around the metropolitan Toronto region. In the late 1970s Calgary froze the development of over 12 square kilometers of land on the south side of the city because of inadequate transportation facilities. Ontario similarly froze most development north of Toronto for a period of 15 years until a major water and sewer trunk servicing scheme was implemented in the early 1980s. The province of Quebec enacted a similar Act to Preserve Agricultural Land in 1978.

1973 *Assisted Home-Ownership Program (AHOP)* was introduced.

Neighbourhood Improvement Program (NIP) was introduced.

Residential Rehabilitation Assistance Program (RRAP) was introduced.

Nonprofit and Cooperative Housing Programs under NHA *Sections 15.1 and 34.18* was introduced.

More unilateral federal and provincial housing measures began in place of joint cost-shared federal-provincial activities.

Canada's first provincial department of housing was established in British Columbia.

Saskatchewan Housing Corporation established. Its responsibilities were formerly held by the Housing and Urban Renewal Branch of the Department of Municipal Affairs established in 1966 (see *Provincial Housing Corporations*).

Urban Renewal Program was terminated.

1974 *Multiple Unit Residential Building (MURB)* scheme was introduced.

Registered Home Ownership Savings Plan (RHOSP) was introduced.

Rural and Native Housing (RNH) Programs were introduced.

Cityhome, in Toronto, became the first municipal non-profit housing corporation established to take advantage of 1973 *NHA* amendments.

Tax deductibility of carrying charges on land awaiting redevelopment was removed.

Survey of Housing Units (SHU) was undertaken.

British Columbia became the first province to deal with the loss of rental units resulting from condominium conversions by amending its legislation to give municipalities the ability to stop the conversion of rental units.

Rentalsman was introduced in British Columbia to mediate landlord-tenant disputes and review large rent increases. Rentalsman was abolished in 1985, in conjunction with downsizing of rent regulation in that province.

FST rate for building materials was dropped to 5%.

1975 *Assisted Rental Program (ARP)* was introduced.

Rent Supplement Program under *NHA Section 44.1(b)* was introduced.

Provinces were requested to impose rent controls as part of the federal government's wage and price control program.

CYC was terminated.

1976 *Home Insulation Program (HIP)* was introduced.

CCURR was terminated.

United Nations Conference on Human Settlements (Habitat) was held in Vancouver.

City of Toronto adopted a special form of zoning (so-called mixed-use districts) for areas of permissible redevelopment within the Central Area in 1976; the zoning provides flexibility in range of uses and room for negotiation of density while at the same time limiting the exercise of discretion.

1977 *Canadian Home Insulation Program (CHIP)* was introduced in Nova Scotia and Prince Edward Island.

Shelter Allowance For Elderly Renters (SAFER) was introduced in British Columbia (see *Shelter Allowances*).

Prince Edward Island became the last province to adopt a Condominium Act.

1978 Nonprofit and Cooperative Housing Programs under *NHA Section 56.1* were introduced.

Report of the Federal-Provincial Task Force on the Supply and Price of Serviced Residential Land (Greenspan Report) attributed rising serviced land prices in part to an increase in scope and extent of land subdivision regulation.

Tax deductibility of carrying charges on land awaiting redevelopment was restored.

AHOP, ARP, NHA Sections 15.1, 34.18, 40, 42 and *NIP* were terminated. *GPM* replaced *AHOP* and *ARP*.

1979 *Community Services Contribution Program (CSCP)* was introduced.

NHA mortgage loan insurance was extended to cover existing rental housing.

CSCP was terminated.

1980 *Canada Oil Substitution Program (COSP)* was introduced.

Building Energy Technology Transfer (BETT) Program was introduced.

NHA insurance was introduced for existing rental buildings.

1981 Canada Mortgage Renewal Plan (*CMRP*) was introduced.

Canada Rental Supply Plan (CRSP) was introduced.

Soft costs were now treated as capital costs in rental buildings.

HIP was terminated.

1982 *Canadian Home-Ownership Stimulation Plan (CHOSP)* was introduced.

Canada Home Renovation Plan (CHRP) was introduced.

Insurance application fees on *NHA* home ownership loans were raised for the first time, and premium schedules were altered to reflect differences in risk among borrowers.

CMHC guideline was introduced stipulating 5% of all units in social housing projects be wheelchair modified.

MURB was terminated.

1983 *NHA Section 56.1* evaluation was produced.

CHOSP, CHRP and *CMRP* were terminated.

1984 *Mortgage Rate Protection Plan (MRPP)* was introduced.

Super Energy Efficient Home (SEEH) Program was introduced.

CRSP was terminated.

Insurance premium structure was altered to reflect differences in risk of borrower on *NHA* loans for rental housing.

1985 Federal Consultation "Paper on Housing" was published.

COSP and *RHOSP* were terminated. A lifetime capital gains exemption of $500,000 was initiated to be phased in over a period of years. In 1987 a cap of $100,000 was proposed under federal tax reform. Principal dwellings remained exempt from capital gains taxation.

Premium schedule for loans for home ownership was changed to reflect further differences in risk of borrowers.

1986 *Mortgage-Backed Securities (MBS)* Program was introduced.

New Federal/Provincial cost-sharing and delivery arrangement for social housing was introduced (see *NHA Section 56.1*)

Federal Cooperative Housing Program utilizing Index-linked Mortgages (ILMs) was introduced (see *NHA Section 56.1*)

"Income Mixing Approach" in Federal Social Housing Programs ended and was replaced by specific targeting to only the most needy households.

BETT Program, CHIP and *HIL Program* were terminated.

Introduction of *NHA* mortgage insurance for second mortgages.

1987 *NHA* mortgage premiums and minimum allowable downpayments were reduced.

Federal tax reform proposals were announced.

CHDC was terminated.

References

Adams, J. 1968. "A Tenant Looks at Public Housing Projects." In Snider, 26-31

Akerlof, G. 1970. "The Market for Lemons: Qualitative Uncertainty and the Market Mechanism." *Quarterly Journal of Economics* 84: 488-500.

Alberta House Cost Comparison Study, 1986. Edmonton: Department of Housing.

Allen, C. 1982. "Refuge from the Storm: Transitional Housing for Battered Women." *Habitat* 25: no.4.

Ames, H.B. [1897] 1972. *The City Below the Hill.* Toronto: University of Toronto Press.

Andrews, H. 1986. "The Effects of Neighbourhood Social Mix on Adolescents' Social Networks and Recreational Activities." *Urban Studies* 23: 501-7.

Archambeault, J. 1947. "Le logement populaire problème capital." L'Ecole Social Populaire, no. 397. Montréal.

Archer, P. 1979. *A Compendium of Rent-to Income Scales in Use in Public Housing and Rent Supplement Programs in Canada.* Ottawa: Canada Mortgage and Housing Corporation.

Armstrong, A.H. 1968. "Thomas Adams and the Commission of Conservation." In *Planning the Canadian Environment,* edited by L.O. Gertler, 17-35. Montreal: Harvest House.

Armstrong, P. 1984. *Labour Pains: Women's Work in Crisis.* Toronto: The Women's Press.

Arnold, E. 1986. "The Measurement of the Affordability of Housing." MA thesis, School of Urban and Regional Planning, Queen's University, Kingston, Ontario.

Artibise, A.F.J. 1982. "In Pursuit of Growth: Municipal Boosterism and Urban Development in the Canadian Prairie West, 1871-1913." In *Shaping the Urban Landscape: Aspects of the Canadian City Building Process,* edited by G. Stelter and A.F.J. Artibise, 116-47. Ottawa: Carleton University Press.

Artibise, A.F.J. and G. Stelter, eds. 1979 *The Usable Urban Past.* Ottawa: Carleton University Press.

Atkinson, T. 1982. "The Stability and Validity of Quality of Life Measures." *Social Indicators Research* 10: 113-32.

Bairstow and Associates Consulting Limited. 1985. *Opportunities for Manufactured Housing in Canada.* Report prepared for CMHC. Ottawa.

Bairstow, D. 1973. *Demographic and Economic Aspects of Housing Canada's Elderly.* Ottawa: CMHC.

––––––. 1976. "Housing Needs and Expenditure Patterns of the Low-Income Elderly in British Columbia." CMHC Regional Office, Vancouver.

Baker, William B. 1971. "Community Development in Changing Rural Society." In Draper, 84-92.

Balakrishnan, T.R. and G.K. Jarvis. 1979. "Changing Patterns of Spatial Differentiation in Urban Canada, 1961- 1971." *Canadian Review of Sociology and Anthropology* 16, no.2: 218-27

Ball, M. 1983. *Housing Policy and Economic Power: The Political Economy of Owner Occupation.* London: Methuen.

Barnard, P. 1974. *Concrete Building Systems in the Toronto Area, 1968-1974.* Report prepared for CMHC.

Barnard, Peter Associates. 1985. *Under Pressure: Prospects for Ontario's Low-rise Rental Stock.* Toronto: Ontario Ministry of Housing.

Bates, P.T. 1983. "An Analysis of Housing Policies in Resource Towns." ME Design thesis, University of Calgary.

Bates, S. 1955. "Five Ways to Better Housing." *Canadian Homes and Gardens,* no. 9 (August): 49-51.

Baum, D.J. 1974. *The Final Plateau.* Toronto: McClelland Burns and MacEachern.

Bernardin-Haldeman, V. 1982. *L'Habitat et les personnes ageés.* Faculté des sciences sociales, Université Laval, Québec.

Berry, B.J.L. 1982. "Inner City Futures: An American Dilemma Revisited." In *Internal Structure of the City,* edited by L.S. Bourne, 555-72. Second Edition. New York: Oxford University Press.

––––––. 1985. "Islands of Renewal in Seas of Decay." In *The New Urban Reality,* edited by P.E. Peterson, 69-96. Washington, DC: Brookings Institution.

Blanc-Schneegans, B. 1982. *Fermont: un nouveau concept de l'habitat bilan pour les résidents et les organismes locaux.* Ottawa: Société canadienne d'hypothèques et de logement.

Bossons, J. 1978. "Housing Demand and Household Wealth: Evidence for Home Owners." In Bourne and Hitchcock, 86-106.

Bourne, L.S. 1986. "Recent Housing Policy Issues in Canada: A Retreat from Social Housing?" *Housing Studies* 1, no.2: 122-9.

Bourne L.S. and J.R. Hitchcock, eds. 1978. *Urban Housing Markets: Recent Directions in Research and Policy.* Toronto: University of Toronto Press.

Bowser, S. 1957. "Row housing: Don Mills, Ontario: James Murray and Henry Fliess, Associate Architects." *Canadian Architect* 2, no.2: 23-6.

Boyle, P.P. 1986. "Aftermath of a Crisis? Mortgage Loan Default Insurance." *Resource: The Canadian Journal of Real Estate,* October 1986: 4-8.

Bradbury, J.H. 1978. "Class Structures and Class Conflicts in British Columbia Resource Towns: 1965 to 1972." *BC Studies* 37: 3-18.

———. 1984a. "Declining Single-industry Communities in Quebec-Labra-dor." *Journal of Canadian Studies* 19, no.3: 125-39.

———. 1984b. "The Impact of Industrial Cycles in the Mining Sector: The Case of the Quebec-Labrador Region in Canada." *International Journal of Urban and Regional Research* 8, no.3: 311-31.

Bradbury, J.H. and I. St-Martin. 1983. "Winding Down in a Quebec Mining Town: A Case Study of Schefferville." *Canadian Geographer* 27, no.2: 128-44.

Bradbury, J.H. and J.M. Wolfe, eds. 1983. *Recession, Planning and Socio-Economic Change in The Quebec-Labrador Iron Mining Region.* McGill Subarctic Research Paper 38. Centre for Northern Studies and Research, McGill University, Montreal.

Bradbury, K. L. and Downs, A., eds. 1981. *Do Housing Allowances Work?* Washington, DC: The Brookings Institution.

Breton, R. 1964. "Institutional Completeness of Ethnic Communities and the Personal Relations of Immigrants." *American Journal of Sociology* 70: 193-205.

Brink, S. 1985. "Housing Elderly People in Canada: Working Towards a Continuum of Housing Choices Appropriate to Their Needs." In *Innovations in Housing and Living Arrangements for Seniors,* edited by G. Gutman and N. Blackie. The Gerontology Research Centre, Simon Fraser University, Burnaby, British Columbia.

Brown, J.C. 1977. *A Hit and Miss Affair: Policies for Disabled People in Canada.* Ottawa: Canadian Council on Social Development.

Brown, P.W. 1983. "The Demographic Future: Impacts on the Demand for Housing in Canada, 1981-2001." In Gau and Goldberg, 5-32

Bryden, K. 1974. *Old-Age Pensions and Policy-Making in Canada.* Montreal: McGill-Queen's University Press.

Bunting, T. 1984. *Residential Investment in Older Neighbourhoods.* Department of Geography, University of Waterloo.

Byler, J.W. and R.A. Gschwind. 1980. "Data Resources for Monitoring Change."

In *Residential Mobility and Public Policy,* edited by W.A.V. Clark and E.G.Moore. Beverly Hills: Sage (Volume 19, Urban Affairs Annual Reviews).

Byler, J.W. and S. Gale. 1978. "Social Accounts and Planning for Changes in Urban Housing Markets." *Environment and Planning A* 10: 247-66.

Canada. Advisory Committee on Reconstruction and Community Planning. 1944. *Final Report of the Subcommittee on Housing and Community Planning (Curtis Report).* No. 4. Ottawa: King's Printer.

———. 1962. *Report of the Royal Commission on Banking and Finance.* Ottawa: Queen's Printer.

———. Task Force on Housing and Urban Development. 1969. *Report (Hellyer Report).* Ottawa: Queen's Printer.

———. Department of Regional Economic Expansion (DREE). 1979. *Single-Sector Communities.* Ottawa.

Canada. House of Commons. 1982. *Obstacles: Report of the Special Committee on the Disabled and the Handicapped.* Ottawa: Ministry of Supply and Services.

Canada. Task Force on Mining Communities. 1982. *Report.* Ottawa.

———. 1985. *Consultation Paper on Housing.* Ottawa: Queen's Printer.

———. Department of Finance. 1986. *Economic Review* April 1985. Ottawa: Ministry of Supply and Services Canada.

Canada Mortgage and Housing Corporation, City Urban Assistance Research Group. 1972. *Urban Annual Review* 2A. Ottawa.

———. 1981. *Housing Affordability Problems and Housing Need in Canada and the US: A Comparative Study.* Report prepared under the Memorandum of Understanding between CMHC and the US Department of Housing and Urban Development. Ottawa.

———. 1983a. *Section 56.1 Nonprofit and Cooperative Housing Program Evaluation.* Ottawa.

———. 1983b. "Social Housing Evaluation." Ottawa.

———. 1984a. *Evaluation of Le Breton Flats by Residents.* Ottawa.

———. 1984b. *Housing in Canada: A Statistical Profile.* Ottawa.

———. 1985a. "Housing Issues in the 1980s and 1990s: Structural Adjustment in the Residential Construction Industry." Ottawa: CMHC Research Division.

———. 1985b. *Research Plan.* Ottawa.

———. 1985c. "Changes to the Existing Rental Apartment Stock in Metropolitan Toronto." Supplement One to the CMHC Local Housing Market Report (December). Ottawa.

———. 1986a. "Evaluation of NHA Mortgage Loan Insurance." Ottawa.

————. 1986b. *Guide for the Planning and Monitoring Committees.* Ottawa.

————. 1986c. *Residential Rehabilitation Assistance Program Evaluation.* Ottawa: CMHC Program Evaluation Division.

————. 1986d. *Wildwood Housing Study.* Ottawa.

Canadian Council on Social Development. 1973. *Beyond Shelter: A Study of NHA-Financed Housing for the Elderly.* Ottawa.

————. 1977. *A Review of Canadian Social Housing Policy,* by J. Patterson and P. Streich. Ottawa.

————. 1985. *Deinstitutionalization: Costs and Effects.* Ottawa.

Canadian Housing Statistics (CHS). 1955-86. Ottawa: CMHC.

Canadian Welfare Council. 1961. *Homeless Transient Men.* Ottawa.

Cape, E. 1985. "Out of Sight, Out of Mind: Aging Women in Rural Society." *Women and Environments* 7, no.3: 4-6.

Capozza, D.R. and G.W. Gau. 1983. "Optimal Mortgage Instrument Designs." In Goldberg and Gau, 233-58.

Carver, H. 1948. *Houses for Canadians: A Study of Housing Problems in the Toronto Area.* Toronto: University of Toronto Press.

Carver, H. and A.L. Hopwood. 1948. *Rents for Regent Park.* Toronto: Civic Advisory Council of Toronto.

Carver, H. and R. Adamson. 1946. *How Much Housing Does Greater Toronto Need?* Toronto: Advisory Committee on Reconstruction.

Central Mortgage and Housing Corporation. 1970. *Housing in Canada 1946-1970.* A Supplement to the 25th Annual Report of CMHC. Ottawa.

Chappell, N. *et al.* 1986. *Aging and Health Care: A Social Perspective.* Toronto: Holt, Rinehart and Winston.

Che-Alford, J. 1985. *Mortgagor Households in Canada: Their Geographic and Household Characteristics, Affordability and Housing Problems.* Cat. 99-945 Occasional. Ottawa: Statistics Canada.

Choko, M.H. 1986. "The Evolution of Rental Housing Market Problems: Montreal as a Case Study 1825-1986." Resource paper for the "Housing Progress in Canada since 1945" study. Centre for Urban and Community Studies, University of Toronto.

Choko, M.H. et F. Dansereau. 1987. *Restauration résidentielle et co-propriété au centre-ville de Montréal.* Etudes et documents 53. Montréal: INRS-Urbanisation.

City of Toronto Planning and Development Department. 1980. "Housing Deconversion: Why the City of Toronto is Losing Homes Almost as Fast as It Is Building Them." *Research Bulletin* 16.

————. 1984. "Toronto Region Incomes." *Research Bulletin* 24.

————. 1986. "Trends in Housing Occupancy." *Research Bulletin* 26.

Clark, S.D. 1966. *The Suburban Society.* Toronto: University of Toronto Press.

Clark, W.A.V. 1982. "Recent Research on Migration and Mobility: A Review and Interpretation." *Progress in Planning* 18, no. 1.

Clayton, F.A. 1974. "Income Taxes and Subsidies to Homeowners and Renters: A Comparison of US and Canadian Experience." *Canadian Tax Journal* 22, no.3: 295-305.

Clayton Research Associates Limited. 1980. "Lender Attitudes to Graduated Payment Mortgages and Social Housing Loans." Report prepared for CMHC.

————. 1984a. *Rental Housing in Canada under Rent Control and Decontrol Scenarios, 1985-91.* Toronto.

————. 1984b. *A Longer-term Rental Housing Strategy for Canada.* Toronto: Canadian Home Builders' Association.

————. Various years. *Canadian Housing Monthly Analysis.* Toronto.

————. 1987. "How Does Your Firm Stack up Financially." *Canadian Monthly Housing Analysis.* 19 March (Special Supplement). Toronto.

————. 1988. *Summary Report: The Changing Housing Industry in Canada, 1946-2001.* With assistance from Scanada Consultants Limited. NHA 6121 02/89. Ottawa: CMHC.

Coffey, W. and M. Polèse. 1987. *Still Living Together.* Montreal: Institute for Research on Public Policy.

Colcord, J.C. 1939. *Your Community.* New York: Russell Sage Foundation.

Colton, T.J. 1980. *Big Daddy: Frederick B. Gardiner and the Building of Metropolitan Toronto.* Toronto: University of Toronto Press.

Community Development Strategies Evaluation. 1982. *Final Report.* Department of Regional Science, University of Pennsylvania.

Community Resources Consultants. 1979. *Housing for Emotionally Disadvantaged Adults.* Position Paper no. 2. Canadian Mental Health Association, Toronto.

Compton, Freeman H. 1971. "Community Development Theory and Practice." In Draper, 382-96.

Connidis, I. 1983. "Living Arrangement Choices of Older Residents: Assessing Quantitiative Results with Qualitative Data." *Canadian Journal of Sociology* 8, no.4: 359-75.

Copp, T. 1974. *The Anatomy of Poverty: The Condition of the Working Class in Montreal, 1897-1929.* Toronto: McClelland and Stewart.

Corke, S.E. 1983. *Land Use Controls in British Columbia.* Research Paper 138. Toronto: University of Toronto, Center for Urban and Community Studies.

————. 1986. "Provincial Housing and Shelter-Support Programs for the Elderly: Ontario." In Gutman and Blackie, 121-32.

Corke, S. and M.E. Wexler. 1986. "Choices: A Revised Approach to Housing Policy and Program Development for Ontario's Aging Population."

"Council Aproves $1 Bllion Plan for Downtown Housing." *Globe and Mail* 13 July 1988: 1-2.

Crenna, C.D. 1971. "Proposals for Urban Assistance Programs." Ottawa: CMHC.

Cullingworth, J.B. 1987. *Urban and Regional Planning in Canada.* New Brunswick, New Jersey: Transaction.

Damas and Smith Limited. 1980. *Residential Conversions In Canada.* Ottawa: CMHC Technical Research Division, Policy Development and Research Sector.

Dansereau F., J. Godbout et J-P. Collin. 1981. *La transformation d'immeubles locatifs en co-propriété d'occupation.* Rapport soumis au Gouvernement du Québec. Montréal: INRS- Urbanisation.

Dansereau, F. and M. Beaudry. 1985. "Les mutations de l'espace habité montréalais: 1971-1981." Paper presented at the ACSALF symposium, "La morphologie sociale en mutation au Québec." Chicoutimi (May). INRS-Urbanisation, Montréal.

————. 1986. "Les mutations de l'space habité montréalais: 1971-1981." In *La morphologie sociale en mutation au Québec,* sous la direction de S. Langlois et F. Trudel. *Cahiers de l'ACFAS* 41: 283-308.

Darroch, A.G. and W.G. Marston. 1971. "The Social Class Basis of Ethnic Residential Segregation: The Canadian Case." *American Journal of Sociology* 77, no.3: 491-510.

Davies, G.W. 1978. "Theoretical Approaches to Filtering in the Urban Housing Market." In Bourne and Hitchcock, 139-63.

Dear, M. and J. Wolch. 1987. *Landscapes of Despair: From Deinstitutionalization to Homelessness.* Princeton: Princeton University Press.

Delaney, J. 1991. "The Garden Suburb of Lindenlea, Ottawa: A Model for the First Federal Housing Policy, 1918-24." *Urban History Review* 19: 151-65.

Denman, D.R. 1978. *The Place of Property.* New York: State Mutual Book.

Dennis, M. and S. Fish. 1972. *Programs in Search of a Policy: Low Income Housing in Canada.* Toronto: Hakkert.

Diaz-Delfino, M. 1984. "The St. Lawrence Neighbourhood: An Evaluative Study of the Social Integration of Its Residents." MSc thesis, Department of Geography, University of Toronto.

Dietze, S.H. 1968. *The Physical Development of Remote Resource Towns.* Ottawa: CMHC.

Dominion Bureau of Statistics. 1948. *Life Tables for Canada, 1945.* Ottawa.

Doherty, E.A. 1984. *Residential Construction Practises in Alberta, 1900-1971.* Report prepared for the Alberta Department of Housing.

Doucet, M.J. and J.C. Weaver. 1985. "Material Culture and the North American House: The Era of the Common Man, 1870-1920." *Journal of American History* 72, no.3: 560-87.

Draper, J.A., ed. 1971. *Citizen Participation: Canada.* Toronto: New Press.

Driedger, L. and G. Church. 1974. "Residential Segregation and Institutional Completeness: A Comparison of Ethnic Minorities." *Canadian Review of Sociology and Anthropology* 11, no.1: 30-52.

Duncan, O.D. and B. Duncan. 1956. "Residential Distribution and Occupational Stratification." *American Journal of Sociology* 60: 493-503.

Economic Council of Canada. 1974. *Toward More Stable Growth in Construction.* Ottawa.

Eichler, M. 1983. *Families in Canada Today: Recent Changes and their Policy Consequences.* Toronto: Gage.

Emery, F.E. and E.L. Trist. 1973. *Towards a Social Ecology.* New York: Plenum Press.

Emmi, P.C. 1984. "Primal/Dual Relationships in a Pair of Multi-sectoral Housing Market Models." *Journal of Regional Science* 24: 17-34.

Fallis, G. 1980. *Housing Programs and Income Distribution in Ontario.* Ontario Economic Council Research Study 17. Toronto: University of Toronto Press.

––––––. 1985. *Housing Economics.* Toronto: Butterworths.

Falta, P. and G. Cayouette. 1977. "Le logement intégré: Facteur d'évolution sociale" ("Social Change through Integrated Housing"). Projet normalisation. Montréal: Canadian Paraplegic Association.

Farge, B. 1986. "Women and the Canadian Co-op Experience: Women's Leadership in Co-ops: Some Questions." *Women and Environments* 8, no.1: 13-5.

Feldman, L. and K. Graham. 1979. *Bargaining for Cities: Municipalities and Intergovernmental Relations, An Assessment.* Ottawa: Institute for Research on Public Policy.

Fillmore, S. 1955. "250 Acres of Homes." *Canadian Homes and Gardens,* May: 26-27, 84-91.

Firestone, O.J. 1951. *Residential Real Estate In Canada.* Toronto: University of Toronto Press.

Fletcher, R.H. and I.M. Robinson. 1977. *Inventory Report 1976 of Canadian Resource Communities.* Study prepared for the Urban Policy Analysis Branch, Ministry for Urban Affairs, Ottawa.

Form, W. 1951. "Stratification in Low and Middle-income Housing Areas." *Journal of Social Issues* 7: 109-31.

Fraser, G. 1972. *Fighting Back.* Toronto: Hakkert.

Front d'action populaire en réaménagement urbain (FRAPRU). 1984. *Pour une politique globale d'accès au logement.* Montréal.

Galloway, M. 1978. "User Adaptations of Wartime Housing." MSc thesis, Faculty of Environmental Design, University of Calgary.

Gans, H. 1961. "The Balanced Community: Homogeneity or Heterogeneity in Residential Areas?" *American Institute of Planners Journal* 27, no.3: 176-84.

Genovese, R. 1981. "A Women's Self-help Network as a Response to Service Needs in the Suburbs." In *Women and the American City,* edited by C.R. Stimpson *et al.,* 245-53. Chicago: University of Chicago Press.

Gertler, L. and R. Crowley. 1977. *Changing Canadian Cities: The Next 25 Years.* Toronto: McClelland and Stewart.

Gietema, W.A. and Nimick, E.H. 1987. "Impediments to the Market Acceptance of Prefabricated Wood Panel Systems." MSc thesis, Real Estate Development, Massachusetts Institute of Technology, Cambridge, Massachusetts.

Gingrich, P. 1984. "Decline of the Family Wage." *Perception* 7, no.5: 15-7.

Glendon, M.A. 1981. *The New Family and the New Property.* Toronto: Butterworths.

Gluskin, I. 1976. *Cadillac-Fairview Corporation Limited: A Corporate Background Report.* Study 3. Royal Commission on Corporate Concentration.

Goldberg, M.A. 1980. "Municipal Arrogance or Economic Rationality? The Case of High Servicing Standards." *Canadian Public Policy* 6: 78-88.

Goldberg M.A. and G.W. Gau, eds. 1983. *North American Housing Markets into the Twenty-first Century.* Cambridge, Massachusetts: Ballinger.

Goldberg, M.A. and J. Mercer. 1986. *The Myth of the North American City.* Vancouver: University of British Columbia Press.

Greenspan, D.B. 1978. *Down to Earth.* Report of the Task Force on the Supply and Price of Residential Land. Ottawa: CMHC.

Greenspan, D.B. *et al.* 1977. *Synthesis and Summary of Technical Research.* Report of the Federal/Provincial Task Force on the Supply and Price of Serviced Residential Land. Volume 2.

Greer-Wootten, B. and S. Velidis. 1983. *The Relationship between Objective and Subjective Indicators of the Quality of Residential Environments.* Ottawa: CMHC.

Grigsby, W., M. Baratz and D. MacLennen. 1984. *The Dynamics of Neighborhood Change and Decline.* Research Report Series, 4, Department of City and Regional Planning, University of Pennsylvania.

Groupe d'intervention urbaine de Montréal (GIUM). 1984. *Patrimoine résidentiel du Grand Plateau Mont-Royal: Sondage des résidents.* Montréal.

Guest, D. 1980. *The Emergence of Social Security in Canada.* Vancouver: University of British Columbia Press.

Gutman, G.M. 1975. *Senior Citizen's Housing Study.* Report 1: *Similarities and Differences Between Applicants for Self-contained Suites, Board-Residence and Non-Applicants.* University of British Columbia, Centre for Continuing Education, Vancouver.

————. 1976. *Senior Citizen's Housing Study.* Report 2: *After the Move: A Study of Reactions to Multi and Uni-Level Accommodation for Seniors.* University of British Columbia, Centre for Continuing Education, Vancouver.

Gutman, G.M. and N.K. Blackie, eds. 1986. *Aging in Place: Housing Adaptations and Options for Remaining in the Community.* The Gerontology Research Centre, Simon Fraser University, Burnaby, British Columbia.

Hallman, H.W. 1984. "Defining Neighbourhood." *Urban Resources* 1, no.3: 6.

Hamilton, S.W. 1981. *Regulation and Other Forms of Government Intervention Regarding Real Property.* Technical Report 13. Ottawa: Economic Council of Canada.

————, ed. 1978. *Condominiums: A Decade of Experience in British Columbia.* Vancouver: British Columbia Real Estate Association.

Hamm, B. 1982. "Social Area Analysis and Factorial Ecology: A Review of Substantive Findings." In *Urban Patterns,* edited by G.A. Theodorson, 316-37. University Park, Pennsylvania: Pennsylvania State University Press.

Hanushek, E.A. and J.M. Quigley. 1979. "The Dynamics of the Housing Market: A Stock Adjustment Model of Housing Consumption." *Journal of Urban Economics* 5: 90-111.

Harloe, M. 1985. *Private Rented Housing in the United States and Europe.* Beckenham, England: Croom Helm.

Harris, R. 1986a. "Homeownership and Class in Modern Canada." *International Journal of Urban and Regional Research* 10, no.1: 67-86.

————. 1986b. "Class Differences in Urban Home Ownership: An Analysis of Recent Canadian Trends." *Housing Studies* 1, no.3: 133-46.

————. 1987. *The Growth of Home Ownership in Toronto, 1899-1913.* Research Paper 163, Centre for Urban and Community Studies, University of Toronto.

Harrison, B.R. 1981. *Living Alone in Canada: Demographic and Economic Perspectives.* Catalogue 98-811. Ottawa: Statistics Canada

Harvey, R. 1986. "Housing Heartbreak." *Toronto Star* 19 April.

Hayden, D. 1981. *The Grand Domestic Revolution.* Cambridge: MIT Press.

————. 1984. *Redesigning the American Dream: The Future of Housing, Work, and Family Life*. New York: W.W. Norton.

Health and Welfare Canada. Various years. *Homes for Special Care*. Statistical Summary. Ottawa.

Helman, C. 1981. "Milton Park: Co-op Housing Goes Big Time." In *After the Developers*, edited by J. Lorimer and C. MacGregor. Toronto: Lorimer.

Henderson, J.V. 1979. "Theories of Group, Jurisdiction, and City Size." In *Current Issues in Urban Economics*, edited by P. Mieszkowski and M. Straszheim, 235-69 Baltimore: Johns Hopkins University Press.

Henderson, J.V. and Y.M. Ioannides. 1983. A Model of Housing Tenure Choice. *American Economic Review* 73 (March): 98-113.

Hepworth, H.P. 1975. *Personal Social Services in Canada: A Review*. Volume 4: *Residential Services for Children in Care*. Volume 6: *Residential and Community Services for Old People*. Ottawa: Canadian Council on Social Development.

————. 1985. "Trends in Provincial Social Service Department Expenditures, 1963-1982." In *Canadian Social Welfare Policy*, edited by J.S. Ismael, 152-63. Edmonton: University of Alberta Press.

Heraud, B.J. 1968. "Social Class and the New Towns." *Urban Studies* 5: 33-58.

Herchak, R. 1973. "Housing in Rural Canada: The Role of CMHC." Ottawa: CMHC.

Hess, E. 1984. "Native Employment in Northern Canadian Resource Towns: The Case of the Naskapi in Schefferville." MA thesis, McGill University, Montreal.

Heumann, L. and D. Boldy. 1982. *Housing for the Elderly: Planning and Policy Formulation in Western Europe and North America*. New York: St. Martin's Press.

Higgins, D.J.H. 1986. *Local and Urban Politics in Canada*. Toronto: Gage.

Hill, F.I. 1976. *Canadian Urban Trends*. Volume 2. Toronto: Copp Clark.

Himelfarb, A. 1976. *The Social Characteristics of One-Industry Towns in Canada; A Background Report*. Ottawa: Royal Commission on Corporate Concentration.

Hirschleifer, J. and G. Riley. 1979. "The Analysis of Uncertainty and Information: An Expository Survey." *Journal of Economic Literature* 17: 1375-1421.

Hitchcock, J. 1984. "Toronto's Post-war Urban Environment." *Environmental Education and Information* 3, no.3: 196-213.

Hodge, D.C. 1981. "Residential Revitalization and Displacement in a Growth Region." *Geographical Review* 71: 188-200.

Hodge, G. 1986. *Planning Canadian Communities*. Toronto: Methuen.

Hodge, G. and M. Qadeer. 1983. *Towns and Villages in Canada.* Toronto: Butterworths.

Holdsworth, D. 1977. "House and Home in Vancouver: Images of West Coast Urbanism, 1886-1929." In *The Canadian City: Essays in Urban History,* edited by A.F.J. Artibise and G. Stelter, 186-211. Toronto: McClelland and Stewart.

———. 1986. "Cottages and Castles for Vancouver Homeseekers." *BC Studies* 69-70: 11-32.

Hooper, J. 1984. "Franklin House: A Haven for Vancouver's Skid Road Women." *Impact* 7: 3.

Howell, S.C. 1980. *Designing for Aging: Patterns of Use.* Cambridge, Massachusetts: MIT Press.

Howenstein, E.J. 1981. "Rental Housing in Industrialized Countries: Issues and Policies." In *Rental Housing: Is There a Crisis?* edited by J. Weicher, K. Villani, E. Roistacher, 99-108. Washington DC: Urban Institute Press.

Hoyt, H. 1939. *The Structure and Growth of Residential Neighborhoods in American Cities.* Washington, DC: Federal Housing Administration.

Hulchanski, J.D. 1984. *St. Lawrence and False Creek: A Review of the Planning and Development of Two New Inner City Neighbourhoods.* CPI#10. School of Community and Regional Planning, University of British Columbia.

Hulchanski, J.D. and J. Patterson. 1984. "Two Commentaries on CMHC's Evaluation of its Non-profit Housing Programs: Is It an Evaluation? *Plan Canada* 24, no.1: 28-36.

Hum, D.P.J. 1983. *Federalism and the Poor: A Review of the Canada Assistance Plan.* Toronto: Ontario Economic Council.

Hunter, A.A. 1982. *Class Tells: On Social Inequality in Canada.* Toronto: Butterworths.

Huttman, E.D. 1977. *Housing and Social Services for the Elderly: Social Policy Trends.* New York: Praeger Publishers.

Institute of Urban Studies. 1986. *The Neighbourhood Improvement Program, 1973-1983: A National Review of an Intergovernmental Initiative.* Research and Working Paper 15, Institute of Urban Studies, University of Winnipeg.

Joint Task Force on Neighbourhood Support Services. 1983. *Neighbourhoods Under Stress.* Toronto: Social Planning Council of Metropolitan Toronto.

Jones, L.D. 1983. *The State of the Rental Housing Market.* Ottawa: CMHC.

———. 1984a. *Public Mortgage Insurance in Canada: Its Relevance to the 1980s and Beyond.* Report prepared for CMHC.

———. 1984b. *Wealth Effects on Households' Tenure Choice, Housing Demand and Housing Finance Decisions.* Ottawa: CMHC.

Keller, S. 1966. "Social Class in Physical Planning." *International Social Sciences Journal* 17, no.4: 494-512.

Kitchen, H. 1984. *Local Government Finance in Canada.* Toronto: Canadian Tax Foundation.

Klein and Sears; Environics Research Group; Clayton Research Associates; Lewinberg Consultants; Walker, Poole, Milligan. 1983. *Study of Residential Intensification and Rental Housing Conservation.* Toronto: Ontario Ministry of Housing.

Klodawsky, F., A. Spector, and C. Hendrix. 1983. *The Housing Needs of Single Parent Families in Canada.* Ottawa: CMHC.

Klodawsky, F., A. Spector and D. Rose. 1985. *Canadian Housing Policies and Single Parent Families: How Mothers Lose.* Ottawa: CMHC.

Klodawsky, F. and A. Spector. 1984. "Housing Policy as Implicit Family Policy: The Case of Mother-led Families." In *Changing Values: Proceedings of the Second National Annual Conference, 162-75.* Ottawa: Family Service Canada.

Klodawsky, F. and A. Spector. 1985. "Mother-led Families and the Built Environment in Canada." *Women and Environments* 7, no.2: 12-7.

Koopmans, T.C. 1957. "Allocation of Resources and the Price System." In *Three Essays on the State of Economic Science, 3-126.* New York: McGraw-Hill.

Kosta, V.J. 1957. *Neighbourhood Planning.* Winnipeg: The Appraisal Institute.

Krauter, J.F. and M. Davis. 1978. *Minority Canadians: Ethnic Groups.* Toronto: Methuen.

Laboratoire de Recherche en Sciences Immobiliers, Université du Québec à Montréal (LARSI-UQAM). 1985. *Impact de la restauration dans les quartiers centraux de Montréal.* Ottawa: CMHC.

Langlois, S. 1984. "L'impact du double revenu sur la structure des besoins dans les ménages." *Recherches sociographiques* 25, no.2: 211-66.

Laska, S. and D. Spain, eds. 1980. *Back to the City: Issues in Neighborhood Renovation.* New York: Pergamon.

Lawton, M.P. 1976. "Housing Problems of Community-resident Elderly." In *Occasional Papers in Housing and Community Affairs,* edited by R.P. Boynton, 1: 39-74. Washington, DC: Dept. of Housing and Urban Development.

Leacy, F.H., M.C. Urquhart and K.A.H. Buckley. 1983. *Historical Statistics of Canada.* Second Edition. Ottawa: Ministry of Supply and Services.

League for Social Reconstruction. 1935. *Social Planning for Canada.* Toronto: Nelson and Sons [University of Toronto Press, 1975].

Lemon, J. 1985. *Toronto Since 1918: An Illustrated History.* Toronto: Lorimer.

Lessard, D and F. Modigliani. 1975. "Inflation and the Housing Market:

Problems and Solutions." In *New Mortgage Designs for Stable Housing in an Inflationary Environment,* edited by Modigliani and Lessard, 13-45. Boston: Federal Reserve Bank of Boston.

Lewinberg Consultants Ltd. 1984. *In Your Neighbourhood.* Toronto: Lewinberg Consultants Ltd.

Ley, D. 1985. *Gentrification in Canadian Inner Cities: Patterns, Analysis, Impacts and Policy.* Department of Geography, University of British Columbia. Prepared for CMHC.

Ley, D. 1986. "Alternative Explanations for Inner City Gentrification: A Canadian Assessment." *Annals of the American Association of Geographers 76,* no.4: 521-35.

Lilla, M. 1984. "Why the Income distribution Is So Misleading." *The Public Interest,* Fall 1984: 62-76

Lithwick, H. 1983. *Human Settlement Policies in Periods of Economic Stress.* Report to ECE. Ottawa: CMHC.

Lorimer, J. 1972. *A Citizen's Guide to City Politics.* Toronto: James Lewis and Samuel.

Lorimer, J. and E. Ross, eds. 1976. *The City Book.* Toronto: Lorimer.

Lotscher, L. 1985. *Lebensqualität Kanadischer Stadte.* Basler Beitrage zür Geographie Heft 33. Basel: Bopp and Schwabe.

Loynes, R.M.A. 1972. *CPI and IPI as Measures of Recent Price Change.* A study for the Prices and Incomes Commission. Ottawa: Ministry of Supply and Services.

Lucas, R.A. 1971. *Minetown, Milltown, Railtown: Life in Canadian Communities of Single Industry.* Toronto: University of Toronto Press.

Luxton, M. 1980. *More Than a Labour of Love: Three Generations of Women's Work in the Home.* Toronto: Women's Press.

Lynch, K. 1981. *Good City Form.* Cambridge, Massachusetts: MIT Press.

Lyon, D. and R. Fenton. 1984. *The Development of Downtown Winnipeg: Historical Perspective on Decline and Revitalization.* Institute of Urban Studies, University of Winnipeg.

Lyon, Deborah. 1986. "Rethinking the Neighbourhood Improvement Program." *Newsletter* (Institute of Urban Studies, University of Winnipeg), no. 17 (April): 3.

MacDonnell, S. 1981. *Vulnerable Mothers, Vulnerable Children.* Halifax: Department of Social Services, Nova Scotia.

Mackenzie, S. 1987. "Women's Responses to Economic Restructuring: Changing Gender, Changing Space." In *The Politics of Diversity,* edited by R. Hamilton and M. Barrett, 81-100. London: Verso.

Macpherson, C.B. 1978. *Property: Mainstream and Critical Positions.* Toronto: University of Toronto Press.

Maher, C.A. 1974. "Spatial Patterns in Urban Housing Markets: Filtering in Toronto 1953-71." *The Canadian Geographer* 18, no.2: 108-24.

Makuch, S.M. 1986. "Urban Law and Policy Development in Canada: The Myth and the Reality." In *Labour Law and Urban Law in Canada,* edited by I. Bernier and A. LaJoie, 167-191. Toronto: University of Toronto Press.

Makuch, S.M. and A. Weinrib. 1985. *Security of Tenure.* Research Study 11. Toronto: Ontario Commission of Inquiry into Residential Tenancies.

Marrett, C.B. 1973. "Social Stratification in Urban Areas." In *Segregation in Residential Areas,* edited by A.H. Hawley and V.P. Rock. Washington, DC: National Academy of Sciences.

Marshall, V.W., ed. 1980. *Aging in Canada: Social Perspectives.* 2nd Edition. Don Mills: Fitzhenry and Whiteside.

Marson, J.C. 1981. *Montreal in Evolution: Historical Analysis of the Development of Montreal's Architecture and Urban Environment.* Montreal and Kingston: McGill-Queen's University Press.

Martin Goldfarb Consultants Limited. 1968. "Public Housing." In Snider, 37-41.

Martin, D.M. 1976. *Battered Wives.* San Francisco: Glide Publications.

Maslow, A.H. 1954. *Motivation and Personality.* New York: Harper & Row.

Mathews, G. 1986. *L'évolution de l'occupation du parc résidentiel plus ancien de Montréal de 1951 à 1979.* Etudes et documents 46, INRS-Urbanisation, Montréal.

Matwijiw, P. 1979. "Ethnicity and Urban Residence: Winnipeg, 1941- 1971." *Canadian Geographer* 23, no.1: 45-61.

McAfee, A. 1972. "Evolving Inner City Residential Neighbourhoods: The Case of Vancouver's West End." In *Peoples of the Living Land: Geography of Cultural Diversity in British Columbia,* edited by J.V. Minghi, 163-82. Vancouver: Tantalus.

McCann, L.D. 1972. "Changing Morphology of Residential Areas in Transition." PhD thesis, Department of Geography, University of Alberta.

———. 1978. "The Changing Internal Structure of Canadian Resource Towns." *Plan Canada* 18, no.1: 46-59.

McClain, J. and C. Doyle. 1984. *Women and Housing: Changing Needs and the Failure of Policy.* Toronto: James Lorimer & Co.

McFadyen, S. and D. Johnson. 1981. *Land Use Regulation in Edmonton.* Economic Council of Canada, Regulation Reference Working Paper 16. Ottawa.

McGrath, D. 1982. "Who Must Leave? Alternative Images of Urban Revitalization." *APA Journal* 48, no.2: 196-202.

McKee, C., S. Clatworthy and S. Frenette. 1979. *Housing: Inner City Type Older Areas.* Institute of Urban Studies, University of Winnipeg.

McKellar, J. 1985. *Industrialized Housing: The Japanese Experience.* Ottawa: CMHC.

McKellar, J. *et al.* 1986. *Technology and the Housing Industry: Developing a Framework for Research.* Cambridge, Massachusetts: Joint Center for Housing Studies at MIT and Harvard University, Working Paper.

Mclaughlin, J., P.Dickson and W.Morrison. 1985. "Building the New Brunswick Information Network: The Property Assessment Component." *Computers in Public Agencies: Sharing Solutions Volume I: Lands Records Systems,* 96-103. Papers from the 1985 Annual Conference of the Urban and Regional Information Systems Association, Ottawa.

McMaster, L. 1985. "Population and Land Analysis (Plans) Information Systems." *Computers in Public Agencies: Sharing Solutions Volume II: Geoprocessing,* 62-73. Papers from the 1985 Annual Conference of the Urban and Regional Information Systems Association, Ottawa.

McWhinnie, J.R. 1982. "Measuring Disability." In *Social Indicator Development Programme* (Special Studies). Paris: OECD.

Mellett, C.J. 1983. *At the End of the Rope: A Study of Women's Emergency Housing Needs in the Halifax-Dartmouth Area.* Halifax: Women's Emergency Housing Coalition.

Merrett, S. and R. Smith. 1986. "Stock and Flow in the Analysis of Vacant Residential Property." *Town Planning Review* 57, no.1: 51-67.

Merrill Lynch. 1982. "Housing Industry: A Merrill Lynch Basic Report." January.

Metropolitan Toronto. 1983. *No Place to Go: A Study of Homelessness in Metropolitan Toronto.* Prepared for the Metro Toronto Assisted Housing Study (January).

Michelson, W. 1977. *Environmental Choice, Human Behavior and Residential Satisfaction.* New York: Oxford University Press.

————. 1983. *The Impact of Changing Women's Roles on Transportation Needs and Usage: Final Report.* Prepared for United States Department of Transportation, Urban Mass Transportation Administration. Springfield, Virginia: National Technical Information Service.

————. 1985. *From Sun to Sun: Daily Obligations and Community Structure in the Lives of Employed Women and their Families.* Totowa, New Jersey: Rowman and Allanheld.

Mills, E.S. 1987. "Has the United States Overinvested in Housing?" *American Real Estate and Urban Economics Association Journal* 15, no.1: 601-16.

Ministry of State for Urban Affairs. 1975. *Urban Indicators: Quality of Life Comparisons for Canadian Cities.* Ottawa: MSUA

Miron, J.R. 1983. "Demographic Change and Housing Demand in the 1980s and 1990s." New Neighbourhood Conference, sponsored by the Ontario Ministry of Municipal Affairs and Housing, Toronto, January.

———. 1984. *Housing Affordability and Willingness to Pay.* Research Paper 154, Centre for Urban and Community Studies, University of Toronto.

———. 1988. *Housing in Postwar Canada: Demographic Change, Household Formation and Housing Demand.* Montreal and Kingston: McGill-Queen's University Press.

Miron, J.R. and J.B. Cullingworth. 1983. *Rent Control: Impacts on Income Distribution, Affordability and Security of Tenure.* Toronto: Centre for Urban and Community Studies, University of Toronto.

Modell J. and T.K. Hareven. 1973. "Urbanization and the Malleable Household: An Examination of Boarding and Lodging in American Families." *Journal of Marriage and the Family* 35: 467-79.

Moore, E.G., 1980. "Beyond the Census: Data Needs and Urban Policy Analysis." In *Philosophy in Geography,* edited by S.Gale and G.Olsson, 269-86. Dordrecht: Reidel.

Moore, E.G. and S.J. Clatworthy. 1978. "The Role of Urban Data Systems in the Analysis of Housing Issues." In Bourne and Hitchcock, 228-60.

Morgenstern, J. 1982. *Environmental Competence Among Independent Elderly Households.* Toronto: Institute of Environmental Research, Inc.

Morrison, P.S. 1978. "Residential Property Conversion: Subdivision, Merger and Quality Change in the Inner City Housing Stock, Metropolitan Toronto, 1958-1973." PhD thesis, Department of Geography, University of Toronto.

Muller, A. 1978. *The Market For New Housing in Metropolitan Toronto.* Toronto: Ontario Economic Council.

Mumford, L. 1938. *The Culture of Cities.* New York: Harcourt Brace Yovanovich.

Myers, D. 1978. "Aging of Population and Housing: A New Perspective on Planning for More Balanced Metropolitan Growth." *Growth and Change* 9: 8-13.

———. 1980. *Measuring Housing Progress in the Seventies: Definitions and New indicators.* Working Paper No. 64, Joint Centre for Urban Studies of MIT and Harvard University.

———. 1982. "A Cohort-based Indicator of Housing Progress." *Population Research and Policy Review* 1: 109-36.

National Association of Home Builders. 1985. *Housing America: The Challenges Ahead.* Long Range Planning Report of the National Association of Home Builders. Washington, DC: NAHB.

National Council on Welfare. 1984. *Sixty-five and Older.* Ottawa: Minister of Supply and Services.

Neil, C.C. and T.B. Brealey. 1982. "Home Ownership in New Resource Towns." *Human Resource Management Australia,* February: 38-44.

Neil, C.C., T.B. Brealey and J.A. Jones. 1982. *The Development of Single Enterprise Resource Towns.* Occasional Paper 25, Centre for Human Settlements, University of British Columbia.

Ng, W. 1984. "Social Mix in Urban Neighbourhoods." MA thesis, School of Community and Regional Planning, University of British Columbia.

Norquay, G. and R. Weiler. 1981. *Services to Victims and Witnesses of Crime in Canada.* Ottawa: Solicitor General of Canada (Research Division).

Northwest Territories, Legislative Assembly. 1984. *Interim Report of the Special Committee on Housing.* Yellowknife.

Nuttall, G. and G. Korzenstein. 1985. "Evolution of a Regional Information System for the Metropolitan Toronto Planning Department." In *Computers in Public Agencies: Sharing Solutions Volume IV: Data Processing, Education, Public Administration, Public Works, Regional Agencies, Transportation,* 120-31. Papers from the 1985 Annual Conference of the Urban and Regional Information Systems Association, Ottawa.

Ontario Housing Corporation. 1983. "Rent Supplement Program: Tenant Satisfaction." Toronto: OHC Operational Planning Branch.

Ontario. Lieutenant-Governor's Committee on Housing Conditions in Toronto. 1934. *The [Bruce] Report.* Toronto: The Committee.

———. Ministry of Community and Social Services. 1986. *A New Agenda: Health and Social Service Strategies for Ontario's Seniors.* Toronto: Queen's Printer.

———. Ministry of Housing. 1986. *A Place to Call Home: Housing Solutions for Low-Income Singles in Ontario.* Report of the Ontario Task Force on Roomers, Boarders and Lodgers. Toronto.

———. Planning Act Review Committee. 1977 *Report.* Toronto: Ontario Ministry of Housing.

Oosterhoff, A.H. and W.B. Rayner. 1979. *Losing Ground: The Erosion of Property Rights in Ontario.* Toronto: The Ontario Real Estate Association.

Page and Steele. 1945. "MacGregor House." *Royal Architectural Institute of Canada Journal,* 141.

Paquette, L. 1984. "Fermont: évaluation de la planification." MA thesis, McGill University, Montreal.

Parker, A.L. and L. Rosborough. 1982. *A Matter of Urgency: The Psychiatrically Disabled in the Ottawa-Carleton Community.* Ottawa: Canadian Mental Association.

Parker V.J. 1963. *The Planned Non-Permanent Community: An Approach to*

Development of New Towns Based on Mining Activity. Ottawa: Department of Northern Affairs.

Patterson, J. 1985. *Rent Review in Ontario and Factors Affecting the Supply of Rental Housing.* Toronto: Social Planning Council of Metropolitan Toronto.

Patterson, J. and K. Watson. 1976. *Rent Stabilization: A Review of Current Policies in Canada.* Ottawa: Canadian Council on Social Development.

Pattison, T. 1977. "The Process of Neighbourhood Upgrading and Gentrification." MA thesis, Department of Urban Studies and Planning, MIT, Cambridge, Massachusetts.

Peddie, R. 1978. "Processes of Residential Change: A Case Study of Alexandra Park." PhD thesis, Department of Geography, University of Toronto.

Pesando, James E. and S.M. Turnbull. 1983. "Retractable Debt Instruments and the Provision of Mortgage Rate Insurance and/or the Introduction of 10-Year Mortgages Callable After 5 Years." Report prepared for CMHC.

Peterson, G. 1974. *The Influence of Zoning Regulations on Land and Housing Prices.* Washington: The Urban Institute.

Phipps, A.G. 1982. *Social Impacts of Housing Reinvestment in the Core Neighbourhoods of Saskatoon.* Department of Geography, University of Saskatchewan, Saskatoon.

Pinfield, L.T. and L.D. Etherington. 1982. *Housing Strategies of Resource Firms in Western Canada.* Ottawa: Canada Mortgage and Housing Corporation.

Poapst, J.V. 1962. *The Residential Mortgage Market.* Working paper for the Royal Commission on Banking and Finance. Ottawa: Queen's Printer.

Poapst, J.V. 1975. *Developing the Residential Mortgage Market. Volume 1: A Residential Mortgage Market Corporation.* Report prepared for CMHC.

Poapst, J.V. 1984. "Pension Reform: An Unexplored Option." Toronto: Ontario Ministry of Treasury and Economics.

Popper, F.J. 1988. "Understanding American Land Use Regulation since 1970: A Revisionist Interpretation." *Journal of the American Planning Association,* Summer: 291-301.

Preston, R.E. and L. Russwurm, eds. 1980. *Essays on Canadian Urban Process and Form.* Department of Geography, University of Waterloo.

Priest, G.E. 1985. "Living Arrangements of Canada's Elderly." *Changing Demographic and Economic Factors* (85-1). Gerontology Research Centre, Simon Fraser University, Burnaby, British Columbia.

Proudfoot, S. 1980. *Private Wants and Public Needs: The Regulation of Land Use in the Metropolitan Toronto Area.* Economic Council of Canada, Regulation Reference Working Paper 12.

Pryor, E.T. 1984. "Canadian Husband-Wife Families: Labour Force Participation and Income Trends, 1971-1981. *The Labour Force,* May. Catalogue 71-001. Statistics Canada. 93-109.

Québec, Gouvernement du. 1988. *Rapport de la Commission d'enquête sur les services de santé et les services sociaux.* Québec: Editeur officiel du Québec.

Ramsey, F.P. 1928. "A Mathematical Theory of Savings." *Economic Journal* 38: 543-59.

Rabnett, R.A. and Associates *et al.* 1981. *Conceptual Plan, Tumbler Ridge, Update Northeast Sector, BC.* Victoria: Ministry of Municipal Affairs.

Ray, D.M. *et al.* 1976. *Canadian Urban Trends.* 3 vols. Toronto: Copp Clark

Renaud, F. and M.E. Wexler. 1986. "Housing the Elderly in the Community: A Review of Existing Programs in Quebec." In Gutman and Blackie, 133-47.

Richmond, A.H. 1972. *Ethnic Residential Segregation in Metropolitan Toronto.* Institute for Behavioural Research, York University.

Roberts, E. 1974. *The Residential Desirability of Canadian Cities.* Ottawa: MSUA.

Roberts, R. and G. Paget. 1985. "Socially Responsive Planning: The Tumbler Ridge Experience." *Ekistics* 52.

Robinson, I. 1981. *Canadian Urban Growth Trends.* Vancouver: University of British Columbia Press.

Robinson, I.M. 1962. *New Industrial Towns on Canada's Resource Frontier.* Chicago: University of Chicago Press.

Robson, R. 1985. "The Central Mortgage and Housing Corporation and the Ontario Resource Town." *Environments* 17, no.2: 66-74.

Rose, A. 1957. "Row Housing: Its Social Significance." *The Canadian Architect,* February: 20-23.

———. 1958. *Regent Park: A Study in Slum Clearance.* Toronto: University of Toronto Press.

———. 1972. *Governing Metropolitan Toronto.* Berkeley, California: University of California Press.

———. 1980. *Canadian Housing Policies, 1935-1980.* Toronto: Butterworths.

Rose, D. 1984. "Rethinking Gentrification: Beyond the Uneven Development of Marxist Urban Theory." *Environment and Planning D: Society and Space* 2, no.1: 47-74.

———. 1986. "Urban Restructuring, Labour Force Polarisation and Gentrification: A Canadian Perspective on Recent Theoretical Developments." Paper presented at the Annual Meeting of the Association of American Geographers, Minneapolis (May): INRS-Urbanisation, Montréal.

Rose, D. and C. LeBourdais. 1986. "The Changing Conditions of Female Single Parenthood in Montreal's Inner City and Suburban Neighbourhoods." *Urban Resources* 3, no.2: 45-52.

Rowe, A. 1986. "Housing in Rural Areas and Small Towns." Resource paper for the "Housing Progress in Canada since 1945" study. Centre for Urban and Community Studies, University of Toronto.

Saarinen, O. 1979. "The Influence of Thomas Adams and the British New Town Movement in the Planning of Canadian Resource Communities." In Artibise and Stelter, 219-64.

——. 1986. "Single Sector Communities in Northern Ontario: The Creation and Planning of Dependant Towns." In Stelter and Artibise.

Safarian, A.E. 1959. *The Canadian Economy in the Great Depression.* Toronto: McClelland and Stewart Limited.

Saint-Pierre, J., T.M. Chau et M. Choko. 1985. *Impact de la restauration dans les quartiers centraux de Montréal.* Montréal: LARSI-UQAM.

Salomon Brothers Inc. 1986. *Home Builders' Attitudes toward Structural Plywood Substitutes – Third Annual Survey.* New York: Stock Research.

Saldov, M. 1981. "A Review of the Social Integration Effects of Social Mix." Faculty of Social Work, University of Toronto.

Sanford, B. 1985. "The Origins of Residential Differentiation: Capitalist Industrialisation, Toronto, Ontario, 1851-1881." PhD thesis, Program in Planning, Department of Geography, University of Toronto.

Sarkissian, W. 1975. "The Idea of Social Mix in Town Planning: A Historical Review." *Urban Studies* 13: 231-46.

Sassen-Koob, S. 1984. "The New Labor Demand in Global Cities." In *Cities in Transformation: Class, Capital and the State,* edited by M.P. Smith, 139-71. Urban Affairs Annual Reviews 26. Beverly Hills: Sage.

Savoie, D. J. 1992. *Regional Economic Development.* Toronto: University of Toronto Press.

Scanada Consultants Limited. 1970. *Industrialized Housing Production: Potential Gains Through High-Volume Programming.* Ottawa.

Schoenauer, N. 1982. "Housing at Fermont." Paper presented at a conference on Northern Housing, at LG2, Quebec.

Schwenger, C.W. 1977. "Health Care for Aging Canadians." *Canadian Welfare* 52, no.6: 9-12.

Schwenger, C.W. and M.J. Gross. 1980. "Institutional Care and Institutionalization of the Elderly in Canada." In Marshall, 248-56.

Sewell, John. 1976. "Where the Suburbs Came From." In Lorimer and Ross.

Sharpe, C.A. 1978. *Vacancy Chains and Housing Market Research: A Critical*

Evaluation. Research Note 3, Department of Geography, Memorial University, St. John's, Newfoundland.

Shaw, Melanie. 1987. "Anticipating the Market for Technological Innovations in the Home Building Industry." Master of Science in Civil Engineering thesis, Massachusetts Institute of Technology, Cambridge.

Shaw, W.G.A. 1970. "Homes: the Neglected Element in Canadian Resource Towns Planning." *The Albertan Geographer* 7: 43-9.

Sherebrin, D. 1982. *Leasehold Sheltered Housing for the Elderly in Britain.* Ottawa: CMHC.

Siegan, B.H. 1970. "Non-zoning in Houston." *Journal of Law and Economics* 13: 71-147.

Siggner, A.J. 1979. *An Overview of Demographic, Social and Economic Conditions Among Canada Registered Indian Population.* Ottawa: Indian and Northern Affairs Canada.

Silver, I.R. 1980. *The Economic Evaluation of Residential Building Codes: An Exploratory Study.* Economic Council of Canada, Regulation Reference Working Paper 5.

Silzer, V. 1985. "Toronto's Low-Income Tenants Hardest Hit by Housing Crisis." *City Planning* 3, no.2: 7-11.

Silzer, V. and K. Ward. 1986. "Meeting Housing Needs through the Existing Rental Stock." Toronto: Social Planning Council of Metropolitan Toronto.

Simmons, J.W. 1986. "The Impact of the Public Sector on the Canadian Urban System." In Stelter and Artibise, 21-50.

Simmons, J.W. and L.S. Bourne, 1989. *Urban Growth Trends in Canada, 1981-86: A New Geography of Change.* Major Report 25, Centre for Urban and Community Studies, University of Toronto.

Simon, J. 1986. "Women and the Canadian Co-op Experience: Integrating Housing and Economic Development." *Women and Environments* 8, no.1: 10-12.

Simon, J.C. and G. Wekerle. 1985. *Creating a New Toronto Neighbourhood: The Planning Process and Residents' Experience.* Toronto: CMHC Ontario Region.

Single Displaced Persons Project. 1983. "The Case for Long Term Supportive Housing." Toronto.

Sirard, G., F. Belanger, C. Beauregard, S. Gagnon, D. Veillette. 1986. *Des mères seules, seules, seules: une étude sur la situation des femmes cheffes de familles monoparentale du Centre-Sud à Montréal.* Montréal: La Criée.

Skaburskis, A. 1979. *Demolitions, Conversions, Abandonments: The Extent and Determinants of Housing Stock Losses.* Ottawa: CMHC.

Skaburskis, A. and Associates. 1984. *National Condominium Market Study, Working Paper 1: Literature Review.* Ottawa: CMHC.

———. 1985. *National Condominium Market Study, Working Paper 5: Survey Methods and Response Rates.* Vancouver.

Smith, L.B. 1974. *The Post-war Canadian Housing and Residential Mortgage Markets and the Role of Government.* Toronto: University of Toronto Press.

———. 1983. "The Crisis in Rental Housing: A Canadian Perspective." *Annals of the American Academy of Political and Social Science,* January.

———. 1984. "Household Headship Rates, Household Formation, and Housing Demand in Canada." *Land Economics* 60: 180-8.

Smith, L.B. and G.R. Sparks. 1970. "The Interest Sensitivity of Canadian Mortgage Flows." *Canadian Journal of Economics and Political Science,* August: 407-21.

Smith, L.B. and P. Tomlinson. 1981. "Rent Controls in Ontario: Roofs of Ceilings? *American Real Estate and Urban Economics Association Journal* 9: 93-113.

Smith, N. and P. Williams, eds. 1986. *Gentrification of the City.* London: Allen and Unwin.

Snider, E.L., ed. *User Study of Low-Income Family Housing* (Interim Report no.1). Ottawa: CMHC.

Social Planning Council of Metropolitan Toronto. 1979. *Metro's Suburbs in Transition – Part I: Evolution and Overview.* Toronto.

———. 1980. *Metro's Suburbs in Transition – Part II: Planning Agenda for the 80s.* Toronto.

———. 1983. *People Without Homes: A Permanent Emergency.* Toronto.

———. 1987. *Social Infopac* 6, no.3 (July).

Social Planning Council of Metropolitan Toronto. 1970. *Alexandra Park Relocation.* Volume 1: *After Relocation.* Volume 2: *Supplementary Report.* Toronto.

Social Planning Council of Winnipeg. 1979. *Housing Conditions in Winnipeg: The Identification of Housing Problems and High Need Groups.* Research Paper Series on Social Indicators. Report 1. Winnipeg.

Social Policy Research Associates, SPR Evaluation Group Ltd. 1979. *An Evaluation of RRAP.* Ottawa: CMHC.

Solow, R.M. 1974. "The Economics of Resources or the Resources of Economics." *American Economic Association: Papers and Proceedings* 54: 1-14.

Spelt, J. 1973. *Toronto.* Toronto: Collier-Macmillan.

Spence-Sales, H. 1949. *Planning Legislation in Canada.* Ottawa: CMHC.

Spragge, S. 1979. "A Confluence of Interests: Housing Reform in Toronto, 1900-1920." In Artibise and Stelter, 247-67.

Spurr, P. 1976. *Land and Urban Development: Preliminary Study.* Toronto: Lorimer.

Stach, P.B. 1987. "Zoning: To Plan or To Protect?" *Journal of Planning Literature* 2, no.4: 472-81.

Stapleton, C. 1980. "Reformulation of the Family Life-cycle Concept: Implications for Residential Mobility." *Environment and Planning A* 12, no.10: 1103-18.

Statistics Canada. 1977. *Survey of Incomes, Assets, and Indebtedness of Families in Canada, 1977.* Catalogue 13-572. Ottawa.

———. 1977-84. *Family Characteristics and Labour Force Activity: Annual averages.* Catalogue 71-533. Ottawa.

———. 1983. *Household Facilities by Income and Other Characteristics.* Catalogue 13-567. Ottawa.

———. 1984a. *The Distribution of Wealth in Canada.* Catalogue 13-580. Ottawa.

———. 1984b. *Fertility in Canada. From Baby-boom to Baby-Bust.* Ottawa.

———. 1984c. *Life Tables, Canada and Provinces, 1980-2.* Ottawa.

———. 1984d. *Canada's Native People.* Ottawa.

———. 1984e. *Fixed Capital Flows and Stocks, 1936- 1983.* Ottawa.

——— (Health Division). 1985. *Canadian Health and Disability Survey: Highlights 1983-1984* Catalogue 82-563E, June. Ottawa: Ministry of Supply and Services.

———. 1986a. *The National Balance Sheet Accounts, 1989.* Ottawa.

———. 1986b. *National Income and Expenditure Accounts.* Catalogue 13-001, July Revisions. Ottawa.

———. 1987. *Construction Price Statistics.* First Quarter. Ottawa.

———. 1988. *National Income and Expenditure Accounts: Annual Estimates, 1926-1986.* Catalogue 13-531. Ottawa.

———. 1989. *Historical Labour Force Statistics: Actual Data, Seasonal Factors, Seasonally Adjusted Data, 1988.* Catalogue 71-201. Ottawa.

Steele, M. 1979. *The Demand for Housing in Canada.* Census Analytical Study. Catalogue 99-763. Ottawa: Statistics Canada.

———. 1983. "The Low Consumption Response of Canadian Housing Allowance Recipients." In *Where Do We Go From Here? Proceedings of a Symposium on the Rental Housing Markets and Housing Allowances,* edited by the Canadian Council on Social Development, 57-64. Ottawa.

———. 1985a. *Housing Allowances: An Assessment of the Proposal for a National Program for Canada.* A report prepared for the Canadian Home Builders' Association. University of Guelph.

———. 1985b. *Canadian Housing Allowances: An Economic Analysis.* Toronto: Ontario Economic Council.

———. 1987. *The User Cost and Cash Flow Cost of Homeownership.* Report prepared for CMHC.

———. 1992. "Inflation, the Tax System, Rents and the Return to Home Ownership." In *Readings in Canadian Real Estate,* edited by Gavin Arbuckle and Henry Bartel, 119-33. Toronto: Captus University Publications.

Steele, M. and J.Miron. 1984. *Rent Regulation, Housing Affordability Problems and Market Imperfections.* Research Study No.9. Toronto: Ontario Commission of Inquiry into Residential Tenancies.

Stelter, G. and A.F.J. Artibise. 1977. "Urban History Comes of Age: A Review of Current Research." *City Magazine* 3 no.1: 22-36.

———, eds. 1986. *Power and Place.* Vancouver: University of British Columbia Press.

Stone, L.O. 1967. *Urban Development in Canada.* Ottawa: DBS Census Monograph.

Stone, L.O. and S. Fletcher. 1980. *A Profile of Canada's Older Population.* Montreal: The Institute for Research on Public Policy.

———. 1982. *The Living Arrangements of Canada's Older Women,* Catalogue 86-503. Ottawa: Statistics Canada.

Streich, P.A. 1985. "Canadian Housing Affordability Policies in the 1970s: An Analysis of Federal and Provincial Government Roles and Relationships in Policy Change." PhD thesis, Queen's University, Kingston.

Struyk, R. and B. Soldo. 1980. *Improving the Elderly's Housing: A Key to Preserving the Nation's Housing Stock and Neighborhoods.* Cambridge, Massachusetts: Ballinger.

Struyk, R.J., with S.A. Marshall. 1976. *Urban Homeownership.* Lexington, Massachusetts: Lexington Books.

Sunga, P.S. and G.A. Due. 1975. *MSUA and the Federal Government.* Ottawa: Ministry of State for Urban Affairs.

Teasdale, P.E. et M.E. Wexler. 1986. *Dynamique de la famille, ajustements résidentiels et souplesse du logement.* Université de Montréal, École d'architecture.

TEEGA Research Consultants. 1983. *An Analysis of Dwelling Repairs and Energy Improvements Based on the 1982 Household Facilities and Equipment Survey: Report to CMHC.* Ottawa.

Thorns, D. 1972. *Suburbia.* St. Albans: Granada.

Tiebout, C.M. 1956. "A Pure Theory of Local Expenditures." *Journal of Political Economy* 64: 416-24.

Toronto, City of. 1973. *The City is for All Its Citizens.* The Mayor's Task Force Report re: the Disabled and Elderly. Toronto.

Townsend, P. 1981. "The Structured Dependency of the Elderly: A Creation of Social Policy in the Twentieth Century." *Aging and Society* 1, no.1: 5-28.

Truelove, M. 1986. "Trends in Daycare in Canada." Paper presented at the Annual Meeting of the Association of American Geographers, Minneapolis (May). Department of Geography, Ryerson Polytechnical Institute.

United Nations, Department of Economics and Social Affairs. 1978. *The Role of Housing in Promoting Social Integration.* New York.

———. Commission Human Settlements. 1987. *Shelter and Services for the Poor: A Call to Action.* Nairobi.

Vancouver, City of. 1986. *New Neighbours: How Vancouver's Single-family Residents Feel about Higher Density Housing.* City of Vancouver Planning Department.

Vergès-Escuin, R. 1985. *Inventaire permanent du parc résidentiel canadien par province.* May. Ottawa: CMHC.

Villeneuve, P. and D. Rose. 1985. "Technological Change and the Spatial Division of Labour by Gender in the Montreal Metropolitan Area." Paper presented at the International Geographical Union, Nijmegen, The Netherlands.

Vischer Skaburskis, Planners. 1979a. *Demolitions, Conversions, Abandonments: Working Paper 1 – Case Studies of Vancouver, Calgary, Toronto, Montreal, Saint John.* Report prepared for CMHC.

———. 1979b. "False Creek. Area 6. Phase 1. Post-Occupancy Evaluation. Part 1. Social Mix." Vancouver.

Vischer, J.C. 1987. "The Changing Canadian Suburb." *Plan Canada* 27: 130-40.

Wade, J. 1986. "Wartime Housing Limited, 1941-47: Canadian Housing Policy at the Crossroads." *Urban History Review* 15: 41-59.

Walker, H.W. 1953. *Single-Enterprise Communities in Canada.* Report to CMHC by The Institute of Local Government, Queen's University.

Ward, J. 1985. "Housing Low-Income Singles: A Community Development Approach." Paper presented to the Annual Symposium of the Canadian Association of Housing and Renewal Officials, Toronto.

Ward, K., V. Silzer and N. Singer. 1986. "Meeting Housing Needs Through the Existing Rental Stock." Policy Paper 4. *Affordable Housing: An Agenda for Action.* Toronto: Social Planning Council of Metropolitan Toronto.

Wekerle, G. 1984. "A Woman's Place Is in the City." *Antipode* 6, no.3: 11-20.

———. 1988a. "Canadian Women's Housing Cooperatives: Case Studies in Physical and Scoial Innovation." In *Life Spaces: Gender, Household,*

Employment, edited by C. Andrew and B. Moore-Milroy, 102-40. Vancouver: University of British Columbia Press.

———. 1988b. *Women's Housing Projects in Eight Canadian Cities.* Ottawa: CMHC.

Wekerle, G. and S. Mackenzie. 1985. "Reshaping the Neighbourhood of the Future as We Age in Place." *Canadian Woman Studies/Les cahiers de la femme* 6, no.2: 69-73.

Wellman, B. 1971. "Who Needs Neighbourhoods?" In Draper.

Wexler, M.E. 1985. "Residential Adjustments of the Elderly: A Comparison of Nonmobile and Mobile Elderly in Montreal."

Wheeler, M., ed. 1969. *The Right to Housing.* Montreal: Harvest House.

Wigdor, B.T., ed. 1981. *Housing for an Aging Population: Alternatives.* Toronto: University of Toronto Press.

Willson, K. n.d. "Residential Form and Social Infrastructure in a North American City: Toronto in the Twentieth Century." Centre for Urban and Community Studies, University of Toronto.

Wojciechowski, M., ed. 1984. *Mining Communities: Hard Lessons for the Future.* Centre for Resource Studies, Queen's University.

Wolfe, J. 1985. "Some Present and Future Aspects of Housing and the Third Sector." In *The Metropolis: Proceedings of a Conference in Honour of Hans Blumenfeld,* edited by J. Hitchcock and A. McMaster, 131-56. Department of Geography and Centre for Urban and Community Studies, University of Toronto.

Wolfensberger, W. *et al.* 1972. *The Principle of Normalization in Human Services.* Toronto: National Institute on Mental Retardation.

Wood, P.H.N. 1975. *Classification of Impairments and Handicaps* (Conference Series 75/13). Geneva: World Health Organization.

Yeates, M. 1978. "The Future Urban Requirements of Canada's Elderly." *Plan Canada* 18, no.2 (June): 88-105.

Zay, N. 1966. "Living Arrangements for the Aged." Canadian Conference on Aging, Toronto (Ottawa: Canadian Council on Social Development).

Index